Science and Fiction

For further volumes:
http://www.springer.com/series/11657

Science and Fiction – A Springer Series

This collection of entertaining and thought-provoking books will appeal equally to science buffs, scientists and science-fiction fans. It was born out of the recognition that scientific discovery and the creation of plausible fictional scenarios are often two sides of the same coin. Each relies on an understanding of the way the world works, coupled with the imaginative ability to invent new or alternative explanations—and even other worlds. Authored by practicing scientists as well as writers of hard science fiction, these books explore and exploit the borderlands between accepted science and its fictional counterpart. Uncovering mutual influences, promoting fruitful interaction, narrating and analyzing fictional scenarios, together they serve as a reaction vessel for inspired new ideas in science, technology, and beyond.

Whether fiction, fact, or forever undecidable: the Springer Series "Science and Fiction" intends to go where no one has gone before!

Its largely non-technical books take several different approaches. Journey with their authors as they

- Indulge in science speculation—describing intriguing, plausible yet unproven ideas;
- Exploit science fiction for educational purposes and as a means of promoting critical thinking;
- Explore the interplay of science and science fiction—throughout the history of the genre and looking ahead;
- Delve into related topics including, but not limited to: science as a creative process, the limits of science, interplay of literature and knowledge;
- Tell fictional short stories built around well-defined scientific ideas, with a supplement summarizing the science underlying the plot.

Readers can look forward to a broad range of topics, as intriguing as they are important. Here just a few by way of illustration:

- Time travel, superluminal travel, wormholes, teleportation
- Extraterrestrial intelligence and alien civilizations
- Artificial intelligence, planetary brains, the universe as a computer, simulated worlds
- Non-anthropocentric viewpoints
- Synthetic biology, genetic engineering, developing nanotechnologies
- Eco/infrastructure/meteorite-impact disaster scenarios
- Future scenarios, transhumanism, posthumanism, intelligence explosion
- Virtual worlds, cyberspace dramas
- Consciousness and mind manipulation

Giancarlo Genta

A Man From Planet Earth

A Scientific Novel

 Springer

Giancarlo Genta
Departmant of Mechanics
Technical University
Torino, Italy

ISSN 2197-1188   ISSN 2197-1196   (electronic)
ISBN 978-3-319-21114-5   ISBN 978-3-319-21115-2   (eBook)
DOI 10.1007/978-3-319-21115-2

Library of Congress Control Number: 2015949388

Springer Cham Heidelberg New York Dordrecht London

Cover illustration: Cover figure by © Michael Boehme (michael-boehme.com)

Printed on acid-free paper

Springer International Publishing AG Switzerland is part of Springer Science+Business Media (www.springer.com)

To Franca and Alessandro

Preface

As soon as humans realized the vastness of the Universe, they began to wonder whether the Earth was an island of life in an ocean of inanimate matter or whether other living beings, perhaps intelligent, conscious ones, dwell in the vastness of the Universe. The question "are we alone in the Universe?" is one of the basic questions about ourselves and our role in the Universe to which every one of us tries to give an answer—or decides that an answer cannot be found.

However, there is a difference: while the other questions about God, our soul, and the essence of the universe are metaphysical questions and cannot be answered by our reason (pure reason, Kant would say), questions related to extraterrestrial life and intelligence are within the reach of science, in particular now that space technology has opened new methods and possibilities for study.

The idea that life is not confined to our planet and that it is likely that it has reached the stage of intelligence in other parts of the Universe as well (or rather, in our galaxy, to limit ourselves to a region we can realistically study) is widespread, and many attempts have been made to at least receive some communications from these hypothetical intelligent extraterrestrials.

If we believe that we are not alone in this universe, a number of questions follow: Are these ETIs (an acronym for Extraterrestrial intelligences) similar enough to us to allow some understanding between us and them? Or rather, if they do exist there are likely to be many of them, is it possible that the various intelligent species are similar enough, at least as far as our mental processes are concerned, to make it possible for there to be some sort of understanding among them?

Modern science assumes that physical laws are the same throughout the whole Universe and don't change with time. If chemical and biological evolution is determined by physical laws, is it possible that only one bio-chemistry, only one way for encoding genetic information, or only one cellular structure is possible? Does life necessarily lead to eukaryote cells, multi-cellularity, differentiation of tissues, and so on? If so, extraterrestrials won't be very different from us, and convergent evolution may lead to striking similarities. And, because there are good reasons for intelligence to have evolved in a bipedal form, with eyes in a frontal position to allow binocular vision, intelligent aliens could well be somewhat similar to us. This is clearly an extreme hypothesis, which may also be generalized to include psychological aspects of the nature of intelligent beings.

The opposite hypothesis is that there are many ways in which living beings can evolve and that life based on a biochemistry very different from ours may exist. Environments that are favorable for life may be very different, and every one could produce beings that have little in common with those evolved in different places. If this is the case, it might be very difficult to recognize it (he/she?) as a living being, and even more difficult to recognize it as intelligent. The very definition of intelligence could be difficult, or even impossible.

In this situation, humans could come in contact with something utterly alien, something we might be unable to recognize even as an intelligent being.

The scenario of the novel is basically that: A number of intelligent species living in our galaxy are similar enough to be able to live together in a more or less peaceful society. They recognize each other to the point that the term "*human*" has taken on the meaning of "intelligent being." They see each other not as aliens but as humans of different species.

At a certain point this society comes across a truly alien species, something so different that nobody has been culturally prepared to face.

The scenario of the novel is this encounter and its main theme is what the meaning of a nonhuman intelligence may be.

The story is followed by a short appendix, summarizing the scientific and technological facts, theories, and hypotheses that are behind the novel. It develops the themes outlined in this preface, with some extension to the subjects of space travel and artificial intelligence.

This feature is a characteristic of this *Science and Fiction* series that has been introduced by Springer.

While the section on astrobiology reflects the current understanding of the subject, the short section on space travel is the most hypothetical, because it is based on ideas which have, up to now, received no theoretical or experimental confirmation. This is because a way of allowing the characters to move at a speed higher than that of light had to be devised, so the wormhole approach was chosen.

The author wishes to express his sincere thanks to Chris Caron (publishing editor at Springer), and to Storm Dunlop (language editor), for their constructive criticism and suggestions, which resulted in a great improvement to the present text.

The gratitude of the author goes also to his wife Franca, for the editing work she performed, as usual, on both the novel and the scientific commentary.

Turin, Italy   Giancarlo Genta

Contents

Part I

The Novel

A Man from Planet Earth

Prologue

The 300.000 years between the time when the various human species started traveling among the stars and the time when the humans in our group of galaxies organized themselves into a single pacific and civil order, are usually defined by historians as the 'Intermediate Period'.

Mostly owing to a literature of dubious quality describing fierce battles between fleets of starships, and the epic feats of adventurers hailing from quarrelsome kingdoms and empires, the general public thinks it was a barbaric and romantic time. This picture is mostly false and people living in those times enjoyed long periods of peace and stability. A number of confederations were started in the different galaxies, which eventually evolved into the present civil society. Even the Qhrun crisis, without doubt the worst disaster to happen in that period, was not as universal as many think, and many human species were affected only to a limited extent.

Many epic tales and novels dealing with that crisis are based on the character of Admiral Thomas Taylor, a mysterious figure most serious historians think never existed. In many legends he plays the role of a ruthless adventurer, barbaric warrior or romantic knight. The last of these traditions, mostly originating in the fifth sector of the Milky Way, the sector many say he came from, sometimes even endow him with magical powers.

Academic debates about Admiral Taylor were revived in the last ten years by the well-known Centauri Finding. As everyone knows by now, a rescue ship attempting to find an automatic freighter, lost en route between the Sun and Gliese 442, spotted a large mass adrift at the extreme outskirts of the Alpha Centauri system. It was only thanks to the captain's great skill and extreme care that the wreck could be recovered and identified as an ancient cargo ship, adrift for hundreds of thousands of years.

When it was transported to the Ancient History Museum on Laraki, archeologists realized that it was the most important discovery in the last

© Springer International Publishing Switzerland 2016
G. Genta, *A Man From Planet Earth*, Science and Fiction,
DOI 10.1007/978-3-319-21115-2_1

several thousand years. Notwithstanding the damage the ship had suffered from the accidental explosion of its engines and hundreds of thousands of years of inactivity, her wreck yielded a great deal of information about the technology and life in those remote times.

The most important findings were those by a group of paleoinformatics specialists who, from the faint traces still existing in the hardware of the ship's computer, painstakingly reconstructed the contents of its memory. With these, and many other pieces of information about space navigation and life on board, they managed to piece together what appears to be a reliable account of the Qhrun crisis and of Admiral Taylor's role in it.

The most important aspect of the finding is that archeologists think the ship was abandoned no more than three centuries after the crisis, and this is therefore the most ancient version of the story. Moreover, the system where the freighter was found is close to that of the Sun, an area from which most of these legends seem to come.

The result of this reconstruction work is here offered to modern readers, without any changes except translation into modern language. The author of this reconstruction warns the reader not to attribute any great historical value to a text that is generally thought to be just an epic tale.

'hu-The hua
Head of the Paleoinformatic Section
Institute for the Ancient History of the Milky Way

1   Blockade

"Fifteen minutes to re-entry, Admiral." The voice of star cruiser CH-23426 was calm and slightly bored, as usual. The screens were completely dark, but soon they would fill with stars.

"Thank you, Twenty-six. Have sensors ready and ensure all shields are at full power on re-entry." Admiral Taylor wished he didn't have to waste time monitoring such routine navigation tasks: this mission had delicate diplomatic implications, but it did, of course, also include routine navigation and military operations.

"Of course, Sir," answered the ship's computer.

Like any good captain, Taylor thought he could understand the smallest nuances of his ship's behavior. The more CH-23426 identified herself as the flagship, the more conceitedly she dealt with everyone, he thought to himself. A self-aware star cruiser was a danger that was not to be underestimated. *Twenty-six must be put in her place*, he thought with a smile. After all, she was just a machine.

Star cruiser CH-23426 was one of the oldest ships in the Confederation Starfleet, and one of the few survivors of the battles fought twelve thousand years earlier, during the Civil War in the seventh sector. Actually CH-23426 had fought on the wrong side: she claimed to have fought at the battle of Teryygil, where the secessionists were finally defeated, but her claim was unlikely, because all their ships were destroyed in that battle. Still, without official documents, lost a long time ago, Taylor felt it unnecessary to contradict her on the matter. If she really believed her version, it was better not to risk trouble with the ship's logic units.

"Five minutes to re-entering normal space, all crew at their stations."

"Susan, just after re-entry, set the fastest course to the third planet, to be on target three hours after dawn, local time." At last the ship was coming to life again, after almost three days of inactivity in hyperspace, the time needed to travel 450 parsecs from Qhra'ar base to the Sun's system.

The computer dimmed the lights on the bridge, and the stars duly appeared on the screens. Then, the instant the computer was aware of the signals from its sensors, warning lights flashed and loud alarms started ringing.

"Three unidentified ships at close range." The computer sounded puzzled. "Unidentified ship on collision course."

"Zoom in on the forward screen." Taylor was surprised too: no incursion in the Solar System was expected for at least two weeks and the Qhruns couldn't know about his mission.

A Qhrun intruder filled the forward screen. The other two ships were standing by a short distance away. A quick glance was enough to see it was a perfect blockade, leaving little space for maneuver.

"Emergency deceleration, all available power to the engines." The cruiser was ready to start the only maneuver she had been trained to perform.

"Negative, Twenty-six. Maintain speed. Collision trajectory to the center of the nearest ship. Forward shields only, batteries ready. Instruct torpedoes one to thirty to home on target straight ahead. Get ready to lower the shield and fire straight ahead with all batteries on my order."

"Order canceled, Sir. The chances the enemy will withdraw are nil and the maneuver exceeds the safety envelope and cannot be authorized." The ship was refusing to obey an order that conflicted with its safety protocol.

"Ashkahan, all safety controls off. Everybody ready for manual control."

"I must ask for confirmation of the last order, Sir." The first mate's voice was uncertain. She didn't like the idea of taking manual control.

"I remind you that manual shutoff of safety controls cannot be authorized and. . ." This was the computer again.

"Twenty-six, this is a formal order. The previous order is confirmed and, if not obeyed, all logical functions of the computer will be shut off," Taylor interrupted dryly. He couldn't allow the bureaucratic attitude of the ship and its crew to waste precious time.

"Safety controls off, collision trajectory underway." The voice of the computer was cold now, and Taylor felt a tone of defiance in her voice.

"Good," said the Admiral. "At my order switch all power to the batteries, switch off all non-essential devices and set all generators at one hundred and thirty per cent."

The cruiser was rushing at top sub-light speed towards the nearest intruder. When they were less than a few thousands kilometers from the intruder, Taylor gave the order to lower the shields, fire all batteries and launch all torpedoes. The order to return full power back to the forward shields followed immediately.

In a fraction of a second it was all over. The orange beams of the batteries struck the central area of the intruder and the surrounding space shone with an intense glare. Immediately afterwards the torpedoes encountered the shields. The target's overloaded generators could only take out a few of them, and the others penetrated straight through to the hull, causing the immediate and total breakdown of the shields.

Then the cruiser hit the unprotected hull of the Qhrun ship, splitting it into two pieces. A series of explosions propagated through the whole intruder.

"Engine room here. I must reduce power, the generators are badly overheating." Del-Nah was right; with safety controls shut off, the computer was no longer able to protect the ship.

"Al right, Del-Nah. Power back to nominal level. Twenty-six, safety controls on again. Take full control of the ship." Taylor relaxed in his seat and stroked the armrest, his behavior letting the computer feel that everything was all right and that the emergency was over.

They had run the blockade so fast that the remaining intruders had no time to lower their shields and fire.

Taylor thought he knew Qhruns well; he had spent the last few years studying their behavior. He had bet on the fact that they would not believe a human would risk a collision and had won. Now he was sure they would hunt him in blind fury, without wasting time to get into a proper formation, with their shields down so that they could use all available power for propulsion, raising them only when in range of the star cruiser's batteries.

"Twenty-six, display distances and speeds on the screen." Some figures appeared near the silhouettes of the ships. "Now slow down very gradually, so that they don't realize we are doing so. As soon as the nearest ship is in range, lower the shields and fire with all batteries."

The maneuver had some chance of success; if the nearest ship didn't realize that they had slowed down, they would come within range a few seconds earlier than expected, with their shields still down. It might work.

"Sir, if the other gates are blocked, the intruders can intercept us on our way to the third planet within 18 hours." The computer was doing her best, anticipating his questions. Clearly she was feeling guilty.

"Thank you. If they manage to exchange messages, we'll be forced to go back to Qhra'ar and come back with a squadron." Those two ships had to be destroyed before they could send a message.

One Qhrun ship was closing slightly faster than she realized. The chase went on for about ten minutes and finally the computer suddenly lowered the shields, fired with all batteries for ten seconds, then raised the shields again. The intruder, hit with her shields still down, blew up instantly.

The third ship got within range of the batteries and, after lowering her shields, opened fire. The cruiser's shields were raised just in time and the space around them lit up, but the ship was still far off and the volley wasn't very powerful.

"Ready to fire. Aim at the shield generators. If they go on firing like this, they will run out of energy and will be unable to raise their shields immediately." Actually, the fire from the Qhrun ship was getting less intense, a sign that their generators were being overloaded. Taylor waited a few more seconds and then gave the order. A few well-aimed shots destroyed the shield generators at the very instant the intruder stopped firing to raise her defenses again.

The third Qhrun ship exploded into a cloud of fragments. A subdued bang and a dimming of all the lights on the bridge made it clear that a few fragments had hit the cruiser.

"Report on ship's state," Taylor ordered the computer.

"Shield generators out, Sir. Sensors detect a group of twenty-four torpedoes homing in." Just when the danger seemed to be over, the situation was taking a turn for the worse. Luckily it was a long-distance launch and there was time to prepare countermeasures.

Suddenly the screens were crisscrossed with the barrage of light beams shot against the torpedoes by the ship's computer. A sudden flare showed that a torpedo had been hit.

"Well done, Twenty-six, keep it up," said an upbeat Taylor. "Engine room, can you fix the shield generators?"

"Ikelkahan is out working on the generators," Del-Nah answered.

"Twenty-six, give Ikelkahan all the help you can." The computer rotated the ship on her axis to put the generators on the opposite side relative to the flight path of the torpedoes.

A sudden lateral acceleration told everyone that evasive maneuvers had started. The batteries blasted ten more torpedoes one after another, but the power drain was again overloading the generators. Lateral accelerations were becoming weaker and the explosions were getting closer.

"I don't think I can avoid the last two" the computer said.

"Concentrate on one and try to be hit by the other on the central seam of the hull, which should be strong enough to sustain a little damage," Taylor told the machine.

A few last swerves enabled them to avoid one of the weapons, which was blasted by a direct hit from the lateral batteries. The computer attempted a last attitude correction, but the power was not sufficient and the torpedo hit the hull about thirty feet above the seam, right in the area of the engine room. The explosion opened a gap in the armor and the fragmentation charge penetrated the hull, tearing an opening a few meters long.

The crash of the explosion echoed throughout the whole hull, the lights went off and the generators stopped. The bridge filled with thin smoke that made their eyes water. Taylor didn't hear the dreaded sound of air leaking to space, which meant that the airtight doors had sealed off properly and there was no danger of a general depressurization.

"Twenty-six, emergency systems on," said Taylor, standing up and extracting his space suit from under his seat.

The computer didn't answer.

"Manual control. All sections report damage." Taylor was still speaking when the emergency lights went on and the screen came back on, although only partially. The sections reported immediately; apart from the engine room, which was depressurized: the damage didn't seem too bad.

Ikelkahan, outside the hull, confirmed that his umbilical had withstood the shock and he was heading towards the airlock. "The shield generators are partially fixed, but there is no power. From here the engine room looks badly damaged; Del-Nah was inside and I don't see any sign of life," he concluded, panting from his efforts to get inside quickly.

Taylor already had his space suit on. He saw that Brelkahad was suited as well.

The two of them in space suits hurried out of the bridge and along the corridors. In less than a minute they had reached the airlock and in twenty more seconds they were in the engine room. The explosion had wrecked everything inside; fragments of the torpedo's charge were imbedded in all the walls and the emergency lights were so badly damaged that the large room was only dimly lit. The damaged engines were so radioactive that, even with the heavy shielding of their space suits, they could remain there for no more than ten minutes.

"I think Del-Nah is there," said Ikelkahan's voice coming through the earphones. Brelkahad moved quickly towards the hyperspace drive. Taylor followed him and soon he could see Ikelkahan. Following the latter's directions, he spotted Del-Nah's body wedged between the supporting columns of the hyperspace engine, which had protected her from the fragments. He could see no bloodstains. She was motionless and her hands and face had a grayish-purple color instead of the deep blue shade typical of Nahaqols.

"She is breathing and there are no obvious wounds," Ikelkahan said, adjusting his radio so that everybody could hear him.

Immediately Taylor ordered his space suit computer to extract a pressurized stretcher from his backpack and a small object fell to the floor, immediately taking the form of a rigid plank. They slid the plank under Del-Nah's body and covered her with a transparent, radiation-proof sheet, which immediately inflated, maintaining a controlled atmosphere inside the stretcher.

"Be careful," Taylor said to the computer in the stretcher, "She was subjected to very low pressure and strong radiation for several minutes. . ."

The small computer had already begun to collect data to begin emergency care. "Vital functions active, no sign of strong traumatic events and no immediate danger, but she got a heavy dose of radiation," the computer reported.

Only three minutes had passed since the explosion and Del-Nah was essentially safe. Now it was vital to get out of the radioactive environment as quickly as possible. They headed towards the airlock, and once through, they moved Del-Nah to a non-contaminated stretcher and got rid of the space suits. Taylor instructed the others to carry Del-Nah to the sickbay, telling them not leave her until all the functions of the ship's computer were completely restored, and hurried back to the bridge.

As soon as Taylor was back at his station, Ashkahan reported on the operations she had performed, from the launch of a hyperspace probe to summon a rescue ship, duly escorted, from Gorkh'ar base, to setting a new course towards the third planet, taking into account the low propulsion power now available. The ship would be fully operational, albeit at reduced efficiency, within a few hours.

The computer's voice interrupted her. "This is Star Cruiser CH-23426. I have now restored all my functions and remember what happened." The voice was almost normal, with only a few hesitations showing a slightly altered state.

Taylor relaxed. The emergency was over, and if no other dangers cropped up, the mission would almost be on schedule.

Admiral Taylor's cabin was fairly large. He lay on the force field in the dim light, thinking of what happened earlier that day. Only now did he realize the danger they had been in and felt scared.

The door opened and Susan came in. She crossed the cabin and settled down on the force field. "Tom, I haven't felt so scared since the first time I was in a battle, in the Kistl system."

"Yes, we really had a close call. If it's any comfort to you, I was thinking the same," said Tom, turning towards her.

"Better not to trust simulations too much. The analysis showed that the chances of the Sun's system being blockaded in anything less than two weeks were less than one in ten million. It's not your fault, Tom. With that result nobody would have taken a squadron with him. I wonder how we could get it so wrong, though." Susan got up, went to the niche where the shower was located and, shedding her clothes, stepped inside. The door closed immediately and she was immersed in a cloud of warm, pleasant-smelling vapor.

Tom considered her last words: it was true, simulations had ruled out a blockade of the Sun's system. Perhaps they were overestimating the effects of the incursions beyond the frontier, or was it perhaps that something in the mathematical models was no longer working properly?

After a few minutes, Susan emerged from the shower as a light gown, warm as if it had just been ironed, emerged from a horizontal slit in the wall. She slipped it on and laid back on the force field with a general feeling of wellbeing.

"Twenty-six, dim the lights in our cabin, please. Call us seven hours from now," said Taylor.

He couldn't sleep. Susan's breathing told him she had fallen asleep almost immediately, tired from the emotions of the last few hours. He was tired, too, but he couldn't stop thinking about the blockade, its possible consequences, and above all their foolishness.

Then he started thinking about the events of the last few years, starting from the day that had so radically changed his life. His mind drifted back to planet Earth. Five years... they had passed so quickly, in such frantic activity. His life had changed so much they might as well have been five centuries...

He finally fell asleep, still thinking of that sunny October day on planet Earth.

2 Contact

It was early afternoon on a sunny October day. Professor Taylor had just returned to his room after a light lunch in the University cafeteria and was looking forward to a quiet afternoon's work. He had spent the morning with some foreign colleagues eager to know more about his research on interstellar space missions: but even though such visits allowed him to talk about his true scientific interests, he felt they were really a waste of time.

Taylor was better known for the few papers he had written on relativistic space propulsion and interstellar flight than for his serious work, as he used to call it, on the dynamics of artificial satellites. A few days earlier he had received an invitation to present a paper on deep space missions to an international conference and now he had a whole quiet afternoon in front of him in which to begin writing.

He booted his computer and the phone rang. *Typical* he thought, adding a mental curse, as he lifted the receiver. *The minute you start doing something interesting the bloody phone rings. . .*

"Reception speaking. A Mr. Singh claims he has an appointment with you," the bored receptionist said.

He barely stopped himself from cursing out loud. It was true, a Mr. Singh had called a week earlier saying he wanted to speak with him about one of his papers on relativistic propulsion and they had agreed to meet that afternoon. There was nothing he could do. As he left his room, he tried to remember where this guy had said he came from. Probably from some Indian university, he thought, visualizing a Sikh wearing a turban.

A few minutes later he was at the main entrance. The receptionist pointed at a man sitting silently in an armchair. He was a short, thin man, apparently in his forties, with dark skin and a thin, well tended, moustache.

"Professor Taylor? My name is Singh, from the University of Madras," said the visitor, handing him a business card. Taylor noticed he had a foreign accent, but different from that of all the Indians he had ever met. He took the card and slipped it into his pocket. In the few minutes it took to reach his room they made the usual small chit-chat ("Is this your first time in the States?", "When did you arrive?", "How was your trip?" . . .)

"You work at Madras University?" asked Taylor as soon as they were sitting in his room, more to start a conversation that he hoped would be as short as possible rather than out of any real interest.

The other looked around, as if to be sure the door was closed and nobody would overhear them. Then he leaned forward and instead of answering the question started speaking in a low voice.

"I think we had better put aside the usual formalities and go directly to the point of my visit." The foreigner's voice was tense and he spoke in what seemed a low and solemn tone. He assumed a conspiratorial attitude that Taylor found rather funny.

"May I ask you to listen to me even if I say things you find hard to believe?"

Here we go, he's a religious fanatic belonging to some sect or other and is here to convert me to his faith, thought Taylor, regretting the wasted afternoon. *Or, worse, some inventor coming to show me a device to do something in a more efficient or economical way. . . or someone in contact with an extraterrestrial entity who, having heard that I work in astrophysics, comes to convince me of the revelations he has received. Damn it, why did he come here, and today?*

As he couldn't help listening to him at least for some time, he tried to cheer himself up by thinking that at least he would have something to laugh about with Susan that evening.

"Go on, I'm listening," he said, perhaps louder than he meant, trying not to smile.

"My true name is not Singh but Sinqwahan, or, as it should be pronounced, . . ." and here the foreigner uttered a series of sounds Taylor was sure he could never imitate. "And I don't come from Madras, but from a far more distant place."

Now I will ask him how many light years away, thought Taylor, but he said nothing, just to be polite.

"My planet belongs to the Asawak system, about 4500 parsecs from the Sun" the supposed Dr. Singh concluded, staring at him, waiting for his reactions.

Taylor was expecting something of the sort. But those words, pronounced in that matter of fact tone, somehow shocked him. He decided to keep silent.

"Clearly you do not believe me," continued the man after a few seconds.

"I think your English doesn't sound Indian, and your name doesn't seem common to any language I am aware of, but from this to believing that you traveled 4500 parsecs to come and see me, that's quite a jump. And then, why? To meet a professor who, in spite of having studied space propulsion for years, wouldn't be able to send a probe to even one thousandth of a parsec?" Taylor replied.

"Oh, no. I have not travelled 4500 parsecs. The base of the fleet from which I come is in the Gorkh'ar system, the star you call Rigel, about 260 parsecs from here." Sinqwahan was less tense now and spoke in a quiet voice, as if what he was saying was completely obvious.

"But that's unlikely. Rigel is a multiple star, spectral class B8, and I don't think any habitable planets can be in orbit around it." Taylor was sure the man was lying.

"You are wrong, Professor. Yes, the star is triple, and doesn't have habitable planets. The base of the fifth sector's fleet is on an asteroid orbiting the largest component," said Sinqwahan.

His visitor certainly did know some astronomy, but the world is full of well-informed lunatics and impostors. Taylor was intrigued by the game, and started looking for a strategy to make the man contradict himself.

"Fifth sector of what?" he asked abruptly.

"Of the Galactic Confederation," Sinqwahan answered. "In any case you are right: I am here to talk with you, by order of the sector Coordinator himself."

"And what can this high galactic authority possibly want from me? Shall I refer to him as His Excellency? What may he want from a person living on a planet that has neither a Starfleet nor bases on asteroids?" Taylor said without the slightest attempt to conceal the irony in his words.

Sinqwahan did not seem to recognize his sarcastic tone and went on baldly: "We will deal with details later. It is enough to say that we are in a difficult situation and our Coordinator thinks you might be of great help."

"Not to contradict His Excellency, but how could an earthling be of help to people who travel through the galaxy like you? And why should I bother about people living hundreds of light years from our planet?"

Sinqwahan smiled, amused. "When I say we are in trouble, I mean all of us. The situation of the Earth is actually worse than that of many others. Our analysts predict that you probably have three to four years left. Personally, I think I could stay out of trouble for the rest of my life, or almost, anyway. So you see, I am not asking you to help the Confederation, but the Earth."

At this point Taylor decided to trigger the trap he had been preparing for a few minutes. "What's threatening you? Perhaps an Imperial attack, personally led by Lord Vader? Then you don't need a professor, but a Jedi Knight."

Sinqwahan stared. "Do not joke, please. We are not in a *Star Wars* movie."

Taylor was tasting his triumph: the trap had fired successfully. "I am pleased to see that George Lucas is doing good business with the galactic Confederation," he said.

Sinqwahan realized the point of the joke. "I have spent more than two years on this planet to prepare for this meeting," he answered without trying to hide the anger he was starting to feel, "and believe me, I did not enjoy it all that much. You are a scientist, used to making assumptions; would it be too much to ask you to accept what I am telling you as a working assumption? Later you can check whether it is believable or not."

It was a reasonable idea, Taylor thought, suddenly realizing he had been impolite. Slightly ashamed, he felt an urge to apologize. "You must understand that what you're telling me is hard to believe. . . but I will play your game, and pretend to be acting in a science-fiction movie."

"All right," Sinqwahan replied, smiling again, "but please remember that this movie is neither *Star Wars*, whose action takes place '*A long time ago in a galaxy far, far away*', nor *Star Trek*, which is set in a distant future. Unfortunately the action of this movie is going on here and now, and we are all involved in it, willing or unwilling."

"Then it could be something from *The Twilight Zone*," Taylor couldn't help adding. As soon as he had finished the sentence, he realized he had started joking again.

"Well, but it is still afternoon...", replied Sinqwahan, puzzled, without finishing his sentence.

The boomerang had returned. The man was aware of recent movies like *Star Wars* and *Star Trek,* but clearly didn't know *The Twilight Zone.* He wasn't too young to remember it, so either the series wasn't popular in his country, or it wasn't included in the training of 'Confederation' agents. For the first time Taylor felt a chill down his back and his tendency to be sarcastic began to fade. "But why are you here? Did you chose someone at random, perhaps from the telephone directory?" he asked after a few seconds.

"We know quite a lot about you, believe me, and the choice was made with care. We know, for instance, that you are 48, that you have a wife named Susan, who has a degree in astronomy and is a scientific editor for a well-known publisher. You live in a house just five miles from here, you have no children and this has caused you much distress, but that you have now accepted it," he added after a short pause.

Taylor was amazed. The man knew too many things about him, personal things he shouldn't know. If he wasn't what he pretended, then what was he? An agent, perhaps, not working for a hypothetical Galactic Confederation, but for a more substantial and frightening earthly power?

Until a few years ago he would have thought he was a KGB agent... but surely the Russians had other things to worry about now than playing the alien with him. So what? China? Iran? *From Asimov to Fleming*, he couldn't help thinking. *I should find some champagne at the wrong temperature to check whether he is a Bulgarian spy.* But he wasn't enjoying the joke any more.

He didn't know anything of potential interest to a spy, but it wasn't so unlikely that a potentially hostile nation would link relativistic propulsion or the stability of satellites with some military application. At any rate it was more likely than an alien having been sent there from a thousand parsecs by the 'Coordinator' of a galactic sector. *I must get rid of him as soon as possible*, he decided.

"When you called you asked for a short talk about relativistic propulsion," Taylor said. "Now, if you don't mind, I have a lot of work to do," he added, with little hope of actually getting rid of him so easily. In any case, even if he left now, he seemed to know enough to find him any time.

"Please, take a few more minutes and allow me to show you something." Sinqwahan was not even considering leaving. He opened a briefcase he had left on the floor near his chair and took out a small black cube, placing it on the table. *From agent to office gadget salesman*, Taylor thought. He noticed that the upper side of the cube had a slightly concave circular zone in the center, with a semitransparent look.

"An image of the galaxy, please," Sinqwahan said, and immediately a three-dimensional image of the galaxy materialized about eight inches above the cube. The image was some four feet in diameter and dense enough to prevent one from seeing the wall behind it. Taylor grasped the armrests of his chair and tried not to show his astonishment. He moved his head sideways slightly and realized the image changed when seen from different angles—a truly three-dimensional image.

"Show the Confederation as it was 450 years ago" Sinqwahan said again, and immediately a large part of the galaxy—most of the part outside the nucleus—turned red. *The nucleus is not habitable,* Taylor thought and immediately realized that the three-dimensional projector trick was making him believe what the other had said. *I must be more careful, this is just a trick,* he thought.

"Show the fifth sector," said Taylor to check his idea. A 30-degrees-wide sector turned green.

"Very good, you can ask the projector to show any part of the galaxy, changing the point of view or zooming in and out, as with a graphic computer utility," Sinqwahan said.

Taylor tried a number of commands and the projector visualized the galaxy according to his instructions. The results were always correct, at least as far as his limited knowledge of astronomy allowed him to see. But most impressive were the computer's reactions to his voice commands: none of the voice recognition programs that he was aware of could be used on machines small enough to fit into the small black cube on his desk, which obviously also contained a power source able to produce the images.

"There is a computer inside the projector. You cannot pretend that the computers used by your 'Confederation' are programmed to understand commands in English," he said. Surprise made him aggressive.

"Yes, a computer of a sort, though it is a very simple computer; it just controls a tri-dimensional projector and its logic is very simple. We had to train it to understand English commands: a simple thing since all our machines are programmed to understand many different languages."

Sinqwahan went on: "Now I would like to give you a brief account of the history of the Galactic Confederation, at least so that you can understand the present situation." After a short pause, he continued: "As far as it is known, the first human species to discover interstellar flight was that from Terkar, in the first sector, about one hundred and forty five thousand years ago." His voice was flat, as if he was repeating a lesson to a child.

Taylor shoved his chair back, almost toppling it over: the projector had suddenly shown the image of an erect being with long, thin arms and legs, and a pair of stubby wings. The thin body was wearing a sort of light orange jumpsuit and its bare parts were covered in short, gray hair. But the truly shocking thing was its head, similar to a lizard's.

"Off!" said Sinqwahan immediately, and the image disappeared. "Sorry, I am so sorry. We are all so used to it, I didn't think. . . You cannot believe that evolution would result in intelligent species all looking like you and me. When we say 'human species' we mean intelligent species, and I can assure you that all of them are intellectually quite close. If you could talk with that Terqhatl in a dark room you would not be able to distinguish him from a human of your species or mine. My species, incidentally, is Aswaqat."

Sinqwahan paused, then resumed his primary-school teacher tone. "The differences between humans are caused by the environment in which they evolved. You and I belong to a type anthropologists call 2-0-2—two legs, no wings, two arms. Terqhatl are 2-2-2s, but some species are 4-4-6s! At any rate, it is considered impolite to speak of these things or to wonder at the physical aspect of your partner in a conversation. Anyway, Terqhatl are among the most pleasant and quiet fellows in the galaxy, apart from their awkward taste in clothing, as you have seen. Their ancestors were reptiles, as yours and mine were primates."

"You can put that image on again, I just reacted out of surprise." Taylor was angry with himself for his reaction and had the impression it had been a test.

The history lecture went on for about twenty minutes, with the projector showing images of different looking species, of the galaxy divided into kingdoms, colonies, spheres of influence and so on. At last the orderly structure of a Confederation, subdivided in sectors, emerged. "For about eighty thousands years the human species lived in peace, in a Confederation in which each species has a large degree of autonomy, organizing its planetary systems following different traditions and different political and social systems. This doesn't mean, of course, that there were not occasional conflicts and even wars, but on the whole there is order and security for all."

Taylor made a mental note to inquire further into this idyllic picture of the galaxy. If everything was going so well and the Confederation was the best of all worlds, why had this big boss decided to look for help? Taylor realized he was getting more and more involved in the stranger's words.

"Please remember that I'm just taking your story as a working assumption," he reminded Singh, and immediately realized that the sentence, spoken without any apparent reason, was a sign of surrender to his words.

"I do not ask for anything more. When I finish, you will decide whether you believe me or not," Sinqwahan replied with a smile. He was being tolerant.

At the end of the lecture, Taylor asked the question that had been in his mind for several minutes: "If your Confederation has guaranteed peace and prosperity for almost a hundred thousand years, why did your coordinator send you to look for our help?"

Sinqwahan hesitated. More or less consciously, he had played for time, wanting to postpone this explanation. "Well, I was sent here to speak about this but, believe me, it is a difficult task. When I showed you the first tri-dimensional image of the galaxy, I said that was the situation 450 years ago. If I were to show you the present situation... In fact, to be honest, I cannot even give you an updated account of the situation."

His voice turned sad and doubtful. "The fifth sector has been cut off from the rest of the galaxy for quite a long time and, for all I know, the Confederation may not even exist anymore. Analysts swear that things are not yet at that point, but sometimes I have my doubts."

There was a long pause, as Taylor waited in silence, then Sinqwahan continued.

"I have better tell it in order. Everything started in 1574, by your calendar, with a small event. The mail ship that every two months reached Arkarak, a planet in the 14th sector at the extreme edge of the galaxy, carrying minerals and news from the Warad system, didn't arrive on time. After a week a rescue ship was sent out, but it also disappeared. When a couple of star cruisers, sent to investigate, did not come back, the government of Arkarak dispatched the only star cruiser left in the system to the nearest Starfleet base, asking for a squadron of cruisers to assist with the search. The squadron, a dozen cruisers, left Wintel starbase, never to be seen again. The only thing to come back was a damaged automatic hyperspace probe, with its computer out of order. From the analysis of the data still stored in the wreck, the specialists were able to decode the name of the cruiser that sent it and an image of a huge spaceship of an unknown type."

"After the 'Arkarak incident', as historians now call it," Sinqwahan went on, "many other systems in the 14th sector lost contact with the rest of the galaxy and all rescue missions disappeared without trace. The coordinator of the sector assembled all the cruisers still at his disposal and sent them into the last system that was cut off. When they entered normal space they found a small group of ships blocking the gate, similar to the one in the image brought back by the probe in the Arkarak incident. They ran the blockade, but when they approached the inhabited part of the system, they were faced by a huge fleet. Out of the one hundred and twenty thousand cruisers, only two hundred came back, mostly in poor condition.

Since then we have lost system after system, and not only in the 14th sector. We learnt to put ships at each hyperspace gate, to warn nearby systems of the arrival of the Qhruns, as these invaders were eventually named, even to predict with a high degree of accuracy the time of their arrival, but nothing more. Any attempt to stop them turned into disaster and whole fleets were destroyed, to no avail. Today no one attempts to stop them anymore; we simply retreat from one system to the next. We estimate that we lose one more system every one of your days. The fifth sector has been cut off from the rest of the galaxy for 82 years and its size continues to shrink. Now you understand why I said that our problem is also yours: our analysts anticipate that the Solar System will be invaded within three or four years, at the latest," Sinqwahan concluded.

Taylor was speechless. The whole thing was plainly absurd. Nothing like that ever happened in history—any invader had his weak points, it was always possible to negotiate, to come to some agreement in the end, or at least to wait until his power was undermined by the dynamics of history. "In so many years, you must have found a way to communicate with them. There's always a way to reach a deal," he mumbled finally.

"That's exactly the point that makes the situation hopeless. No attempt to communicate has ever succeeded, we know nothing about them, we have no images or other clues. No transmission from their ships has ever been interpreted, the very name Qhrun was given to them by us."

Sinqwahan paused, then in a low voice, pronouncing the words one by one, he said: "The point is they are not human—we don't know *what* they are, but they are not human."

Taylor thought back to what Sinqwahan meant by human and realized the implications of the sentence. Those overgrown winged lizards in orange jumpsuits had been defined as human, like all intelligent beings in the galaxy, but Qhruns were something completely different, something Sinqwahan spoke about with a sense of shock and even awe.

Minutes went by in silence. It was hard for Sinqwahan to speak of these things. Taylor was shaken by what he had just been told and almost unable to think. If everything the supposed extraterrestrial had said was true, the situation was, as he said, hopeless. And if it was all fantasy, what could possibly be its purpose? Why should someone bother to tell him such a tall tale? A part of his mind still nursed the idea that the guy would suddenly start laughing and tell him it was all a joke and he was on Candid Camera. He would have loved a conclusion like that, even at the cost of having played the fool. But there was that three-dimensional projector in front of him, with that bloody voice-controlled computer.

"All right, so you are—or rather, we, are—in trouble," he said out of desperation, just to break a silence that was becoming unbearable. "But why come here? What does your coordinator want from me? I know nothing about space navigation, or human and non-human beings. He needs more soldiers, a better fleet, more powerful weapons. Space specialists, I mean, not an earthling who knows nothing of these things."

"You've said it." Sinqwahan replied. "That is exactly what the coordinator had in mind. Where our admirals and our technology have failed, perhaps an unbiased outsider, with a more open mind, can find a solution or at least suggest something useful. You must understand. . ."

If they are at this point, they must be really desperate. They are coming to me for a sort of cosmic brainstorming session. Everyone sitting around a table saying silly things and someone recording it all, hoping that something useful will come out. And they want me along because they think I will say even more stupid things than the others, thought Taylor.

"Your coordinator has gone nuts out of despair, if he thinks that this may work," he said.

Sinqwahan pretended not to have heard Taylor's opinion about the coordinator. "Consider that after so many millennia of peace our admirals do not have much inclination to fight and the efficiency of our fleets is low. Perhaps an outsider, coming from a less peaceful world, with a more combative mindset, can think or do something useful."

Now Taylor started to understand. "I see. So someone said 'Let's find someone from a barbaric planet, where people still kill each other readily, and see whether he can think of something', isn't that it?" Without realizing it, Taylor had raised his tone of voice.

"Let's say that we need people from a younger civilization, more prepared to fight, less resigned. Perhaps a few thousand years ago things would have been different," Sinqwahan said, vaguely amused by Taylor's reaction.

"But why me? Why not a professional soldier? Or a weapons expert?" Taylor went on, as if he hadn't heard the other's answer.

"Believe me, we were careful in our choice. We examined thousands of candidates. We need a scientist, a person more used to thinking than to action. We do not want to put you in command of a fleet, we just want you to study the situation and bring in new, unbiased ideas. And we need a technically minded person, and one who knows astronomy, like your wife. For us you are the ideal choice, and in addition you do not have that many links on this planet. You can come with us for some time, as long as you think necessary, even for good, should this attempt fail and this planetary system be invaded soon."

"What on Earth do you mean by involving Susan in this? Up to now you were just speaking of me. . .," Taylor spluttered. Sinqwahan's last words had really worried him.

"Would you come alone? Do you think she would miss this opportunity to have a close look at what she has studied for so many years?"

Taylor had not thought about that aspect of the matter. Apart from the immediate danger for everybody, if the story were true he was being given a unique opportunity. He had worked for years on technologies enabling humans to travel in space and now he was offered the chance to reach worlds everyone considered outside the reach of Earthly humankind, at least for many hundreds of years. And of course his visitor was right about Susan: she would not miss an opportunity like that.

"You're right. At this point we ought to go back to my home. Let me call Susan," he said.

He picked up his phone and called her at work. It wasn't easy to convince her that what he had to tell her was so urgent that she should go home immediately. Without explaining anything—and what could he have explained, since he was still unable to make any sense out of what he'd heard?—he mentioned the possibility of a long journey, which was after all not so far from the truth.

After Sinqwahan had put the projector back in his briefcase, they left. Taylor was relieved to see that the corridor was empty; he didn't need to give explanations for leaving so early, but he wasn't in the mood for talking. And above all, how could he introduce his guest? He laughed at the idea of introducing him to some colleague as an agent of the Galactic Confederation, just to see his reaction. He reached his car and couldn't refrain from asking "Is your flying saucer parked at the main gate?"

As usual, Sinqwahan pretended not to get the joke. "Yes, my car is parked outside the gate."

Taylor drove him to his car, a white Taurus with no car rental sticker, and while Sinqwahan was getting out, said "You obviously know where I live."

"Obviously," answered the other. "See you there in twenty minutes."

Taylor drove off. It was clear they knew everything about him. He wondered if it was wise to go home with him, then he realized that it couldn't make things any worse; whatever Sinqwahan was or represented, he could only continue to act out the comedy. He had no choice.

The twenty-minute drive home felt like eternity. He kept on thinking, with mixed feelings, of the supposed alien's words, swinging from sheer terror to moments of excitement. Often he looked in the mirror, but the road behind him was empty. From the moment he left the university Sinqwahan had disappeared. He was on the point of deciding that everything was just a dream when, approaching home, he saw the white car, parked in front of the house with Sinqwahan standing nearby.

He stopped the engine and got out. "Susan will be here in a few minutes," he said leading his guest into the house. Sinqwahan went through the door and headed for the sitting room, as if he knew the place. He opened his briefcase, took out the projector and summoned up an image of the galaxy. "Susan will like this, and it will take us straight into our subject."

While Sinqwahan was talking, Taylor heard the noise of a car and looked through the window to see Susan arriving. He went out, followed by Sinqwahan. When they reached the front garden Susan was already walking towards the door. "What's going on? Your voice on the phone made me feel it was something important and I came straight home," she said.

"I'll tell you everything. First let me introduce Dr. Singh, from Madras University. Actually this is only an alias," he added in a conspiratorial tone, "his true name is Sinqhan, an agent of the Galactic Confederation at the direct orders of a big boss, nothing less than a sector Coordinator."

Susan laughed. "Please, stop kidding me," she said as she shook hands with the stranger.

"I fear your husband is not joking, even if he has not yet learned my name. It's Sinqwahan, not Sinqhan," said the latter in a calm voice.

"Well, if you both want to go on with this silly joke, please come in, Mister Galactic Agent. I hope you'll enjoy the hospitality of an earthly home," she added, laughing. "But I sure hope you didn't interrupt my work just to listen to these childish games," she said entering the house.

"In the sitting room there is something that might be useful for your work," Taylor said.

"Wow, it's beautiful. What's it? It looks like a galaxy. What's it made from?" she cried out as soon as she saw the three-dimensional image.

"It doesn't just look like a galaxy: *it's* a galaxy—or better, a tri-dimensional image of our galaxy generated by a projector," explained Sinqwahan as he asked for a number of changes in the image.

Susan collapsed onto a chair. Her eyes were staring alternately at the image and the stranger.

"You had both better find an explanation for all this, and one without galactic confederations and sector coordinators. Because there is an explanation, isn't there?" she said, with a quiver in her voice.

"Let's all sit down and you start your tale from the beginning. I'll just bring something cold to drink and then I'll listen to your stuff again," Taylor told the stranger.

Sinqwahan started with the lesson he had given Taylor an hour earlier. This time it took longer, however, as Susan frequently asked for explanations, mainly when he touched some controversial or unknown aspect of astronomy or science in general. It was about half past six when Sinqwahan, took a look at his watch and said abruptly, as if making an obvious suggestion: "Why don't we continue this discussion on the star cruiser that has been orbiting your planet since this morning? All your remaining doubts will rapidly disappear."

The two stared at him in disbelief. Their curiosity was strong, because that would call all bluffs. However, wasn't it dangerous to trust the stranger in that way?

"I promise I will bring you back to Earth as soon as you ask; nobody will force you."

Since they didn't react to his suggestion, Sinqwahan pressed on: "Do not ask me why you can trust me: the only realistic answer is that you cannot avoid doing so. If I am telling the truth, you have nothing to fear and you are right to trust me; if on the contrary what I am suggesting is a trap, I would have the means to make you follow me anyway. But if my intention were to force you to follow me, why stage all this comedy in the first place? I would not need high technology to kidnap two people," Sinqwahan concluded with an encouraging smile.

Taylor wasn't at all encouraged by these words, which could be interpreted as a veiled threat. Susan had the same impression, and replied "So yours is an offer we cannot refuse."

"No, let's say it's an offer it would be stupid to refuse," Sinqwahan retorted in his most reassuring tone of voice.

"I think he's right," Taylor concluded. At this point curiosity was overcoming his caution. "I think we have no choice, now. How will you transport us on board? Will you take a transmitter out of your pocket and order someone to beam us up?"

Sinqwahan burst out laughing. "No, that's science fiction. In the real world, if you want to board an orbiting starship you either use a shuttle or you land the ship. In the present case we have better avoid the second solution, otherwise the captain and myself will get into real trouble when someone reads the ship's log. To land on an inhabited planet not belonging to the Confederation is one of the worst breaches of the rules a captain can commit. A small shuttle, disguised as a helicopter, is waiting for us not far from here."

"All right, so to come down with a cruiser is illegal, but to land with a fake helicopter is not. You have strange laws in your world," Susan said.

"No, actually everything we are doing here is strongly illegal," Sinqwahan countered, "but our Coordinator thinks that sometimes it may be necessary to violate the letter of the law. You see, this law is aimed at preventing interference with human species that have not yet achieved interstellar travel. In this case, however, the end is so important..."

"...that it justifies illegal means," Taylor stepped in. "Your coordinator would like Machiavelli. It looks like politicians are the same all over the galaxy."

"We have little choice. It is a desperate attempt, but we have to try. For this operation we used a cruiser that is not a regular unit of the Starfleet, but a ship used for covert operations by security forces under direct orders from the Coordinator. But to land with the cruiser would attract too much attention, here and in other places," Sinqwahan concluded.

Remembering the joke he had made to himself a few hours earlier, Taylor thought there was actually little to choose between Asimov and Fleming—both would do. At any rate he had to agree that, if the situation described by Sinqwahan was true, the Coordinator couldn't be blamed for acting in this way.

"Let's leave out political philosophy. If we must go, let's take that chopper and see what's going on," Susan interrupted him nervously. At that point she was ready for the truth, any truth.

"You are right," Sinqwahan said getting up and moving towards the door. "Let's go and you'll see that I told you the truth and that nobody will force you against your will."

3   On Board

The helicopter was waiting for them, its engine running and its landing lights on, in a field beside the small road where Sinqwahan had left his car. It was an absolutely normal Bell 222 with a civil registration.

Again Taylor started wondering whether the whole story was just a hoax. On their way there Sinqwahan had called the helicopter's pilot, using a very earthly cell phone, asking him to land. He then called someone else who was instructed to retrieve his car. No 3-D projectors or other outlandish gadgets, just plain old earthly gadgets.

At any rate it was too late to reconsider. Ducking under the rotor, they reached the machine and, after getting into their seats, buckled up automatically. Sinqwahan sat down in the front right-hand seat, near the pilot, without fastening his seat belt, and turned back. Pointing at the pilot, Sinqwahan introduced her as "Ashkahan, the cruiser's first mate. Most of the crew is of my species." The pilot turned to shake their hands. Slightly surprised, Taylor saw that she was a woman. His surprise was far greater when he shook her hand and realized it was webbed. He tried to hide his amazement, unsuccessfully, as the other started laughing.

"Yes, our hands are webbed, as you can see. To work on Earth Sinqwahan had to have some skin removed, but when it is all over I think he will have everything restored."

Sinqwahan was ill at ease and Taylor thought he had reddened, though it was too dark to be sure. "Anthropologists say that when our ancestors came down from the trees our planet was all a swamp and they adapted to a semi-aquatic life," he said quickly, as if to change the subject.

"We came close to growing fins," Ashkahan said with a smile, starting takeoff procedures.

"Do you always use such an earthly greeting?" Susan asked Sinqwahan with a suspicious look.

"No, but we always try to behave like the natives on any planet. Do not forget that Ashkahan is a spacer in our Starfleet and she is used to meeting humans of all species," Sinqwahan explained to her.

While they were speaking the helicopter took off in an absolutely conventional way. If the machine's appearance was a disguise, every detail had been taken care of.

Any doubts Taylor still had quickly disappeared during the short journey. Ashkahan looked earthly enough, perhaps somewhat oriental, but her skin was too dark and her hair unusual, although both still fell well within the normal span of human characteristics. He was sure she would go unnoticed in any earthly situation.

They flew through a thin layer of clouds and suddenly they could see a dark sky full of stars. Sinqwahan said hastily "Don't be afraid now. Ashkahan will stop the dummy rotor and switch off the device that is making all this noise."

An unreal silence suddenly fell on them and all they could hear was the soft hiss of air against the helicopter's fuselage. *No, against the hull of the shuttle*, Taylor thought, relieved. He was so excited by the adventure that the thought it might all just be a trick and the vehicle be just a helicopter had disturbed him. That thought had been in the back of his mind up to that very moment.

"So it was true!" exclaimed Susan, who had clearly come to the same conclusion. "What's our altitude?"

"Twenty thousand feet. Now that the rotor has been folded we can climb faster. We will dock with CH-23426 in fifteen minutes" said Ashkahan. "Start docking procedures," she ordered the shuttle's computer. "I have already established the radio link. Docking in 14 minutes, 26 seconds," was the machine's answer.

All the maneuvers were performed automatically. The earthlings stared at the sky and at the stars that were becoming brighter and brighter, trying to see the silhouette of the starship which had to be somewhere in front of them.

"What's the ship's orbit?" Susan asked of nobody in particular.

"Almost circular, perigee 320 kilometers" the shuttle's computer immediately answered, and it took them several seconds to realize whose voice it was.

The minutes seemed to pass more and more slowly, until a small zone of the sky in front of them began to look completely devoid of stars. The dark patch was wedge-shaped and was growing fast. Then suddenly a thin slice of light appeared in the middle and started growing, taking the shape of a bright rectangle. The shuttle aimed towards the light and eventually entered a large bay through an opening in the hull. It stopped moving forward and slowly landed on the floor.

Ashkahan immediately opened the door and gestured for the others to get out.

Taylor jumped to the floor, realizing he could walk normally. Concentrating on his own gait, he decided that local gravity was slightly lower than on Earth, but not by much.

Sinqwahan obviously anticipated his thoughts, for he said "The gravitational field on this ship is kept at the value on Laraki, about ninety percent of that on Earth."

Taylor was about to answer, when a voice came from a loudspeaker: "Welcome on board. I am star cruiser CH-23426. If you want to communicate with me, just speak and I will hear you in any part of the ship. I will try to make your stay on board as comfortable as possible."

They walked along a corridor until, after about twenty yards, they reached a door that opened automatically and entered a small, well lit, chamber. When they were all inside, the door closed and they felt the whole small room moving: it was a sort of elevator or some internal transportation system.

"In a few seconds we will reach the bridge," Sinqwahan said.

The feeling of motion didn't last long and the door opened again, revealing a large circular room. The walls were covered with screens showing the space around the ship, to the point of giving the illusion of being in a glass cylinder floating in space. Long consoles were sited under the screens and six people were sitting on chairs in front of them. A seventh person was sitting in a seat in the center of the room, a large console at her side. *That must be the captain, assuming one exists,* Taylor thought. Ashkahan moved swiftly towards a console under the largest screen and Sinqwahan nodded for them to take two empty seats.

The captain, an Aswaqat woman, brusquely gave her orders to the computer "Twenty-six, leave orbit. Get behind that large satellite." She was speaking very bad English, far worse than that of the other two Aswaqat.

"Order being processed, Madam," the synthesized voice answered. It took Taylor a little while to realize what the order meant—and then he started worrying.

Sinqwahan explained: "We must leave this orbit. We have already recorded some contact by radar on Earth and they are trying to identify us; better for us to hide behind the Moon, just to give us time to speak."

After a short time Sinqwahan added: "As soon as deorbiting is completed, I would like to introduce you to the crew."

Only then did Taylor look at the other people on board. Of the eight humanoids, five were clearly Aswaqat, two were of a different species, apparently shorter, with larger heads and gray skin that Taylor thought was similar to the skin of elephants. The last one would have had an earthly appearance, had it not been for his deep blue color.

"Five are Aswaqats, two are Irkhans, a species from the fifth sector like ourselves, and the other one is a Nahaqol" explained Sinqwahan, who had noticed that the two were staring at the people on the bridge. Two more crew members, an Aswaqat and a Nahaqol, are currently in the engine room," Sinqwahan concluded. The computer immediately displayed another room on one of the screens, where two humans, one with a deep blue skin, were sitting in front of a long console covered with instruments.

"Come, I will introduce you to the captain," said Sinqwahan, standing up and walking towards the center of the room. The earthlings followed him. Taylor, remembering Ashkahan's words on the shuttle, offered his hand but she ostentatiously ignored him.

"Welcome on board," she said dryly, "though you should be aware that I don't agree with this illegal use of a unit of the fleet. As soon as I am back, I will have this ship assigned to other tasks—I don't like these illegal and absurd operations."

"Captain, I am acting under direct orders from the Coordinator and you know the importance he attaches to this mission. Let me remind you that you can ask to be assigned a different task, but this ship is not a regular unit of the fleet. Moreover, what is going on in this mission is to be considered fully classified," Sinqwahan answered.

"Don't remind me of my duties, Sinqwahan. You aren't even a spacer. And just remember that I cannot be requested to follow illegal orders," the captain retorted. The last sentence was pronounced with contempt.

Rats! We've just came on board and they start quarreling, Taylor thought. "We are sorry to cause you problems" he said, "As soon as Sinqwahan has told us what he has to say, we will return to Earth," he said in his most courteous tone of voice.

"People from backward planets always cause problems," she answered. "The sooner you are back, the better for everybody, but I fear this fellow will try to convince you to stay with us for a long time." The captain sat down and subsequently completely ignored them.

Sinqwahan introduced them to the other members of the crew and they found themselves taking a close look at the two human species they had not yet encountered. Taylor realized that, after everything he had seen and heard in the last few hours, the presence of such creatures seemed strangely normal. *If we spend a couple of days here, I shall find it boring to see just earthlings*, he thought.

After they had been introduced to everyone, Sinqwahan glanced at the earthly watch he was wearing and said: "For you it is nine, and you haven't eaten anything yet. Why don't we have some dinner while we talk?" Without waiting for an answer, he got up and headed towards the elevator they had used to reach the bridge, followed by the two earthlings. Clearly he wanted to minimize the impression given by his minor confrontation with the captain and show that he was in full control of the situation.

As they walked to the mess hall, Taylor realized how hungry he was and thought Sinqwahan had a good idea, not only in leaving the captain. This time the way was a little longer and they walked the length of another corridor, leading to a room with a large table in the center and a number of chairs around it.

As they sat down, the computer asked "Madam, Professor, what would you like for dinner?" Clearly there was a single computer providing for all the ship's functions, from space navigation to the preparation of food.

"The kitchen can synthesize food suitable for dozens of species, but I fear it is not yet familiar with Earth cuisine," Sinqwahan said. Having noticed Taylor's reaction to the word synthesize, he added "What I mean is that food is reconstructed, practically atom by atom, and you cannot tell it from the real thing. It is only a matter of supplying the right recipe, at a molecular level."

"Then we must do with a cheeseburger, French fries and a milkshake, like in a fast food place," ventured Susan. In a few seconds a small portion of the table in front of them slid to one side, and a tray with dishes bearing the food requested and cutlery of an earthly type emerged.

Sinqwahan too asked for something to eat in his own language and another tray emerged from the table with kitchenware and cutlery so different from those used on Earth that if they had seen them in another context they would have had a hard time understanding what the objects were meant for.

They started eating in silence, waiting for Sinqwahan to start talking.

"Yes, it's time to talk business" he said after some time. "First of all, our law forbids contact with planets inhabited by an intelligent species that has not yet reached interstellar flight. We call them class B planets. Can you imagine what would happen to a civilization that doesn't know space flight if it came into contact with the Galactic Confederation? At best, it would accept lifestyles from outside and lose its original characteristics, a bad thing since its peculiar contribution to global civilization would be lost. But a far more likely outcome would be the fate of many aboriginal civilizations on your planet after they came into contact with Europeans. One of the basic laws of the Confederation is to avoid premature contact with civilizations on class B planets, leaving them free to develop autonomously. A number of nearby class C and D planets, habitable but without intelligent life or at least terraformable, are also closed to colonization and left available for the future."

"SETI has little chance of success, then," Susan observed.

"I would dare say, none. We are very careful to avoid transmissions towards Earth. But note that the whole thing isn't motivated by altruism. It's just wise policy to plan for civilizations leaving their star system to find nearby planets to colonize, without having problems with other species. Otherwise the delicate equilibrium the Confederation is based on would not last for long. This is the first and most important point," Sinqwahan said with a note of finality.

Taylor had to agree that the explanation was logical and he had no reason to object.

"Let's move on. As I told you, my mission is to recruit you both as consultants for this crisis. You will have a starship at your disposal, to go where you like in the parts of the galaxy that can still be reached, which, unfortunately, are not many. You will be in a crisis committee directly chosen by the Coordinator and will be given all the information you need, even the most confidential. Your only duty will be to study and try to understand the problem and suggest solutions."

The Earthlings silently looked at one another. The prospect was terribly intriguing; in practice they were being offered nothing less than a whole galaxy, or at least what remained of it, to study, with unlimited resources allocated for the work.

"And if we fail?" Taylor asked, rather realistically.

"I do not want to sound melodramatic," Sinqwahan answered, "but that would mean the end of civilization as we know it. In such a case you will be able to chose between going back to Earth, to which only a few years remain, or coming with us, retreating from system to system, for as long as there is one available."

"You have seemed pretty sure from the beginning that we would accept. Why are you so sure? If we refuse, will you force us?" asked Susan.

Sinqwahan bent forward and, lowering his voice, went on "You would be of no use if we took you by force. We want the most complete and voluntary cooperation from you. I will tell you four good reasons for me to be at least confident that we will obtain it."

He paused shortly then started again: "First, there is your curiosity: you are both scientists and I am offering knowledge you have never even dreamt of. A second reason is your sense of duty. Perhaps you do not care much for the galaxy and the millions of inhabited worlds, but you cannot avoid doing your best to save your planet and yourselves."

"What if we don't believe what you told us about the Qhruns? After all, we only have your word for it," Tom interrupted him.

"Yes, but here I have a third ace up my sleeve. Even if you do not believe me now, won't you have doubts in the days, months and years to come? Think about the hell of living with the thought that you may have lost the opportunity to contribute to an attempt to save all of us. How will you feel four years from now? When you will be looking at the sky fearful of seeing Qhrun intruders!"

"And this is what you call not being melodramatic!" exclaimed Taylor, unable, however, to find any weak points in the speech.

"Now let me come to more personal motives," Sinqwahan went on. "I am sure you know many scientists say that human life will be far longer in the future, even in the near future. Think of the advances in medical sciences achieved in the history of the galactic civilization. How old would you say I am?"

"How can I say, if you are not human. . . sorry, if you are not from Earth? If you were an Earthling, I would guess you are forty–forty-five," answered Tom.

"My species wasn't originally any longer-lived than yours. I was born eighty six years ago and, were it not for the Qhruns, I would have a good chance of living at least three more centuries. Before coming here we discussed this matter with our biologists and they feel sure that, though they cannot work miracles with persons of your age, they can give you a good chance of living at least two and a half more centuries, perhaps more."

"You are offering us an elixir of long life!" Susan exclaimed.

"Something like that. A long life for you, if the Qhrun allow it, and an even longer one for your offspring who, being treated from an early age, could have the same life span as ourselves."

Tom stared at him. " Sinqwahan, you are a darned son of a gun, you and that politician of yours, with whom you must have planned that little speech" he said, but his tone was softening.

"You think you've made an offer we cannot refuse, don't you?" Susan added, as if talking to herself.

"No, just an offer which it would be stupid to refuse, but I think I have already said that. I am offering you a life you never dared to dream of and all I ask in return is that you study the problem, so you can help us, but above all yourselves," Sinqwahan concluded, and as he spoke he realized he had won the argument.

4 Departure

Those words were followed by a long silence. Then Sinqwahan got up saying "I have better leave you alone to decide. You have about twenty minutes, before we are back in Earth orbit."

Tom looked at Susan's eyes and she quickly nodded assent. "There is no need to go on playing this comedy, Sinqwahan. You knew from the beginning that it would end this way, but if you want a formal acceptance, here it is. Do we have to sign something or is an oral contract enough?"

Sinqwahan, who had already reached the door, turned back and solemnly shook their hands "This is customary in your country, I believe. I promise you that you will never regret this decision," he said in a solemn tone that seemed slightly funny to Tom.

Sinqwahan sat down again and they started to discuss details. At that point Susan was pressing for leaving straight away and she almost felt she didn't want to go back home.

They decided to go back quickly to collect a few things, and then to leave immediately. "Actually," Sinqwahan concluded, "you don't need to carry anything. We will provide all you need, because the resources for this project are practically unlimited. You should take only what you need from a psychological point of view." This could have meant carrying nothing at all or the whole of their house. Or the whole planet.

The cruiser had already closed in on the planet, and they went to the shuttle bay and got into the helicopter. Ashkahan was again the pilot. The descent was quick and eventless, and they landed at the same point that they had left a few hour earlier.

Sinqwahan drove them home, where they filled two suitcases with more or less randomly chosen things, wrote two letters of resignation, stating that they needed a long holiday, and left. Susan thought that perhaps they ought to write other letters to relatives and friends, but what could they say? That they were engaged in a secret mission at the orders of a Coordinator of the Galactic Confederation to deal with the invasion of potentially non-human aliens?

They loaded everything into the helicopter and in a very short time they were back on board.

As soon as the shuttle had landed on the floor of the bay, Ashkahan headed for the bridge, to start the maneuver to leave the Solar System, saying: "Tomorrow you will start your first hyperspace journey; it's nothing exciting, but the first time, the idea of traveling hundreds of parsecs is always rather dramatic."

Sinqwahan pointed out that the only noteworthy feature of interstellar travel was boredom. "Two or three days of nothing: nothing happens, you see nothing, no feeling of motion. But on this occasion you will not get bored: you will have plenty of things to study." Then he led them to a fairly large, empty cabin and asked the computer to prepare a bed. Suddenly, a rectangular block about two meters long and wide, and half a meter high emerged from the floor. When Tom observed that it looked quite hard, Sinqwahan threw himself on it, remaining suspended about a foot above the upper surface. "Just a repulsion force field; it supports you with a uniform pressure. The feeling is similar to that of a waterbed, but it is more stable," he explained. Then he added that the air over the force field could be kept at any desired temperature, and concluded "If you prefer to use a blanket, the computer will supply one: many human species like to sleep under a blanket, even if it is not needed. Anthropologists say it is due to a need for protection, a need as old as the species itself."

They tried to lie down and found that it was comfortable, even though they realized that it would take a few days to get used to the strange feeling of floating. Sinqwahan moved towards the door to leave and ordered: "Dim the lights and call them after nine hours, at seven ship time," then, turning towards them, asked: "Are nine hours enough? Tomorrow will be a busy day." Then he wished them good night and left.

When Sinqwahan had left, they asked the computer what the Aswaqats used to wear when sleeping. Two jumpsuits made of a light and soft fabric came out a slit in the wall. Susan took one and, out of surprise, let it fall on the floor: it felt chilly and wet. "Why did we have to end up amongst these semi-aquatic beings? We are lucky we don't have to sleep in water!" she exclaimed, half worried and half amused.

They were already thinking that they would have to sleep with what they had on, when the computer understood the situation and supplied them two other jumpsuits, this time warm and dry. "If I gave these to an Aswaqat he would call a maintenance team," the computer stated ironically.

They were so tired that they fell asleep immediately, notwithstanding the computer, the force field, and all the strange things around them.

The voice of the synthesizer woke Tom "Professor, it is seven, ship time. You have slept for nine hours; Sinqwahan is waiting for you for breakfast"

Without really understanding what was going on, Tom woke up. He turned on one side. expecting to take his watch from the bedside table and, realizing at once that nothing was supporting him, let out a shout.

Susan was already up, wearing a pair of trousers and a light soft jacket of a rusty red color, of the same type as all the crew were wearing the previous evening. She looked at Tom and started laughing "Don't worry, you won't fall down: don't you remember the force field? Get up and start your new life as a spacer."

Suddenly the events of the previous day flooded back into his head. "So it was not just a dream." *Dream or nightmare?* he asked himself. At any rate it was all real, the starship, the Aswaqats and the rest. And that rest included the Qhruns, therefore it was more of a nightmare than a dream.

He dressed quickly, and they walked to the door, which opened. "I will show you the way to the mess hall," the computer said.

Sinqwahan was waiting for them, sitting at the table and eating small chunks of food of an indistinct color from a hemispherical cup, using a long pin. He greeted them and then explained immediately that the presence of people from a planet outside the Confederation had to be kept secret. "It will be easy for you to play the role of Aswaqats, even if you will need a few small accessories, like a wig and gloves with webbed fingers, and new names.

What about Tayqhahat and Susakahan? The first sounds fine and, as for the second one, I have an aunt called that. . ."

"The names may sound very Aswaqat, but I fear you overlooked one small detail: sooner of later we will have to open our mouths or to understand what they are saying. . ." Tom interrupted him.

"All settled. You will be two spacers from Kirkyssi, a small planet that was lost 45 years ago. There were no survivors, except yourselves, obviously. Your cruiser was badly damaged but was able get to the closest hyperspace gate. You were then taken aboard a rescue ship, and here you are. There is no problem about making a few corrections in some of the archives and to build a credible past for you. The best thing is that you, people from Kirkyssi, speak in an awful way, to the point that sometimes it is difficult to understand what you are saying. In a week, using a mental projector, you will become perfect in Kirkysakh, you can bet on that." It was clear that Sinqwahan was enjoying plotting this type of intrigue.

"And all with the approval of your friend the Coordinator, I guess" Tom concluded.

"Sure, it's his idea, and Quanslyaq's," Sinqwahan answered.

"Another of those politicians of yours, all disciples of Machiavelli?" Susan asked.

"Oh, no. He is not a politician, he is an Iktlah, a venerable Iktlah of the Tteroth. A person of infinite wisdom. You will meet him soon and you will see," Sinqwahan concluded.

From his solemn tone it was clear that he was talking about a very respected person.

"Right now," Tom said "maybe you should suggest a breakfast for us, Aswaqat style, possibly not too weird."

"You Kirkysakh really like some small crabs living, or rather, that lived, in the lagoons of your planet. You eat them in a seaweed soup."

"Are you sure that we don't eat something better at breakfast?" Tom asked with a grimace, as two bowls full of a greenish liquid came out of the table, just in front of them. Susan picked up the pin that was near the bowl and, after stirring the liquid, speared an indefinable object, pale brown in color.

"You should eat them as they are, shell and all. The soup should be taken directly, sipping it from the bowl, but only after having eaten the crabs, if you do not want to be impolite," Sinqwahan remarked.

The two slowly ate the small crabs and then sipped the soup.

"How was it?" the Aswaqat asked.

"Not too bad," Tom replied, thinking that perhaps sooner or later he would get used to that sort of breakfast.

"I am glad. But don't try to get me to taste that. No Kirkysakh ever succeeded," he answered. Suddenly the two Earthlings realized that not only did the various species have different habits, but also within the same species the inhabitants of the various systems had widely different ways, lifestyles and traditions.

"Before starting your work, I had best show you some parts of the ship you need to become familiar with," said Sinqwahan after they had finished their breakfast.

First of all he led them to where the shuttle 'lifeboats' were launched, and explained the evacuation procedures in detail. Then, going through the engine room, he led them to a large hall, which he called a gymnasium. "Spacers should exercise at least one hour every day, but nowadays nobody follows the rules anymore. These spacers are getting lazy," he explained with a smile.

The next compartment that they entered Sinqwahan called 'the shooting range'. Actually it was a sort of closed corridor, with a three-dimensional projector able to create moving targets. "Another spacer tradition is their skill with handguns," Sinqwahan explained, opening a locker that contained a number of black objects with a striking similarity to earthly handguns. Tom thought that the shape of many objects is so completely dictated by their use that they must be similar throughout the whole galaxy. He noticed that dust had accumulated on the weapons, noting signs of a lack of cleanliness on board for the very first time.

"It looks as if the crew do not do much training here," was his comment.

Sinqwahan took one of the guns. "You are right, many traditions are disappearing. Most spacers would be unable to hit a shuttle from one hundred feet. But the dust on this is caused by the static electricity that remains after firing."

He explained that the weapons were thermal guns, with a useful range of a hundred yards. "They cause intense heating in the impact point and work on any organic material, vaporizing it in less than a second. If you wear one, when it is on it produces a weak force field around you, preventing you from being hit by a similar weapon. But the instant you shoot the field is switched off."

The Aswaqat switched on his gun. Looking carefully, Tom realized that the air around him had become slightly cloudy. The Aswaqat ordered the computer to start and the corridor became a narrow clearing in a forest of trees with an alien appearance. Suddenly a large four-legged animal emerged from the wood, with a huge head and a mouth full of shining teeth. It jumped in their direction. Although they realized that it was just a three-dimensional image, the Earthlings jumped backwards, while the Aswaqat raised the gun, aiming it at the head of the beast. A black spot appeared on the forehead and immediately the head of the animal burst, as the body fluids boiled. The image disappeared and a fraction of a second later a second animal appeared from a different point. Sinqwahan hit this second target as well and then ordered the computer to stop the simulation.

"Very realistic," Tom said, still impressed by the beasts.

"It is possible to have sounds as well, but I like it better without all the noise and special effects—I don't like videogames," Sinqwahan explained. "By the way, the landscape was from a forest of some planet in the eighth sector; I do not remember the name. Now you understand why spacers had to train with handguns, particularly when the exploration of untamed planets was their main task," he said.

Clearly the galaxy was a much wilder place than the civilization of the Confederation, as it had been described by Sinqwahan, had led them to imagine. Tom decided that he had to learn to use those guns as soon as he could find some time.

Finally the Aswaqat led them to a room he called 'the library'. "It is time to start work now. The first thing you should do is to learn how to use a mental projector." He led them to one of the helmets lined up along a wall—Susan thought that they could be mistaken by hair dryers—and explained that the machine produced a mental image corresponding to the initial screen of a program, with various menus that could be entered simply by focusing the attention on one of the items. However he warned them not to exercise with the projector for too long, and instructed the computer to monitor their condition carefully and to stop the session in the event of any unpredicted reactions.

"You should not let this worry you, the point is only that we do not know how Earthlings may react to this device. This is the first time in many thousands of years we have grown-ups starting to use a mental projector. If you cannot focus your mind on anything, or manage to control the machine in your first session, don't worry. It might take time. See you on the bridge," he said as he left.

"He said we must start work and then went away leaving us here with these darned things," Susan said, sitting down and donning a headpiece.

Tom did the same, and instantly the room around him disappeared from sight. In the darkness, a number of sentences written in unknown characters slowly emerged. He tried to visualize their meaning, and with considerable surprise he realized that, while not being able to read them, he was aware of their meaning: it was a sort of a menu. In fact, he was not receiving words or sentences from the projector, but direct concepts. He tried to concentrate on the first one and, after a few seconds, the image disappeared and another one appeared: a sort of a submenu.

He realized he had started to control the machine, when he found a lesson on the Aswaqat language. He entered that and a voice saying a few words formed in his mind. When he tried to repeat them, the projector took control of the muscles of his throat, tongue and lips to help him to pronounce the words correctly.

Tom carried on with his language class for some time and then, feeling a bit tired, went back to the main menu to search for details on space technology, only to realize that he lacked some of the basic mathematical and physical concepts. And where he lacked the right concepts, everything became so fuzzy that he could not understand what was meant, which made him feel uneasy. *There is no way out, I must proceed step by step, starting with the basic concepts.* Realistically, he decided to go back to his language class.

Time soon passed. He realized that his progress was incredibly fast and for the first time he conceded that playing the role of two Aswaqats was not so absurd, after all.

After more than four hours of work with the mental projector, Sinqwahan appeared and suggested taking a break. "Let's go to the sick bay, so that you can have a quick check up and proceed with the anti-degenerative treatment. The earlier we start, the better."

They did that, then went back to the bridge where the screens were full of stars, one of which was brighter than the others. When Tom was told it was the Sun he realized how far from the inner Solar System they were now.

As soon as she saw the Earthlings, the captain said to Tom, without even greeting him: "I see Sinqwahan succeeded in keeping you on board. Rascals like him always get what they look for, particularly when it is against the law." Then she turned towards Sinqwahan "We need to set course to Ytl, isn't that right?"

"Yes, we must follow instructions. But at reduced speed, the journey must last at least six days," Sinqwahan answered, in an authoritative voice.

"As Mr. Coordinator wants. We will proceed with two hyperspace crossings then" the captain said, defiantly. "Twenty-six, set course for Gorkh'ar" she ordered the cruiser.

"Negative, Twenty-six. We must not get close to any Starfleet bases; our mission is covert. Set course for Lakatl. And first of all we must get rid of that shuttle. Twenty-six, let me talk to the computer of Shuttle Number Two." From Sinqwahan's voice it was clear that he wouldn't allow any objections.

"I would remind you that I am in command of this ship, and that you are not qualified to give orders," the captain stated coldly. Then, facing the screen, she gave her orders: "Twenty-six, course for Lakatl. Establish contact with Shuttle Two: Mr. Sinqwahan has something to tell it."

Sinqwahan instructed the shuttle to enter a circular orbit around the Sun, and then to power down so that it would remain undetected, until they needed it again."

One of the screens showed the shuttle moving away, and Tom realized how absurd the presence of that helicopter was there, in the outermost reaches of the Solar System. "Ten seconds to entering hyperspace," the computer said. Tom held his breath, but he felt absolutely nothing, no impression of movement. Ten seconds later all the stars just disappeared from every screen: their long hyperspace crossing had started.

"Disappointing, isn't it?" Sinqwahan asked. "Interstellar journeys are like this: dark screens and boredom." All the spacers got up, one by one, and left the bridge; there was nothing to do for two days but wait. The bridge was empty by the time they stood up.

They spent a few hours wandering through the ship and then Tom insisted on spending some more time at the mental projector. After trying some more Aswaqat food, Kirkyssi style, at dinner, and taking a short walk, exploring new corridors and small compartments whose use they couldn't guess, they retired to their cabin. They were far less tired than the previous evening, even if all those hours at the mental projector had taken their toll.

They laid down on the bed and Tom, after having exchanged a glance with Susan, said with a not too loud a voice "Twenty-six, can you please dim the lights and exclude this room for an hour-and-a-half?"

"Immediately," was the answer. It seemed the computer was not reacting badly to this exclusion. But after a moment she said, with a hint of uncertainty in her voice "I wanted just to warn you to be careful: the repulsion field ends abruptly at the edges of the bed and, if you go beyond it, you will fall on the floor. I say this just because you are not used to it yet," she added with a tone that was even more uncertain.

"Thank you Twenty-six," Susan answered.

Tom tried to call the computer, without any reply. The machine was not listening.

"What a shameless machine. What do you think she understood what we are doing?" Susan asked laughing.

"The insight that these computer have is terrific. Or perhaps terrific insight is not required to predict this," Tom answered, starting to open her jumpsuit.

They made love for the first time on a repulsion field. Initially, it was not a very satisfactory experience, with the feeling of falling down and the fear of moving outside the field. But soon they started to get used to it and to appreciate the freedom of movement given by the way they floated in the air.

When the computer came on again, they were already dressed and lying on the bed, hand in hand. "I am back. I hope it was a new, pleasant experience."

The voice of the computer didn't betray any emotion. *Obviously,* Tom thought, *how could a computer have emotions?* At any rate he was not in the mood to talk about certain subjects with a machine. "It is none of your business, Twenty-six," he answered, trying not to be rude.

"You are wrong, Professor, this is also my business. I am in charge of the psycho-physical wellbeing of everybody on board and. . ." the computer was speaking slowly, as if she was carefully choosing the words. The machine was trying to touch as lightly as possible on the subject. ". . .what you have just done is very important from that point of view."

"I am sorry, Twenty-six. We are fine then, aren't we Susan?"

"You need not worry, Twenty-six; it was beautiful," Susan immediately confirmed.

"I am glad" the computer went on. "But perhaps I can help."

Tom didn't have time to wonder what the machine meant by those words, when the room disappeared, and a reddish light filled the space around them. They were on a rusty-colored beach, only a few feet from an almost motionless sea, sending out reddish sparkles. Looking around, Tom realized that the sand was made of microscopic quartz crystals, reflecting the light of a reddish sun, low on the horizon. Two moons, one the size of the Earth's Moon and the other smaller, where shining in a slowly darkening sky.

Tom realized that, apart form the image, generated by three-dimensional projectors, he could hear the soft noise of the water and many low sounds coming from the forest behind them. He could even smell a sweet scent and feel a soft breeze, slightly cooler than the air in the cabin.

"Sunset at lake Aw, on Yjloq," the computer said, with the tone of a tourist guide. "Do you like it?"

"It's wonderful Twenty-six. Very romantic." Susan said. "Someday we must go to Yjloq."

"Why?" asked Twenty-six. "The image is far better. Here I can modify it just as you wish. I can add objects, people, everything you want. I can create images, sounds, scents of any type."

"Virtual reality, and of the most sophisticated type," Tom said.

"Virtual reality and virtual fiction" the computer added.

"Twenty-six, could you please disconnect this room for one more hour" Susan asked, putting her hand on the lacing of her jumpsuit.

"Well, as you like. Do you want me to leave that image?" the computer asked. "It will remain static, and you cannot interact with it, but it can stay."

"So leave the image. See you then" Tom concluded, realizing he had used a greeting to a machine.

"See you in sixty minutes," the computer replied.

Tom turned towards Susan and got nearer "Don't tell me that for your psycho-physical wellbeing, you. . ." he began, rather amused.

"Don't be silly," Susan interrupted him with a laugh. "I wanted to talk you without that sort of Big Brother—or rather, Big Sister—listening. Nothing better than making her believe that her romantic landscape did its job."

"I am impressed by the fact that all these computers are programmed for the wellbeing and, apparently, the happiness of humans. Too nice to be true. It is impressive how much everybody depends on computers. On the bridge they are all sitting at their posts doing nothing, waiting for a *deus ex machina* to lead them through the galaxy. But then, what are they there for? Why not use automatic ships? You want to study something, and a mental projector teaches you. Anything you want to do, there is a computer to satisfy your desires. Rather like Disneyland, don't you agree?" Susan continued.

"Perhaps it is all in the natural course of things. Even on Earth we totally depend on our machines, even if we don't realize it since we are used to machines of that sort."

"From a certain point of view you are right, but our machines do not think, or at least are not programmed to pretend they think. Is it only a quantitative difference or something more? You are the expert in technology, what do you think?"

Tom was silent, thinking, for a couple of minutes, then went on to say: "Think about a man only one century ago dealing with an automatic lift or a computer-controlled machine. Wouldn't he have the same feeling? Here they talk about machine psychology, but is it so different from the theory of automatic controls, only far more advanced? We always come back to the same question: where do quantitative differences end and qualitative ones start?"

"Perhaps the difference is just in how you feel about them at a subjective level. At any rate their attitude towards human beings remind me Asimov's

Laws of Robotics. Are you able to quote them for me, please? As literally as you can?" Susan was lying on her back, her eyes closed, to concentrate better.

"I do not know whether I can quote them exactly. Let's see. . . First: a robot cannot harm a human or allow, by not acting, a human to be harmed. Second: a robot must obey orders from humans, except when they are in conflict with the First Law. Third: a robot must protect its own existence, except when it is in conflict with the first two laws."

"They seem very reasonable: something like that must apply to any form of artificial intelligence, even to our Big Sister," Susan decided.

"Don't forget that this cruiser is a military unit and that Sinqwahan spoke about wars. Clearly in war, the Second Law must override the First and the whole theoretical construction collapses." He paused and then went on. "Wait a minute. Asimov himself formulated a sort of Zero Law, in the sense that it comes before the First. It is something like this: a robot cannot harm humanity or allow, through inactivity, humanity to be harmed. Obviously this law has the highest priority, but is very difficult to apply, since it may be unclear what is good for humanity.

"That's the point," Susan exclaimed. "Just imagine that one of these Big Sisters decided that humanity needs regeneration. Then they decided to stage a powerful enemy, separating larger and larger portions of humanity from the others. The whole thing for the good of humanity itself."

"You imply that these Qhruns don't exist, they are only figments of computers?"

"No, I don't mean it is actually like that. I mean it might be. After all, who has seen these Qhruns? Who has succeeded in contacting them? Nobody has ever established contact with occupied systems, nobody has proof that people living there died. Is the whole thing virtual reality, like this wonderful landscape surrounding us?"

Tom remained silent, meditating on what she had said. "In that case the Coordinator would be right: they are so used to depend on their computers that they would never suspect something like this. Perhaps it is a possibility, but I think it is unwise to become sold on any theory before we have studied the situation in more detail."

"Perhaps we should get into the habit of excluding the computer periodically to discuss such things. She will be happy, thinking of our psycho-physical wellbeing," Susan concluded.

"Well, I feel we should not disappoint her, with all the care she has put into this three-dimensional image. . ." Tom said, snuggling up close to her.

On the third day they re-entered three-dimensional space in the Lakatl system and this maneuver was as little eventful as their entry to hyperspace had been. No warship was detected; just a pair of freighters on their way between

Lakatl and two mining bases on small asteroids, and a large cargo ship leaving the system for an unknown destination.

They entered hyperspace again near Satq, another system without any significant features. At re-entry into three-dimensional space the sensors detected a star cruiser at about twenty-eight milliparsecs and the two ships exchanged identification signals, but Sinqwahan ordered that no specific message regarding their destination, nor the presence of passengers should be broadcast. Then they immediately started the procedure for the third crossing, finally, towards Ytl.

After more than six days their journey ended in the Ytl system.

5   The Iktlah

The yellow star in the Ytl system had just appeared on the screens, when the computer announced an incoming message for Sinqwahan. "Switch the communication to the small screen," he said.

The main screen was showing three star cruisers standing close to the hyperspace gate and a smaller screen soon came on, showing the bridge of a starship. The captain was a creature with a large head, practically without a nose and a short beak instead of a mouth. He was clearly a Tteroth, as the Aswaqat had called the people living in the Ytl system.

"Captain Ertlaq presents the warmest welcome from the venerable Iktlah Quanslyaq to the honorable Sinqwahan and to his distinguished guests Tayqhahat and Susakahan from Kirkyssi," the captain of the cruiser uttered slowly. "May the air sustain your wings," he ended.

Tom looked more closely at what at a first glance he had taken for a hump on the back of the Tteroth, and realized that a pair of wings were folded behind his body. *Rats! That must be a 2-2-2,* he thought, almost expecting to see him unfold his wings and flit across the bridge.

Sinqwahan bowed deeply and with a solemn voice answered "Sinqwahan from Asawak greets the famous Captain Ertlaq and thanks the venerable Quanslyaq for his welcome. May the air be favorable to you." He turned towards the Earthlings and nodded to them to answer the greeting. First Tom and then Susan said a few words, trying to imitate the solemn and formal tone of the Aswaqat. However Tom omitted the last part fearing he might start laughing. *Better to be impolite than make things worse by laughing in front of a famous leader,* he thought, and immediately realized how provincial that thought was. *Any greetings, by any human species, are perhaps quite funny, when taken out of context.*

"The venerable Quanslyaq sends me to escort you to Ytl. We will follow a trajectory that will take us to Ytlatq early in the morning tomorrow." The four ships moved in close formation towards Ytl, with two of the cruisers flanking their starship while the third preceded them.

Sinqwahan nodded to the two Earthlings to follow him, and headed towards the library. "It is a great honor to find an escort at a hyperspace gate, particularly because Tteroths are extremely formal," he said as soon as they were in the corridor. Then he explained that Tteroths were one of the youngest human species in the sector, because they had attained starflight no more than twenty thousand years earlier, and that in the past they had a highly stratified society, ruled by a close-knit aristocracy.

"I believe that this is the reason why they are so formal. But do not misjudge them: they are usually far more open and intelligent than you might imagine. You will realize this as soon as we reach the surface of their planet," he ended.

Once in the library, Tom could at last see the outside of a star cruiser: the Tteroth ship to starboard was a few hundred yards away and filled the whole window. They remained there fascinated, even though the light was not enough to see all the details clearly.

When Tom moved away from the window he realized that he needed to go back to the mental projector to learn more about Tteroths. "Let's start studying; I don't want to give a poor impression to the venerable Quanslyaq. I think we will undergo what amounts to an examination."

"Don't make fun of him for having that title," Susan answered, "from the way everybody speaks about him, he must be a remarkable figure."

"He may well be. Perhaps it is simply that I have developed an allergy to high-sounding titles," Tom said, donning the projector's headpiece.

Their session with the projector gathering information about Tteroths lasted more than four hours; then they spent the rest of the day in the usual way. The following morning they finally got ready for their first visit to an alien planet, something that had them very excited.

The cruiser soon entered a circular orbit around the planet, and Ertlaq instructed their shuttle to land at Ytlatq, wishing them auspicious entry to the planet, in his usual formal and solemn style. The shuttle quickly emerged into space and started losing altitude, heading towards the surface.

When the shuttle entered the thick layer of dark clouds covering the surface under them, it was caught by the strong turbulence and started bouncing around. The two Earthlings were not greatly affected by this motion, which was no worse than the turbulence that often occurs when flying in aircraft, but the two Aswaqat were uneasy. "It is not only me, who am no traveler at all, but even spacers are not used to this," Sinqwahan said after a short time, adding that he had not been at all happy on the few occasions he had to use an aircraft on Earth.

Eventually they broke out of the clouds and could see the surface of the planet. They were flying over a dark gray, rough sea. It was early morning, but the light was dim and the windows were covered by drops of rain. In a few minutes they came in sight of a tall, gray cliff.

After reaching the cliff the shuttle landed on a clearing on top of it, where a few other shuttles and atmospheric winged aircraft were parked. A small group of Tteroths led the Earthlings and the Aswaqats to a nearby vehicle, after a complex welcome ritual. Despite the dim light due to the clouds and the thin rain, Tom was able to observe them from a close distance, confirming his impression that they were facing large birds.

From the documents he had studied, it was clear that Tteroths had kept their ability to fly throughout the evolutionary stages that had led to intelligence, and Tom was eager to see how large the wings would be, that would be capable sustaining such heavy creatures.

The vehicle floated at a few inches from the ground and started moving across the flat, well-kept lawn without making any noise as they proceeded along the top of the cliff. The edge was however not visible, because there was an interrupted line of low buildings between the vehicle and the cliff. "The dwellings are mostly below the surface and receive light from windows in the cliff," Sinqwahan explained. "Tteroths like to live as close to the edge as possible." Tom wondered whether this preference could be a result of an ancestral memory of nests on ledges of cliffs.

"The venerable Quanslyaq lives in what was formerly the Ytlatq royal palace. We will be there in a few minutes," one of the Tteroths explained.

Tom was unable to refrain any longer from asking him whether they still used their wings. "Flying is tiring; let youngsters use their wings, I prefer driving!" he stated, with a chuckle.

The vehicle entered a narrow tunnel sloping downwards and stopped at the entrance to a corridor leading in the direction of the cliff. The Earthlings and the two Aswaqats, accompanied by a Tteroth, walked through it to a huge hall, well lit by large windows directly opening on the cliff. Tom cautiously went over to one of them and looked down: he could see the white foam of the waves breaking against the rocks several hundred meters below.

At last they reached a door, leading into another large hall and Sinqwahan nodded at Tom and then Susan to go in. The room was almost devoid of furniture and along one of the walls Tom could see some Tteroths—about fifteen, he guessed—in a line. They were crouching on the floor, their arms down alongside their bodies, with their half-opened wings wrapping them like a cloak. Against the opposite wall there was a couch, on which another Tteroth was lying. The venerable Quanslyaq, Tom guessed.

Tom went out to the center of the hall and then crouched down, keeping his arms down, along his body. *I cannot wrap my wings around my body,* he thought, remembering how, as he learned the day before, one must behave in front of a venerable Iktlah.

"Tayqhahat from Kirkyssi humbly salutes the venerable Quanslyaq. May the air sustain your wings," he uttered slowly. He bowed deeply and after a few moments said again "May the air be propitious to the venerable Iktlah and to his disciples," he concluded and remained there, without moving.

The Venerable slightly bowed his head and answered "Welcome, Tayqhahat. Welcome you too, Susakahan, and you Aswaqats. May the wind of the cliff sustain you." Then he paused and added smiling "Although I do not see how it possibly could, anyway." He paused again and then added "I see that you have studied our ways. It is from the minor details that you can see a thoughtful man, and this makes me think that Aintlhad and myself were right when we decided to call on you."

Tom thought he had to answer with the modesty a young man was expected to show when talking to a Venerable Iktlah. "Your praise embarrasses me. I do not dare to contradict your words, but I cannot either agree with what you have said about me, humble servant of your greatness." Even as he was speaking he realized that he had gone rather too far.

Quanslyaq burst into a laugh. "You speak like the royal governess in *The Flight of Princess Hethl.* Come here, so we can talk more easily." The Iktlah was right: among the texts Tom had studied with the mental projector, one was *The Flight of Princess Hethl*: a long historical novel, dating from pre-space times, in which Princess Hethl, imprisoned with her father the King, by a hideous usurper, plunged from the cliff and, after a flight lasting several days, reached a group of followers who, after hundreds and hundreds of pages, succeeded in restoring dynastic rights.

"Well, the royal governess was not an unpleasant character," answered Tom, who wanted to show the Venerable that he knew at least something about classic Tteroth literature.

"No, not unpleasant," the Iktlah concluded, "but so boring. . ." He gestured with one hand towards the silent figures crouched in a row along the wall and the fifteen disciples silently got up and, in a slow procession, headed towards the door. "Ruklyaq, you can stay. Come here," he added, and one of the figures left the procession and crouched near the couch.

When all the disciples were out of the room, Quanslyaq, talking directly to the Earthlings, said "Tom, Susan, please sit down in a comfortable way: to crouch in our way is surely painful for you. And call me just Quanslyaq, leaving out the 'venerable' and the royal governess manners. You see, we Tteroth are somewhat provincial, because we seldom leave our planet. But I have been a spacer, I saw more planets and strange humans than Sinqwahan, whose only longing is to go back to his home on Laraki, and also more than these inner-planet spacers who nowadays man Starfleet ships."

He was talking in a friendly and amused tone, but Ashkahan reacted with a cold glance to that 'inner-planet spacers', which she felt was a gratuitous insult.

Quanslyaq realized it and spoke again, this time directly to her "My dear Ashkahan, don't get upset; I didn't mean you are a bad spacer. If you were chosen for this mission you are undoubtedly up to your task. I just mean that nowadays the Starfleet is no longer as it was in the old days, not even as it was when I was serving on a star cruiser."

He looked round at everyone and read the disbelief in their eyes. "Don't think that these words are the usual sort of complaint from an old man. I do miss the times I stretched my wings in the wind of the cliff and let the up-draft carry my body for hours. But I don't miss only that. What I mean has nothing to do with my past. The point is that for centuries our civilization has become too static: the number of the newly colonized planets has been decreasing steadily from century to century, and the number of new human species bringing new ideas and energy into the Confederation has been in sharp decline."

"Then the Qhrun crisis started. I remember when a messenger arrived here with news of the Arkarak incident: I was just a boy and I was not very impressed. And even years later, many came to the conclusion that the Qhrun were a blessing from God: a healthy shock to revive a civilization that had become too static. Well, if the Qhruns are a remedy, someone messed up the dosage and the patient is dying."

He paused, and then went on with a lower voice "But old I am without doubt. I am the oldest person on this planet: in Earthly years my age is four hundred and seventy-two, and I should have gone to meet my ancestors several decades ago. Here the saying is that after spending all my life studying the Qhruns and trying to understand how it is possible for us to survive this crisis, I cannot leave this world before seeing at least a hint of a solution. But this is just superstition; one of these day Ytl will lose its oldest dweller, Qhruns notwithstanding."

Tom took advantage of the old man's pause to say "You must have come up with a number of explanations in all those years. Perhaps you should tell us your ideas and give us a starting point for our work."

"No, that is exactly what I must avoid. I can tell you all my life, the events I have witnessed, but not my theories. If I did this I would influence your reasoning and all the work of calling you from a planet outside the

Confederation would turn out to be useless. You are right in that we have made thousands of theories. Simple ones, complicated ones and even absurd ones. Believe me, we even thought that Qhruns do not exist, that they are just a delusion produced by our computers to revive our civilization. It is easy to say such things here, in the quiet atmosphere of this palace. But when you are out there, on a star cruiser under fire from the batteries of the intruders. . . No, that's not some virtual reality."

When he paused, Tom said: "I must say that Susan had exactly the same idea. We decided not to mention it to anybody, but now, after you have said the same thing. . . Can you please tell us why you think it cannot be true?"

"So you liked that explanation too, did you? It's a soothing idea, and above all it agrees with our views of the universe: the universe is essentially good and in such a universe evil cannot prevail. For millennia, philosophers have discussed the problem of evil, of how divine goodness can allow it to exist. It is a problem that has been present in all cultures from before their space ages: it's a problem all human species have encountered individually. Many Iktlahs thought that this is not a problem but The Problem. But no civilization believes that evil will prevail in the end. And Qhruns are exactly that: just destruction, an evil that prevails over everything and everybody. So it is nice to create a theory in which they don't exist, and are just a delusion created for our own good by something we have built to serve and protect us."

"There could be however a similar, but less acceptable, explanation," Susan interrupted. "What if the Qhruns are a delusion created by the computers, perhaps with the best intentions, but which has then spun out of control? You say that when you are under the fire of their batteries you realize that it cannot be just virtual reality. But cannot the computer simulate the attack, perhaps to the point of destroying the ship and itself, if it thinks that such an action serves some overall good? Or, worse, because someone ordered it?"

"No, Susan. I don't think this is possible. I am no computer scientist, but I have dealt with computers all my life, and I assure you that star cruisers have a strong instinct for self-preservation. I have been a starship captain, and I know only too well that sometimes they behave like cowards. But no more arguments on this subject: if you think that it is worthwhile your working on this theory, do it. Isn't it true that sometimes when working on wrong theories you come across useful discoveries?"

"But above all, we asked you to come here because it seems that we are unable to deal with this situation, so it is up to you to decide the direction you need to take. I would just warn you not to get fond of some arbitrary explanation."

"For what it may be worth, I would like to tell you about my most dramatic encounter with Qhruns," Quanslyaq went on. "It was about 250 years ago. I was no longer young, but I was not the old cripple you see now. I was the captain of a star cruiser and Aintlhad was my first mate. Our friendship had started several years before, when he was a young officer at Qhra'ar base and I was teaching space propulsion to rookies like him. At that time Qhruns had just entered the sector and the danger seemed to be a distant one. We convinced an admiral to give us a squadron of a dozen cruisers and started wandering through the galaxy, looking for contact."

"We searched for years, from system to system, from one base to another. We were looking for anybody who had had contact with Qhruns. We questioned hundreds of automatic probes, and when I say question a probe I mean dismantling all the memory cards, piece by piece and then have them decoded by a team of computer scientists. With humans you cannot do that, but you can get close: we had with us a team of psychologists who squeezed out even the deepest memories and those removed from the conscious mind."

"My worst experience occurred about thirty years later. Still looking for a contact, we entered the system of Klatq, a system with no habitable planets because its star is triple. As soon as we were out of hyperspace, and while we were still admiring the stars, the sensors warned us that a group of three ships was at 50 milliparsecs."

Quanslyaq paused, then went on: "You should have seen the sight. . . I was staring out in awe; there is nothing in the universe more striking than a multiple star seen from a close distance. The primary was an enormous red giant. The second component was yellow, a standard G0 class star; if the other two stars had not been there, I bet it would have had habitable planets. The smallest one was a white dwarf, but it was far closer to the point where we entered the system so it was almost as bright as the other two. The red and yellow components were so close that there was a plasma bridge joining them. The plasma was almost as bright as the stars themselves, and everything was continuously moving, with flares and streaks of fire all around. . ." Quanslyaq stopped abruptly.

"Sorry, I let description of the view make me lose the thread. What was I saying? You see what it is to get old! Well. . . I decided to send messages and to cautiously close in. The Qhrun ships dashed towards us at top speed, with no attempt to answer our messages, and when they were almost in range, I realized that the data from the sensors were wrong."

"We had nine ships in front of us, not three. When the computer took the readings they were moored alongside one another so that the computer saw just three masses. When I say that Starfleet is no longer the Starfleet of old: the rules state that the mass of each object detected by sensors should be evaluated, but the computer didn't check. Three objects were seen and it assumed that there were three ships."

"We tried to regain the gate, but they were on us in seconds. When they neutralized our shields, I gave the order for everyone to don space suits.

I had not completely closed my helmet when I was thrown against a bulkhead and the last thing I remember was our engines blowing up."

"I regained consciousness a few hours later. I was in space, in the middle of a sea of fragments. Aintlhad was floating close to me; he was holding my arm and pushing me slowly towards a larger fragment of a starship. I tried to talk to him but he gestured to me he had switched off my transmitter. When he pointed to one of the Qhrun ships that was slowly drifting through the wreckage, using its batteries to destroy anything that moved, I understood we had to avoid betraying our presence. We got inside what looked like a large fragment of a storeroom and waited there in hiding. One of my wings was broken in several places; actually since then I have not been able to really fly, just glide from the top of the cliff."

"After two days the Qhrun ships went away, one by one, through the hyperspace gate. Aintlhad had found a small thruster and was using it to move around. He found a few air bottles, a container full of water, and some food. I was motionless, attached to an eyebolt protruding from the wreckage. After three days Aintlhad arrived, exultant, with a hyperspace probe. It was badly battered, but he was able to fix it and gave it instructions by operating directly on its computer. We tied ourselves to the probe and he launched it towards the hyperspace gate."

By now the Iktlah was talking slowly and painfully, and seemed even older than his actual age. The memories were still making a deep impression on him. "It was the worst nightmare of my life. First, there was the acceleration, without being protected by the artificial gravitational field of a starship. Then hyperspace itself. You crossed it in a star cruiser: it's boring, and that's all. But unprotected outside, it's a different thing. You are suspended in nothingness: not a light, nothing except yourself. You realize that your body still exists only because you feel it, and, believe me, I could feel my broken wing. But apart from that, only your mind still exists for you. You are not in contact with anything. I knew that Aintlhad was there, close to me. But what does 'close' mean in hyperspace? Even the topology is different, all your normal concepts of three-dimensional space do not hold anymore. Physicists can write equations to describe it, which nobody understands. I mean, nobody can give a truly meaningful physical explanation in words. The only thing I knew was that those equations work: they have been understood at a physical and mathematical level for thousands of generations; that probe was leading us somewhere."

"Before entering hyperspace Aintlhad had programmed the probe to reach Istlot, a two-days' journey. To me it seemed like eternity: two days of nothing.

Then suddenly the stars appeared around us. and the universe, the probe, Aintlhad and everything were there again. Fifty-two hours later, an automatic cargo ship, homing on messages from the probe, picked us up."

The Iktlah remained silent for a few seconds, then he started again, in a somewhat stronger voice, as if emerging from his nightmare. "You now know what links me and Aintlhad. When our story became known, we were considered heroes. He became a hero, I mean. Nobody had ever traveled through hyperspace outside a starship, riding a probe, nobody had ever survived such an experience. And I was already quite well known. I was a young Iktlah then, perhaps the youngest, but I was known in the whole sector for my stubborn search for knowledge about the Qhruns. Nowadays when the only stories speak of defeat and of systems abandoned one after the other, without even fighting, for doing far less than what he had done, you become more famous than Admiral Vertearis, the man who defeated the Xartians."

Quanslyaq paused once more and then concluded, with a smile in his voice "I think that the political career of Aintlhad and his Coordinatorship has a lot to do with the stupid computer that put us in that predicament." The stress caused by his story was now vanishing and the Iktlah didn't look so old.

After a long silence, perhaps more to hear a voice than to obtain any useful answer, Tom said: "What do you want from us then? You tried everything that could be attempted; you saw them as closely as a human can do; you studied them for centuries. . . What can two Earthlings do in such a situation?"

Quanslyaq looked in their eyes, first Tom's and then Susan's. "Call it insight, call it divine inspiration, call it a delusion of a senile old man. I think that there should be a different approach to this problem, perhaps simpler and more indirect. Don't let anything we might tell you influence your search. Don't try to think like us; remain two Earthlings, under your Aswaqat disguise." Here he gave an amused look at Sinqwahan. "You will have all my support, for all that my small moral power can be of use, and above all the support of Aintlhad. After swimming through hyperspace, he became the most respected coordinator the sector has had for many centuries."

He turned towards the crouched figure who had remained motionless and silent for the whole time. "Ruklyaq will come with you. God knows I would like to come, but for years I have not left this planet. I rarely leave this hall, now. Ruklyaq is a clever young man, eventually he will be an Iktlah. He has the quickness of mind and the humility of heart to understand. I am sure he will be of great help, he and all those others that Aintlhad will appoint to the crisis committee."

Ruklyaq moved slowly, bowing his head. "I will do my best not to disappoint you, Iktlah," he said and then resumed his silence.

Quanslyaq spoke again. "I must warn you. You must be careful, very careful. Obviously when you try to contact the Qhruns. You must not trust the experience of anybody, human or machine. Everything must be checked and checked again. If only I had ordered that computer to evaluate the mass of those ships... I would have saved my squadron and avoided a nightmare for myself, but perhaps Aintlhad would not be coordinator now and you would not be here. Who knows, the infinite wisdom of God can use an idiotic and lazy computer."

"But the danger does not come only from the Qhruns. These are hard times, when humans are put to test. And they often react in unpredictable ways. Expect anything from all of us, from the highest heroism to the worst treason. Don't even trust yourselves; you don't know how you may react under extreme conditions. You must have the greatest intellectual courage, pursuing bold ideas, but be very cautious when taking any action. Before doing anything, simulate your actions on computers over and over again. Include in your simulations possible treason and your own weakness."

The voice of the Iktlah was very low now, just a whisper. The old man paused and then concluded, with a calmer voice "But now go, and don't lose hope. It's impossible that God created the various forms of humankind in this galaxy only to have them end this way." Quanslyaq was tired now, and was barely able to take the hands of the Earthlings in his hands, concluding: "May God bless you and be with you in this mission." Then, with a lighter tone, added "May air sustain your wings, distinguished Tayqhahat and Susakahan from Kirkyssi."

He dismissed Ruklyaq in a similar way, who was nervous at having to leave his home planet—apparently for the first time—and then, rather more quickly, the two Aswaqats.

They got up one after the other, and left, followed by Ruklyaq. The way back to the orbiting starship was quick and without problems. The weather was now better and the sea, still rough, was shining in the light of the yellow star. The shuttle moved quickly upwards and in just a few minutes Ytl became nothing more than a gray planet, with blue shading.

"Now we will set course for Laraki, where the Coordinator is waiting for us, and looking forward to meet you," Sinqwahan said jumping out onto the floor of the star cruiser's shuttle bay.

The captain was nervous, and from the few sentences that she spoke it was clear that she didn't like Tteroths, and Quanslyaq in particular. "Those Iktlah always speak as if they were inspired by their God, just to cover the stupid things that come out of their bird-brains." It was clear that Sinqwahan was upset by that barbed allusion to the physical aspect of a particular human species.

That evening, after retiring to their cabin, Tom called the computer and asked it to synthesize a piece of soap. Susan was amazed by the request, but a quick glance from Tom convinced her that she had best say nothing.

"Of course," the computer answered. "Do you want a particular color or scent?"

Susan was even more amazed to hear him answer, in a studied, casual tone "As for the color, I like dark soap. . . something like brown or black. Yes, black is best, if possible. And for the scent, no scent at all will do."

Four small black rectangular blocks came out from the usual slit in the wall. Tom took them and hid them in one of the suitcases. When Susan asked him why he made that weird request, Tom just answered quickly: "Just little Earthly habits," and never touched upon the subject again.

The following day the cruiser entered hyperspace, setting course towards the Satq system, which they had already visited on their way to Ytl.

6 Treason

After reaching Satq they performed the usual routine checks and quickly exchanged data with the automatic station located at the hyperspace gate. Then the captain suggested setting course for Kistl, a little-inhabited system on the route between Satq and Laraki.

Sinqwahan insisted that they ought to stop in the Urt system, but the captain, with uncharacteristic kindness, pointed out that two inhabited planets orbited Urt's star, and so they were likely to encounter units of Starfleet who might ask questions about their flight plan.

Sinqwahan had to agree that she was right and that Kistl was a better choice. "Twenty-six, set course towards Kistl," she ordered, ending the discussion with a triumphant note to her voice. Tom thought that such an attitude was clearly disproportionate, relative to the importance of such a decision. *We need to understand her; Sinqwahan made her look a fool in front of the crew and now that she won something this reaction is the result,* he thought, trying to control the uneasiness he felt growing in the back of his mind.

It was the longest hyperspace crossing the Earthlings had ever experienced: six long days without re-entering three-dimensional space.

Time was passing slowly and Tom was using the mental projector for longer and longer periods. He felt distressed, anxious to learn, and felt that any time he spent away from the projector was irremediably lost. It seemed as if everything might depend on some information that was available, buried somewhere in the machine's memory, but which he had yet to learn. He felt restless, and was relieved only when he was in the shooting range training—Susan said playing—with the thermal guns.

He discussed this feeling several times with Susan, but she tried to appease him by observing that it was a natural consequence of Quanslyaq's words. Ruklyaq was also uneasy, but he ascribed that to the fact that he had never been on a starship before.

Everybody was already on the bridge when the computer warned them, in accordance with the rules, that they would re-enter three-dimensional space in ten minutes. The captain also seemed nervous. It was as if everybody was waiting for something unexpected to occur.

As the stars re-appeared on the screen all the alarms went off. *That's it,* Tom thought, almost relieved that at last the wait was over. *Now we will see what the hell is going on.*

"Maximum alert. Five unidentified star cruisers straight ahead, at close range," the computer said, with a note of surprise. A cruiser, not dissimilar from the one on which they were traveling, filled the forward screen.

At least they are not Qhruns, Tom thought, relieved. The captain rose quickly and started to move towards the door leading to the corridor. She looked nervous, but not at all surprised by what was happening. Glancing to one side, Tom could see that Ulkahan had also got up and was slowly heading towards the door.

Everybody else was still sitting at their posts, looking at the screens, now showing all the unidentified starships.

Tom got up and, trying to be inconspicuous, sneaked towards the door.

"Captain, I have succeeded in identifying one of those ships. . ." the voice of the computer started to say.

"Shut up, Twenty-six," the captain interrupted. Then, turning her head to one side, she saw Tom, who was already midway between his station and the door. "Get back to your seat immediately, Earthling. Everyone remain at their stations," she shouted in an excited voice.

Instead of obeying, Tom rushed to the door and operated the manual locking device. Then he leaned his back against the closed door and remained there motionless, without knowing exactly why he had acted in that way.

"Twenty-six, get a shuttle ready and open that door." The captain, followed by Ulkahan, was now close to the door. Suddenly Tom understood the situation and realized he had to remain where he was.

The captain stopped three feet from him. "Out of there, Earthling," she shouted. As he didn't budge, she shouted again, in a louder voice "Out of the way, that is an order."

She was becoming more and more excited and Tom realized he had to make her betray her intentions.

"Twenty-six, that order is canceled. I charge Captain Istlahad with negligence and misconduct. After the alarms went off she omitted to activate the shields and now she is attempting to leave the bridge during an alert," Tom, said defiantly, trying to keep cool. "I formally ask for her to be removed. . ."

"Step back, you wretched Earthling!" Istlahad shouted, taking a thermal gun from the inner pocket of her jacket.

Tom froze against the door. "I formally ask for her to be removed from command," he said, trying not to show how tense he was.

Everybody froze: none of them had ever seen a thermal gun aimed at a human. Then Susan suddenly got up.

"Stop, Susan, don't move. Everything is under control," Tom said, halting her in her tracks.

"Step back, you bloody fool!" Istlahad repeated. A few instants later, seeing that he didn't move, she fired.

A beam of light shot from the thermal gun Istlahad had in her hand. At about four inches from Tom's body the light diffused in a faint glow, and everyone realized he was protected by an energy shield.

Tom had stiffened, with all his muscles contracted, as he anticipated feeling the shot against the shield. He had imagined it would be something like a bullet hitting against a bulletproof jacket, but all he felt was just the warmth radiated back from the shield when the energy from the weapon was absorbed by it.

"I formally repeat my request, now adding the charge of treason. Istlahad has attempted to take the life of a person under the personal protection of the sector Coordinator. Twenty-six, the first mate should be given command duties and the captain must be arrested," Tom went on, now in a quieter voice. Then he extracted a thermal gun from his pocket and aimed it towards Istlahad. "Throw down your gun and surrender. When we reach Laraki, you will have a fair trial."

The computer did not react; it was clear that she didn't realize what was going on; however she raised the shields of the starship. Everybody on board was motionless; only Ulkahan was standing behind Istlahad, looking uncertain of what to do.

Istlahad turned to one side and aimed the gun at Ashkahan "You have a shield protecting you, but she has none. Step back and open that door, otherwise I will fire."

Tom was expecting that move, but he knew it would give her little advantage. "To shoot you must lower your shield. You can kill one of us, but then it will be your turn."

"What do you want to do, Earthling? Can't you see we have a deadlock?" Istlahad said, keeping her gun aimed at Ashkahan.

"Order four of those ships to move away to two milliparsecs. When they are far enough, I will let you go," Tom said, hoping nobody would interfere.

"You cannot hope I do such a thing. You have been quick to understand the situation, but you cannot do a damned thing." She moved the gun sideways, aiming it at Susan. "That's where I will shoot, if you don't step back!" she said defiantly, with an air of triumph.

Tom was waiting for that move, the only one Istlahad could do to carry out her intentions. Now was the moment to try his bluff. Clearly Istlahad hated and despised him: he was an Earthling, the ultimate cause of the way in which the law was being flouted and which the Coordinator had involved all them in. She thought him capable of any wicked action.

Tom burst into a loud laugh. He was staring Istlahad in the eye, trying to guess her thoughts. He waited, hoping that the fury he could see in her eyes would overcome her common sense. Then, abruptly, he ventured on his bluff "Go on, shoot, you are going to do me a favor. I can find as many females like her as I wish, on Earth," and here he stressed the word females, hoping that this violation of the Aswaqat code of behavior would increase her rage and contempt. But as he pronounced his last words he thought. *Hell, I've gone too far. You must never raise like a fool when bluffing.*

But by then Istlahad was blinded by her rage and was ready to believe anything. "Bloody philistine," she hissed. She turned towards the screen, while keeping her gun aimed at Susan, and said "Twenty-six, put Admiral Ratlqal on the main screen."

The wrinkled face of an Irkhan appeared on the main screen. "Admiral, we must change our plans. Please, order four of the ships to move back to two milliparsecs."

"Impossible. Try to get out of that ship," the Irkhan answered. Clearly the computer had only partially obeyed the order and the video was only one way: the admiral could not see what was going on.

"No Admiral. Do as I told you." Now Tom could see that Istlahad was scared. He had to help her, otherwise. . .

"Admiral Ratlqal, here Tayqhahat from Kirkyssi speaking. You had better follow her suggestion. She has important information and you should not destroy this ship before we are on a shuttle," Tom said, trying to use his best Kirkysakh accent.

"Who are you? I can't see anything. Start broadcasting images," the Irkhan said.

"We cannot, Sir. The video system is gone. Istlahad and myself had trouble with the crew, and some of the systems got damaged. I'm on your side, even though I was under cover. We have vital news for you." He was bluffing again:

What if the admiral had no followers in Starfleet? That 'I'm on your side' was perhaps a fatal error.

"All right, the plan is cancelled. Four ships will leave, but you come on board immediately." The Irkhan's face disappeared from the screen.

"Hell, I have to thank you, you barbarian!" Istlahad hissed at Tom.

"I don't accept thanks from a traitor. Now tell us what the plan was that you were speaking about, otherwise I won't let you leave." Now that the immediate danger was over Tom felt the recent tension draining away and had an urge to start laughing. He hardly succeeded in controlling himself.

He looked around. Everybody was still: they clearly didn't understand what had just happened. Only Sinqwahan was following his actions: he gestured for him to go on.

Istlahad said nothing. "Can you deny that if I had let you out those five ships would have opened fire against us as soon as your shuttle was at a safe distance?" Tom said. And that that traitor would have blown this ship apart with you still on board? How long would you guess our shields would have withstood the fire of the batteries of five cruisers?"

"Perhaps five seconds. And I think I have already thanked you for saving my life. And now get away from that door and let me out," was Istlahad's answer.

Only then did they realize what danger their ship had been in.

"Twenty-six, tell me when those four darned ships are at two milliparsecs," Tom ordered the computer.

"Certainly, Professor," the computer answered. "I have succeeded in identifying those five starships. They belonged to a small squadron commanded by admiral Ratlqal: they were based at Qhra'ar and were listed as missing six years ago, after they attempted to contact the Qhruns"

"A few minutes and you will be allowed to go, Istlahad," he told her.

She ignored him, but told Ashkahan, who was again under the aim of her gun: "I have your word that you will not try to blow my shuttle as soon as I am outside, captain?"

At first, Ashkahan didn't understand whom she was talking to and didn't answer. Tom took advantage of her delay to say: "We don't promise anything to traitors. You have the batteries of that cruiser protecting you, and this is all that you can count on. However, I warn you that if they make the slightest mistake and leave us the smallest chance to fire, we will not miss it."

"Fair enough. I couldn't expect any more from a barbarian like you! Believe me that if you make the slightest error the captain of that cruiser will not miss the chance and you will become interstellar dust!"

The computer announced that the four star cruisers had reached a distance of two milliparsecs. Tom moved away from the door, which opened immediately. Istlahad, followed by Ulkahan, hurried out. "My best regards to traitors," Tom said as they passed in front of him.

"Qhruns will destroy you, barbarian," was her answer.

After less than a minute, the screen showed the shuttle racing away.

"Twenty-six, move slowly towards the gate, but remain within range," Tom said.

"I am sorry, Professor," the voice of the computer actually seemed sorry. "I must take orders only from the captain."

All had happened so quickly that Ashkahan was confused and didn't know what to do. "Twenty-six, Tayqhahat is in command now. You must obey his orders," she said.

"Full power to the shields, be ready to switch power to the batteries," Tom ordered.

"Star cruiser CH 23426 ready at your orders, Captain Tayqhahat. All hands to battle stations, all systems battle-ready." The excited tone in the computer's voice cause a shiver to run down Tom's spine. *This computer is now going to play the hero,* he thought half-amused. And then, realizing the situation: *And myself, what am I playing myself? I left Earth no more than a fortnight ago and now, after having taken control of a star cruiser, I am engaging in a fight with a squadron of five cruisers, under the command of an admiral with years of fighting experience. And all these fellows, some with webbed hands, birds, blue- and gray-skinned chaps are following me without a word. Either this is a dream, or we are all a bunch of lunatics.*

Then he suddenly realized that the gray-skinned 'chaps' were of the same species as Admiral Ratlqal, so he turned towards the two Irkhans. "You know that admiral, don't you?"

The two Irkhans looked at each other and then Raoqal spoke for both. "Yes, we know him well. I don't know what the hell has gone wrong with him. He was a good and respected officer. Now, if we have to fight against him, we are ready. Irkhan or not, what matters to us is that he has turned out to be a traitor."

He looked sincere. The words of the Iktlah, warning him not to trust anybody, had shaken Tom deeply, and what was going on was an impressive demonstration of how right he had been.

He realized he had to develop a plan for the battle. If he could prevent Istlahad from getting onto that ship the presence of Earthlings could still be kept confidential. The screen was showing the four cruisers slowly getting closer.

"Twenty-six, how much time will elapse between the shuttle entering the nearest starship and when the other four will be close enough to fire?"

"One minute and fifty seconds, Sir."

Far better than he expected. Perhaps it was possible to do something, provided that the captain of that cruiser was crazy enough not to have the shuttle wait in space until the other starships were able to assist the recovery maneuver. "Tactical screen on, zoom on the shuttle bay of the nearest ship." The main screen changed to show the cruiser and, in front of it, the shuttle moving towards her. "Identify that ship and tell me if it is of a type much different from yourself."

"She is CH 471233; She is far newer than myself. I cannot say that we are equal, but there is no significant difference."

"Check whether the central-shield generators will be aligned with the door of the shuttle bay when they lower the shield that protects the entrance."

The computer remained silent for a few seconds, then she cried out, sounding elated "Yes, Sir! They will be exactly in line!."

"Don't get excited, Twenty-six. If that captain is not a fool, he will wait to lower his shields until the others ships arrive, or at least will turn his ship round to let the shuttle in without danger. By the way, the rules state he must do that way... But if they lower the shield without modifying their attitude we will have a few seconds to score a good hit."

"I bet all you like that he will lower his shields without thinking twice," Sinqwahan interrupted the computer, who was about to answer.

"He will not have a chance to repeat that mistake again, then," Tom stated, grimly. "Twenty-six, if they let the shuttle in, lower the shields, fire with the front batteries through the opening of the shuttle bay. Without checking the result of the first volley, after a second open fire again, aiming at the center section." Tom looked around and realized that none of the crew had ever been into a battle, except in simulations.

"I think it is better to wait one second and a half. The circuits of the shields have some inertia," the computer answered.

"It is up to you to chose the best timing", Tom concluded. The shuttle was now close to the cruiser. It didn't stop and the ship had not started to make any attitude-correction maneuver. *Is it possible they don't understand the danger they are facing?* Tom asked himself. And was also wondering whether he had the right to destroy a starship, and probably also the lives of the twelve humans who were in space in front of him. *If we let Istlahad out as she wanted, they wouldn't hesitate to kill us,* he thought, realizing that everyone else was asking themselves the same question.

Then everything happened so quickly that Tom just watched the scene as if it were a movie. The shuttle homed in, there was a slight change of color of the hull showing that the shield had been lowered; then light streaks from the front batteries; a blinding glare in the central part of the cruiser, where the shield generator was located; and then, a second and a half later, an explosion of light.

The central part of the cruiser vaporized in a glare so intense that the screen darkened automatically, not to blind them. The cruiser's hull divided into two large fragments, which were then rocked by secondary explosions.

The cruiser was quickly reduced to a cloud of incandescent small fragments, cooling and disappearing in the darkness of space within a few seconds.

"Twenty-six, move towards the hyperspace gate, remaining out of the range of their batteries, and with the shields down." Tom's voice sounded in the deep silence and shook all of them. Nobody was able to articulate a word.

Tom realized he had to go on relying only on himself and the computer. The four cruisers were closing in. At a certain point two of them headed towards the fragments of the first starship, while the other two continued the chase. *This is another mistake we can exploit,* Tom thought.

"Twenty-six, identify the ships chasing us and tell me where the admiral is."

"They are CH 43198 and CH 43176, the latter with the admiral."

"Be ready to release thirty two torpedoes from the rear tubes. They will remain inactive in space until the two ships pursuing us are within their range. At that precise moment they will split into four groups of eight. The first three groups will attack CH 43198 in three waves, spaced thirty seconds apart. The first wave will attack the whole ship, the second will concentrate on the center section and the third will concentrate on the generators of the central shields. The last group will simulate an attack against CH 43176 distributed on the whole ship and then will concentrate against the shield generators at the same time as the third wave arrives against CH 43198. You will slow down so as to get into firing range at the same moment as the torpedoes reach CH 43176. Is that all clear?"

"It is quite complicated. I fear the computers of torpedoes cannot behave in such a complex way," the computer answered.

"You give them detailed instructions. You oversee them to ensure that they will follow them," Tom ordered.

"We will do our best. When shall I raise the shields?"

"Shields down for the whole maneuver. Raise them only if you see that the torpedoes fail to perform their task."

Ashkahan turned towards Tom. "It's all too complicated. Starships aren't used to performing such complex maneuvers," she said in an alarmed tone.

"They will get used to it. If we want to have a chance against the Qhruns, we must try new things, and trust computers more," Tom answered.

"Thank you, Sir." the computer broke in. *Now she is committed to the idea of demonstrating that she can succeed. We made it a point of pride and all her systems will perform at their best,* Tom thought.

They were all staring at the screens showing the two starships chasing them. Abruptly one started swerving, without much success. One after the other the torpedoes of the first wave exploded against the shields. The second wave was stopped by the shields too, but now much closer to the hull, a sign that the shield generators were badly overloaded. After about thirty seconds, the third wave went in. The first torpedoes saturated the shield generators but the last two couldn't be stopped and caused a general shield failure. A few seconds later, the fourth torpedo group met the second cruiser.

"Well done, Twenty-six. That three seconds delay was a masterpiece." The torpedoes aimed at Admiral Ratqal's starship didn't go directly against the shields, but performed a number of simulated attacks preventing the cruiser from taking any offensive maneuver.

"Thank you, Sir." If she weren't a machine, you would have said the computer was beaming. The commendation didn't distract her: a set of shots from all the batteries blew the hull of CH 43198, which was now defenseless, to pieces.

"Concentrate all fire on the other ship, but be careful not to overdo it: I want them all alive, Admiral and starship included. I want the Admiral on the main screen."

The shield generators were destroyed first, followed by the batteries: now the cruiser hung there, motionless and defenseless. The face of the Irkhan, looking grayer than usual, appeared on the screen.

"Admiral, you must surrender, immediately. First, order your two ships to remain close to the wreck of the first cruiser, with their shields down," Tom dictated.

The admiral didn't move: he had no intention of complying.

"Twenty-six, punch a pair of holes in their hull, just behind the bridge." Two beams of light homed in on the cruiser, and their bridge was invaded by smoke, before the airtight doors shut.

"Admiral, don't think I will wait until your friends rescue you. Either you order them to stay where they are or next volley will be aimed straight at the bridge."

This time the admiral gave the order, and the other two starships, which had already started moving closer, went back to the point were the wreck of the first ship was.

"Now switch off all your computer's higher functions and surrender your ship. My computer will take control." The admiral obeyed with a quickness that seemed suspect to Tom. *All these stories of treason are making me paranoid,* he thought. "Twenty-six, enter all memories and check whether there are any residual orders. Look specifically for self-destruction orders," he added.

It took about ten minutes to perform all the checks, but in the end she declared that there was no danger. Only at that point did Tom give the order to start docking, connecting the two ships directly through an airlock. He then asked Sinqwahan to go, with four armed crew, to the lock to take care of the prisoners.

"Twenty-six, can you please create ten separate compartments in one of the storerooms, to use them as cells? I do not want the prisoners to be able to communicate with one another," he stated. The transfer of the prisoners took more than half an hour, because they were carefully searched individually and then accompanied to their cells.

"Sinqwahan, which is the nearest system on the route to Laraki?" he asked.

"Telqtl, only three hours from here," the Aswaqat answered.

Tom ordered the computer to enter immediately hyperspace towards Telqtl. "Immediately after re-entry into three-dimensional space, we must enter hyperspace again. The choice of Telqtl is so obvious that they will follow us," he added.

They carefully chose their next destination: Lot'lat, a system with a double star and no inhabitable planet.

Now it was time to relax, thinking of what had happened. Sinqwahan was the first to speak. "Congratulations, Tom. The Kistl battle, as this fight will be known, will be mentioned in Starfleet textbooks. You saved this ship, stopped Istlahad and captured Admiral Ratlqal without getting a single scratch. Really remarkable!"

"I am sorry for Istlahad and Ulkahan, but I couldn't allow them to go around to say who Tayqhahat is, am I right?" Tom asked by way of return. He was really sorry for Istlahad: she was not a pleasant person, but he certainly didn't like having her reduced to interstellar dust.

"You did what you had to," Sinqwahan interrupted him. "She knew the danger she was facing when she decided to lure us into that trap."

"I don't think she thought she was in any real danger," Ashkahan said. "If it were not for him, none of us would have understood in time what was going on, and she would have been able to carry out her plans." She paused and stared briefly at the other crew members, one by one, and then continued: "Tom, I think I speak on behalf of everybody on board. I'm asking you to take command of this starship, not only for this emergency but for good." Then, looking at Susan, she went on: "Now we need a navigator. You know astronomy, I believe that you can do that job, with the help of the computer. So the ship will be fully crewed again. Everybody would be glad if you do, Twenty-six as well, I think"

The voice of the computer broke in immediately "I would be pleased, Ashkahan. During the battle I really enjoyed working with Captain Tayqhahat."

"But you cannot. I am an Earthling, I do not even belong to Starfleet. . . How do you think an appointment like this would be endorsed by Headquarters. . .?"

Sinqwahan interrupted him "Why, Captain Tayqhahat? The Coordinator will be most upset if you, a spacer from Kirkyssi, refuse to obey an order to take command of this unit."

These words caused everybody to start laughing, a long, relaxing laugh that helped to dissipate the tension of the last hour.

"I think that now they have really framed us, Tom," Susan commented. This was the first time she had shown any reaction since Istlahad had aimed the gun at her. For the whole battle she felt as if she were watching a show in which she was only marginally involved.

"I fear so," he commented, with a sad look, that wasn't really genuine. The idea of taking command of a starship was very alluring. "If I must play the captain, I must start by grilling our prisoners."

He turned towards the two Irkhans "Raoqal, Ilkalt, please tell us all you know about this admiral."

"I already said he had always been a respected officer; I don't understand how. . ." Raoqal started.

"No, Raoqal. Not your impressions, please. I need facts: how old is he, where he was recently, and so on," Tom stopped him, to be in turn interrupted by the computer: "I have prepared a short report about him, using all the relevant information that I have found in my memory banks."

Tom thanked the machine, who projected the documentation onto the main screen. It was quite complete, and they took more than a quarter of an hour to go through it all; the result was the portrait of a good officer, like many, efficient without being a genius, with no great achievements but with nothing that would cause anyone to anticipate treason.

When they got to the end, Tom asked the two Irkhans whether they had anything to add. "That portrayal is doubtless correct. But something is missing; he was eager to act, to try to get in contact with the Qhruns, to save the Confederation," Raoqal said.

"No, it was not just the desire to save the Confederation as a political entity. He was motivated by a moral, I would even dare say, a religious, urge. He tended to use prophetic tones when he spoke about the need to save the various human species," Ilkalt corrected him.

"He spoke like an Iktlah?" Tom asked.

"No, I have seldom listened to the speeches by Iktlahs, but I think they are more tolerant, more flexible in their ideas. He was much more of a bigot, a narrow-minded person," Ilkalt added.

Turning towards Sinqwahan, Tom asked: "You ought to know how to examine a prisoner."

"Not at all, I know no more than you, about that. We need a psychologist— let's wait until we reach Laraki," he answered.

"No, I want to squeeze out any possible information from them as soon as possible. I fear that someone else will make another try to prevent us from reaching Laraki."

"Then I would like to talk privately with you, Susan and Ruklyaq," Sinqwahan stated.

They moved to the library and the Aswaqat had the computer isolate the room.

"Now that nobody can hear us, you must tell me something," Tom began. "I believe you understood what was going on before I did. When I was against that door, with that witch aiming her gun at me, you didn't look very surprised. Why didn't you do a damned thing? If I hadn't been wearing a shield, she was going to roast my guts."

"I knew you had a shield. First, because if you had none, you would not stay there, waiting to be roasted. Second, because you had a faint glimmer around you. Only overexcited people would fail to notice that. Third, because I think you are too clever to steal a thermal gun from the shooting range, taking all the trouble of substituting a poor imitation made from soap for it, and then to leave it goodness knows where at the most critical moment, re-entry into three-dimensional space." Sinqwahan stood calmly, looking around and enjoying everyone's amazed expressions.

"How did you come to realize that?" Tom asked, coolly.

Susan burst in laughter "So that's why there was that silly request for soap. Where did you get such a stupid idea, from a Mickey Mouse cartoon? A gun made from soap!"

"That idea was not so stupid, since it worked. And I didn't get it from a Mickey Mouse cartoon, but from the story of an escape from Alcatraz. If it was enough to cheat professionals like those prison warders, I told myself, it would also work with our computer," answered Tom, slightly annoyed by that word 'stupid'.

Sinqwahan went on "I am also sure that you noticed that one of the thermal guns was free from dust, and that it was not the one we used. You also concluded that someone on board was practicing with guns."

"You bet I did. It was before reaching Ytl. After what Quanslyaq said I grew suspicious, I made that fake gun and I stole a real one. I am not so naive, Susan, I checked that the guns are not charged within their box and that the computer cannot see exactly what is inside."

"I must confess that when I found the fake weapon I got scared," Sinqwahan said. "Then I scraped some of the material and I had it analyzed by Twenty-six. When she told me it was a fragment of the soap she had prepared for you, I felt relieved. Since then, at any rate, I always kept one of my thermal guns in my pocket."

"Then you too had a gun on the bridge! Why didn't you speak out, to prevent Ashkahan and then Susan from being in such danger?" Tom exclaimed.

"On the contrary, there was a very good reason not to move. I wanted you to be the one to get us out of that mess. Recordings of all that went on, obviously after all reference to Earth and Earthlings have been wiped out, will be seen in the whole sector. Within one month Tayqhahat will be a hero. Apart from providing you with a past, we will also provide you with a future, and a very bright one; everybody will expect that you, after defeating those traitors, will deal with the Qhruns."

"I said once before that you were a son of a gun. You knew everything from the first moment and waited to see how I would deal with the whole mess, just so that you could exploit the situation. . ." Tom said, not knowing whether he should get angry or laugh.

"You are wrong. I knew earlier than that, just as you did. Don't tell me that Istlahad's consideration, when she suggested setting course towards Kistl, wasn't highly suspicious."

"And then why did you accept?" Tom asked. He could not believe that Sinqwahan had anticipated everything, but now, thinking about it, he realized that he had had his suspicions too, and had got ready, although he didn't know exactly what to expect.

"Why? Because I wanted to make her to come out, into the open. And about you calling me a son of a gun, don't forget that you will be the one who in the end will benefit from the situation. You will be the hero, not me. And there is one thing I don't like: you called me that, but you used the same words for Istlahad. What do you really mean by that?"

"Well, we mean a shrewd person, who is ready to act cunningly. But said in a friendly way it is not an insult, is almost a way of acknowledging the ability to solve a problem in an. . . unconventional way, on the border of illegality."

"Considering the way you manipulated Istlahad's feelings to make her to betray her intentions and what you told the admiral to save the ship, I daresay that you are an expert in solving situations in an unconventional way, as you say. You son of a gun, too." They both started laughing.

Susan didn't laugh at all. "You're both very clever. Two clever sons of a gun, as you have admitted. But next time you think of being so clever, please do at least tell me beforehand about your beautiful clever ideas. Do you think it is nice to find a thermal gun aimed at you, to be close to getting roasted, and then possibly reduced to interstellar dust in a battle without the least understanding of what is going on?"

Tom realized that Susan was right. "I am sorry, you have every reason to complain. The point is that I didn't want to frighten you with what could be just fantasies. Then when the crisis actually broke we did not even have time to exchange a single word."

Sinqwahan felt guilty too, so he added: "Actually there was little danger, at least from Istlahad. If she really put her threat into practice, I would have jumped in front of you with my shield on. Tom knew perfectly well she could do nothing but threaten you, but I was ready to protect you from the thermal gun. As for the dangers of the battle, there was nothing we could do. We had to face them and I fear that we will have to do it again."

"Now we must try to get as much information as we can from our prisoners. We must first find out whether they succeeded in contacting the Qhruns," Tom interrupted him.

"You actually believe that they were able to establish a contact?" Sinqwahan asked.

"I don't know, but I wouldn't rule it out. Let's summarize: we have an admiral, reliable and proficient, as everybody said. He left in quest of a contact with the Qhruns. He disappears and nobody hears from him for six years. Then he comes from nowhere to ambush a starship performing a secret mission ordered directly by the Coordinator. And, what is even worse, does it in collusion with a Starfleet captain and believes me when I said that I was on his side. This can only mean than he is part of a conspiracy together with spacers in the fleet. Don't you agree, Sinqwahan?"

"It seems so and, if it is like that, the situation is far worse than you think. I can swear that I never heard of any conspiracies. And if there is anyone who should know, it's me."

"When did you last contact your people?" Tom asked.

"As the computer is not listening, I can tell you: in the Satq system I received and sent back a hyperspace probe. At my base they know where I am or, rather, they knew before you decided to go to Telqtl and Lot'lat without letting me inform them. We use hyperspace probes that no one in Starfleet can identify and classified contact procedures. I can tell you that that eight days ago, when the last probe that I received was dispatched, Security wasn't aware of any conspiracy."

"Whoever they are, their information network must be better than that, as they tried to destroy only this ship, among all those that are moving around in this sector." Tom added. "Let's assume that our admiral succeeded in his contact. His actions could be explained if he was acting on behalf of the

Qhruns." At this point Tom fell silent, seeing the expressions of disbelief on the faces of the others.

"Well," he went on after a paused, "you can be lured to the other side in many ways. They could convince him by argument, or force him somehow, or they could even control him in some ways. Let's consider bold ideas: the man in a cell on this ship might not be an Irkhan admiral at all, but a Qhrun in disguise."

This time they all reacted in utter disbelief, everybody trying to speak at once.

"Impossible. Let's be serious: a Qhrun on board?" Ruklyaq shouted.

"Why not? Who has ever seen a Qhrun? What does one of them look like? And is there a single person who can say for sure that Qhruns cannot capture a human being, use his body and go around this galaxy pretending to be Irkhan admirals?"

"You read too much science fiction!" Susan exclaimed.

"Are you sure that what you have seen in the last week is not more unbelievable? And then, to quote again Quanslyaq," and here with a nod towards Ruklyaq, to which the latter didn't react, "we are here to combine this galactic civilization's way of thinking of with a different culture." Sinqwahan waved his hand as if he wanted to speak, but Tom went on. "And our Earthly culture also includes the weirdest aliens of science fiction."

He looked around and realized that nobody understood what he meant by his last words. "I mean that if the Qhruns are not human, as you, Sinqwahan, suggested the day you recruited us, any hypothesis might be correct, even that it is possible to produce copies of human beings."

"What do you suggest then? If anything is possible, if the admiral might be a Qhrun, capturing him might have been a fatal mistake," Sinqwahan said.

"It might, but also not to have captured him might have been a mistake. We must be extremely careful. Now I suggest that all prisoners undergo a comprehensive medical examination, particularly a psychic one. Then we question each one of them carefully. Nobody, not even the computer, must be informed of what has been said here. I also suggest giving thermal guns to all the crew, just in case something turns sour," Tom answered.

"I will tell Twenty-six to take control of this room again and then we can start." Sinqwahan got up and after reaching the corridor, gave a few orders to the computer.

Tom checked that the prisoners were still in their cells and that they could not communicate with each other. "Twenty-six, can you check whether any non-human life form was carried on board, in particular by our prisoners?"

"Certainly, I have already checked that no dangerous microbes or viruses were been carried on board. And there are no other life forms on board, except bacteria."

Tom glanced at the others. "Now we will lead each prisoner to the sick bay, one by one. You must check that their vital parameters are within the standard range for Irkhans. I want a thorough check, including scanning the electrical activity of their brains and a full set of psychological tests."

"Is there any danger that they have carried some disease on board?" the computer asked in a worried voice.

Tom thought that perhaps such a suspicion might prevent even worse fears from spreading. "No, not a great danger, just one that we cannot exclude. We need an accurate check."

Sinqwahan and Ashkahan took the first prisoner from his cell to the sick bay. They carried guns and, as a further precaution, and also to strengthen the idea that an epidemic disease might have been carried on board, they had protective overalls and masks.

The check-up took more than half an hour and then the prisoner was escorted back to his cell. The computer stated that all species parameters were within standard limits and that no disease was detectable.

Over the next five hours all the prisoners were checked in the same way. During this procedure, the starship re-entered three-dimensional space in the Telqtl system and started the next crossing towards the Lot'lat system. In the few minutes between the maneuvers, Sinqwahan was able to send a probe to inform his headquarters on Laraki about the latest developments.

Shortly after the computer had finished checking the last prisoner (finding nothing out of order), the cruiser re-entered three-dimensional space in the Lot'lat system, which was completely empty. This time Tom ordered the cruiser to stay in the system for a longer time to check whether any ship was following them.

Then, after starting the hyperspace crossing towards Laraki, Ashkahan, Raoqal and Ilkalt went to fetch the admiral. Tom wanted the two Irkhans to be present for the whole time, as they were the only ones on board who could understand, from small details in the behavior of the prisoners, whether they were other than what they pretended to be or whether their minds had been manipulated.

They sat around a table and Tom tried to assume an air of self-possession. Ashkahan came in, saying that the admiral was present outside the door. "Bring him in," Tom said.

The admiral entered, followed by the two Irkhans, who remained besides the door, their guns ready. Tom gestured for them to move to a position from which they could observe the prisoner's face.

Tom stared the admiral straight in the eyes for some time. Then he decided to play it easy and asked "Your name and your rank, please."

The admiral stared back at him and, after a few seconds, answered "Ratlqal from Ikrat. Starfleet admiral for the fifth sector."

Tom felt better after that beginning, because he had feared that the admiral would refuse to answer altogether. "I must warn you that I intend to charge you with treason, for attacking a star cruiser performing a mission directly ordered by the Coordinator of this sector. How can you explain your behavior?" He refrained from looking around to see the reactions of the others.

The admiral didn't answer, but started questioning: "You are Tayqhahat from Kirkyssi, aren't you? And now you have been appointed captain of this ship?"

The question found the Earthling unprepared. He wondered whether he should react in annoyance—"you have no right to ask questions"—or follow a more flexible line. "Yes, I am Tayqhahat from Kirkyssi, and as the captain of the ship you intended to attack treacherously, I am asking you again to explain your behavior," he answered, opting for the second solution.

"Captain Tayqhahat, you lied to me. You are unworthy of the command someone has entrusted to you." The admiral's voice was clearly starting to reveal his anger.

"Admiral, you would have destroyed my ship if I had not acted in that way. My goal was to save this starship... and I succeeded, as you can see," Tom stopped abruptly. He had fallen in his opponent's trap. *I do not owe him any explanation*, he thought, and added, even at risk of interrupting that dialogue he had perhaps started, "At any rate it is not for you to discuss my behavior."

"You are right, captain. You saved your ship, and that was your main duty." The admiral was apparently satisfied by his answer.

"You disappeared six years ago, after leaving in search of contact with the Qhruns. What happened?"

The admiral didn't answer. The silence became quickly unbearable, so that Tom repeated his question: "I asked you to tell me what happened during those years. Did you succeed in your attempt?"

Since he persisted in his silence, after a while Tom added: "Your silence can be considered as an acknowledgement of guilt. I ask you whether you had a contact with the Qhruns and, in that case, whether you attacked this ship by their order."

The admiral didn't move. *I should have studied the criminal laws of the Confederation. Now he will say that I cannot question him unless his lawyer is present*, Tom thought, aware of the comical side of the situation.

Not having the slightest idea on how to go on, he decided to resort to some melodramatic effect and, raising his voice, he said solemnly "Admiral Ratlqal from Ikrat, I formally charge you with acting in collusion with our enemy and attempting to destroy a Confederation starship. If you do not answer for these charges here, you will do to the Coordinator himself."

The admiral remained motionless for another minute, then he suddenly raised his head and stared at those present, one after the other. At that motion Sinqwahan, betraying his uneasiness, took his gun from his pocket and aimed it at the admiral.

The latter raised an hand and said, loud and solemnly "You abominable heathen, you are trying to stop the actions of the Angels of Wrath! Have no doubt, the Almighty will punish you and all those who oppose his holy will. Don't you realize that this society is foul from its foundations and deserves to be destroyed? His Angels are here, among us, and your foolish attempts to stop them will be defeated." His voice had become louder and louder and his last words filled the room. His eyes stared at Tom and his raised hand seemed to reach up towards the ceiling.

A great deal of confusion followed these words. Ruklyaq, who had witnessed the whole scene while remaining quietly seated, sprang to his feet, backed against a wall, wrapped his wings around his body and covered his face with a hand. "He is an Iskrat-is-Thn. It's impossible, an Iskrat-is-Thn!" he shouted hysterically.

Sinqwahan remained on his seat, but his hand holding the gun was shaking. The two Irkhans backed away, staring at the admiral, and loudly spoke some sentences in their own language, which Tom was unable to understand. Ashkahan also moved back, taking out her gun. Only Susan, although impressed by the admiral's words, didn't move, more amazed than frightened.

Tom realized the danger. *Now someone, in panic, will roast the admiral, and we shall loose any chance of getting any further information*, he thought. He stood up and, trying to use a calm and reassuring tone, firmly stated: "Silence, please, and above all, put your weapons away." Then, pointing to the two Irkhans and Ashkahan, went on: "Now take the admiral back to his cell, while we stay here to discuss the situation. Provided everyone can keep their nerve and speak calmly."

The first one to recover her composure was Ashkahan. "Sorry, we were not prepared to hear such words. None of us could imagine the he might be an Iskrat-is-Thn," she said, with a rather contrite voice. She was clearly ashamed of her reaction. "Admiral, we must escort you back to your cell," she concluded.

The admiral got up and moved towards the corridor, followed by the three spacers. When the door closed behind them, Tom turned towards Sinqwahan and Ruklyaq and said "And now, please, is there anyone who can explain to me what in this galaxy is an Iskrat-is-Thn?"

7   The Coordinator

"And now, please, is there anyone who can explain to me what in this galaxy is an Iskrat-is-Thn? Provided we can maintain a reasonable degree of self-control," Tom added, aiming his words at Ruklyaq, who was still standing at the back of the room.

"Please, Ruklyaq, explain things to him. You are almost an Iktlah and you should know," Sinqwahan said, addressing the Tteroth.

Ruklyaq slowly came up to the table and sat down on a chair. He was more at ease now, but looking carefully at him Tom noticed that his wings were shaking.

"Well. . . Actually I don't know where to start. . ." Ruklyaq paused for a while, as if he was collecting his ideas, then continued: "You know that the various civilizations have widely differing religious creeds: the majority eventually developed monotheistic religions, or ones based on a single divine principle, expressed in a wide variety of forms. The role of religion in everyday life has even greater variations: you can find civilizations that developed a true theocratic structure and others where the role of religion is only marginal. Moreover, the same civilization may swing between different attitudes over time, and such changes can be quite quick, with times of the order of a few centuries."

Seeing the disbelieving expressions of the Earthlings, he added: "Don't be surprised, the historical dynamics of a society involving a striking number of planetary systems is extremely slow. And the increase of the average life-span slows it down further."

Ruklyaq paused for some time, and then went on: "But back to the point. Groups with highly characteristic and sometimes extremist ideas periodically came into being in the more isolated systems, particularly in times of hardship. Iskrat-is-Thn is one of them. Historians date them back to the Xartian crisis, the crisis ended by Admiral Verterais"

"But these are things that took place 18,000 years ago," Sinqwahan interrupted. "I meant the Iskrat-is-Thn of today, or rather of two hundred years ago. They have nothing to do with what you are saying!"

"On the contrary they have everything to do with it! You cannot understand what is happening, if you don't start from them!" Ruklyaq had become excited. "Xartians were a fairly young species which, after entering the Confederation, developed hostile feelings towards nearby civilizations. Some of their leaders preached their right to expand in zones assigned to other species, instead of looking for free planets in more distant parts of the galaxy. The crisis broke out when they occupied a few systems that were already inhabited."

"The Coordinator of the relevant sector put the case to the Central Coordinator, and before the fleet could do anything, the crisis had expanded, and involved a couple of sectors. The Xartians found some allies among other species that had expansionistic inclinations and a coalition opposing the Confederation was created. The galaxy was practically split into two factions, and a war followed. After some centuries, the Central Coordinator unified all the loyal fleets under the command of Admiral Vertearis, who eventually defeated the Xartian Alliance in a number of battles, which restored the authority of the Confederation. In all this turmoil, a self-appointed prophet started to preach that the end of the universe was approaching. The Xartians were the instruments of the anger of the Almighty, the Angels of Wrath, as Admiral Ratlqal said."

"There are different traditions about the meaning of the term Iskrat-is-Thn by which the members of the sect that he founded were known, and even about the name of the prophet. He preached that only a small number of individuals, the Elects of God, would be saved from the impending destruction of the universe, while the Xartians would disappear, once their purifying role was over."

"If things had stopped there, there would have been no actual danger," Ruklyaq continued. "The point is that the crews of some starships defected, taking their cruisers and joined the sect. They started helping the Xartians, destroying everything and everybody they found unworthy. As long as their targets were the starships of the Confederation, the Xartians treated them as allies. Soon, however, their aim of destroying everything that was not worthy of entering their purified universe started to be in conflict with the goal of the Xartians, which was essentially to create a large zone of influence under their control. Then that bunch of fanatics started attacking the Xartians too, they charged them with betraying the holy duty of purifying the galaxy, whenever their behavior was not fierce enough."

"In the end the sect was chased by both the Confederation fleet and that of the Xartian Alliance. Before the war was over it was completely destroyed. However, the destruction they caused was heavier than that due to the Xartians themselves. And this is all ancient history."

"The thing was practically forgotten for thousands years and the Iskrat-is-Thn ended, along with thousands of other sects, in the pages of those books that are of interest to only a few specialists in ancient history. However, about a hundred and eighty years ago, an authority on ancient history, from the museum of Laraki, came out with the idea that the ancient Iskrat-is-Thn were right, although ahead of their time. The Angels of Wrath were not the Xartians, but the Qhruns, and the elects had a duty to help their work, destroying the sinful establishment of the galaxy, and hastening the advent of a new universal order. The elects could thus live in a universe purified by the Qhruns, who, after having performed their task, would disappear. The sect expanded rapidly, above all among the military and academics, but quick reaction by the Coordinator succeeded in wiping it out quickly, and nobody heard of the Iskrat-is-Thn anymore. Or rather, nobody heard about them until to you questioned our admiral."

Ruklyaq sat silently and nobody said so much as a single word for a sometime.

"But why, in all the recordings I studied, is there not even the slightest mention of all this?" Tom said after a while.

"There is nothing, because this is just a small, insignificant detail in a history lasting tens of thousands years in a galaxy made of millions and millions of inhabited planets. The complexity of the galaxy is such that nobody can know any more than an infinitesimal part of it: I know about this only because Quanslyaq used to tell me that the present crisis is a fertile culture medium for this type of fanatical sect," Ruklyaq said.

The situation was serious, and they needed to know more, and in a hurry. As nobody put forward any ideas, Tom asked Sinqwahan whether it was possible to use a mental projector the 'other way around', so to speak, to extract information from a mind.

Ruklyaq reacted in a scandalized way. "That would be illegal, outrageous!"

"As illegal and outrageous as our presence on board of this ship?" Tom asked, rather amused by the Tteroth's sudden legalism.

"I don't see why it might be illegal to entrust a starship to a spacer from Kirkyssi, even if I must admit that he obtained this responsibility in rather an unconventional way," Sinqwahan stated with an amused tone of voice. "But at any rate to attempt such a thing we would need specialists and equipment that we don't have now." The Aswaqat seemed to enjoy seeing the scandalized air of the others.

There was nothing else that could be done, for now, but continuing to question the other members of the crew of the captured ship. They were brought one by one to the library, but none of them answered any questions. Just one, losing his temper, admitted belonging to the sect.

Ten hours later, when the last one of them had finally been accompanied back to his cell, they realized that they had wasted all that time to no avail. They were aware of a danger, but they knew little more than that.

"It seems we must keep our doubts at least until we reach Laraki," Sinqwahan stated by way of conclusion, getting up and walking towards the door.

Suddenly Susan got up too, exclaiming "Wait, don't go away, we have not yet finished. We have someone else to question!"

Everybody turned towards her, without understanding what she meant.

"We have not yet questioned that star cruiser's computer. If there are recordings of the course of that ship, we can gain useful information. And then, surely it is unlikely that the computer never recorded any conversations among crew members," Susan explained.

Tom suddenly realized that he too had come to accept ship computers to the point that he was no longer consciously aware of their presence. "You are right! The computer should have been questioned first," he exclaimed. But looking around he realized that the others were not very enthusiastic about the idea, and that their expressions were almost frightened.

All except Sinqwahan, who explained "About the course you are right, but as to the other information, nobody can have access to the recordings of a computer regarding the actions or words of a human, without explicit permission from the latter. If this rule were to be infringed, there would be no privacy. . ."

"Well, the privacy of a bunch of criminals can go to hell, when the Confederation had so much at stake! I am sure Aintlhad would agree," Tom interrupted him.

Sinqwahan nodded approvingly, while at the same time gesturing to him not to say any more. The others didn't understand what he meant and Tom realized that the Aswaqat was using gestures typical of Earthlings. He quickly decided to speak only about the information regarding the flight path of the starship.

"Twenty-six, before docking with Admiral Ratqal's cruiser, you entered the memories of that ship to check whether there were any dangers, didn't you?" Tom asked.

The voice of the computer broke in immediately: "Certainly, I did, Sir. It was an explicit order, and a safety precaution I would have taken any way. I carefully checked the memories for accuracy and downloaded the contents into my archives, but I noticed that all those regarding recent years had been erased."

"Accidentally or by purpose?" Sinqwahan asked.

"The last recording was about the departure from Qhra'ar and the following one started when the Admiral started chasing us. As the ship had not yet been hit, it couldn't be due to an accident. Moreover, the Iskrat-is-Thn had the habit of erasing the memory of their computers before entering a battle, to prevent information about their sect from falling into unworthy hands, as they would say. They downloaded all information onto a material object, that could be easily destroyed in the event of capture."

"Did you notice anything hinting that information had been destroyed?" Tom asked.

"Nothing certain, but during the docking maneuver there was a short energy discharge on that starship, as if a very small quantity of matter was being vaporized."

"And so we have lost any chance of obtaining useful information from that computer," Tom concluded. *At least as far as official information is concerned.*

"We still have a chance: once at Laraki, we can have the computer of CH 43176 thoroughly checked. It is practically impossible to completely wipe out all information from a computer with distributed memories and processors. A good computer scientist should be able to retrieve at least something" Sinqwahan suggested.

When the star cruiser re-entered three-dimensional space, the sensors identified four star cruisers approaching them, although still at a distance of about two astronomical units. The ships exchanged the usual automatic messages, confirming that they were four units the Coordinator had sent to escort them into the system. After the few minutes needed for this exchange of messages, a welcome address from the Coordinator was relayed by the commander of the small group of starships.

The Laraki system had two inhabited planets, plus settlements on two more planets and several satellites. More than twenty hours were needed to move into the system as far as the fourth planet, which was the one housing the sector capital.

After a short stay in a parking orbit around the planet, the small convoy began planetary entry. The descent through the atmosphere was performed almost entirely on the dark side of the planet and they could see lights scattered irregularly on the surface. At dawn they were already low over an almost motionless ocean and in view of the spaceport, which was just a large grassy area, where a multitude of starships of all types were lying. Lines of trees divided the various zones and a few low buildings completed the scene.

The group of star cruisers flew for a few minutes over an area scattered with cargo ships and finally reached an empty zone. The four escort ships landed forming a square pattern and the two cruisers landed between them. As soon as they touched the ground, they were surrounded by men armed with heavy hand lasers, apparently all Aswaqat, who came out of one of the low buildings.

"They are here to take care of our Admiral and his comrades," Sinqwahan said, noticing the worried glance Tom gave him as soon as he saw the men.

"What are you doing with our prisoners?" asked the Earthling, who up to that moment had not thought about it.

"That's a good question. I think it is almost a century since we have arrested any spacer, probably from the time Iskrat-is-Thn were creating problems, and even then it was not common, because it wasn't easy to capture them alive. At that time they were deported to a small habitable, but still unsettled, planet. In all probability, our admiral will also end up there", Sinqwahan answered. Then he added in a very low voice "By the way, the Coordinator liked your idea about using the mental projector, and I think he will keep them here for some time, to see whether it is possible to do something of that sort."

After handing the prisoners over to the security agents, they left the cruiser and headed towards a vehicle that was waiting not far away.

Susan asked about the native species of the planet and Sinqwahan explained that originally there was no habitable planet in the Laraki system. The two planets now inhabited had been terraformed thirty thousand years earlier by the Aswaqat.

After getting on a vehicle together with a few guards, they crossed various zones of the port, most of which almost empty. Tom observed that the port had an air of little activity and Sinqwahan confirmed this impression. "Only a hundred years ago you would have seen a different scene. But now the sector is isolated, communications are becoming increasingly difficult and the number of systems one can reach decreases day by day."

Outside the spaceport the feeling of desolation became stronger. The vehicle was moving along a path covered by grass, lined by low buildings: they passed few vehicles and did not see many people around. Laraki was the second planet they had landed on, but they felt that the whole atmosphere was quite different from that of Ytl. Perhaps that was to be expected: Tteroths had not yet been touched directly by the crisis, while Aswaqats had already lost some of their systems and the capital of the sector was affected more than any other planet by the increasing isolation.

Tom saw a building much larger than the others, with the unmistakable aspect of a public building. As they came closer he realized that the facade was decorated with geometrical patterns, giving it an aspect that was both alien and familiar at the same time. The vehicle stopped in front of a flight of steps, leading to a broad entrance. It was lined on both sides by armed guards. From Sinqwahan's amazed expression, Tom realized that their presence was unusual. "It seems that our encounter with the Iskrat-is-Thn has stirred up fears in someone here," he said.

"I have never seen so much security around here," the Aswaqat agreed.

They climbed the steps and then walked along wide corridors. The differences between the style of the ancient royal cliff palace on Ytl, which was simple and austere, and the solemn architecture of the Coordinator's palace was striking. Followed by their escort, they entered a large empty hall. Tom noticed it was full of seats, many with an unusual shape: they would not be able accommodate either Earthlings or Aswaqat. He tried to figure out what strange creatures they had been made for.

The Aswaqat gestured for the Earthlings to follow him and started walking down a side corridor. After about fifty meters, he turned towards a door that led to a smaller hall, with a large table in the center and a number of chairs around it. The walls were covered by large screens, all dark except for one, showing a star field in slow motion, which was probably a map of the sector. A man, an Aswaqat, was standing in front of the screen. He seemed to be concentrating on the map.

As soon as he realized they were in the room, the Aswaqat turned and walked towards them, stretching out his hand: "You are here, at last. I have been looking forward to this very moment for years," he said emphatically.

Tom put his right hand on the shoulder of the Aswaqat, hoping that his imitation of the Aswaqats' greeting was reasonable. "We were looking forward to meet you too, Mr. Coordinator."

"Just Aintlhad, please. I studied you for such a long time that I feel as if I have known you both for a lifetime. But now let's sit down, I believe we have lots of things to talk about." Then, noticing that Sinqwahan had moved towards the door, he said to him "Sinqwahan, please remain. Your presence may be useful."

They sat round the table and Tom opened the briefcase he had carried from the ship. Taking out a tri-dimensional projector, he laid it on the table. He looked again at the Coordinator and realized that now he could easily tell one Aswaqat from another. Aintlhad was older than the others he had met and spoke more slowly, almost solemnly, as if he was well aware of his role. He had however no external signs of his rank.

The voice of the Coordinator put an end to these thoughts. "Well, Tom, having been with us for just twenty days, you have done a good deal. You took a weapon and used it to gain control of a starship. You engaged in battle with a group of five star cruisers, destroying two of them. Another one you captured, with the whole crew; then you questioned an admiral—violating a good number of written and unwritten laws—and made him so mad that he confessed to belonging to the Iskrat-is-Thn. You enthused the crew of your ship to the point of inducing them to put you in command. I feel as if I have let loose a force that God-only-knows-who will be able to control."

"Just a moment, Aintlhad. Things are not quite as you say. Actually I came here to do what you asked me to do, and I found myself involved in a battle between starships. If I hadn't done what I did, we would now be just interstellar dust somewhere on the outskirts of the Kistl system, and your whole strategy would have been history even before it could start."

"No, Tom, you didn't understand what I meant," Aintlhad smiled. "You have nothing to apologize for. You have performed far better that my best hopes. I sent for a scholar who could help us to understand, and instead I found a captain who seems able to became a sort of Admiral Vertearis. Instead of helping us to study the Qhruns, you may help us fight them."

Tom was disconcerted at those words. He had accepted being involved in what was a work of an unusual type, but which was essentially not very different from his own work, and for which he felt he was fit. Now they were forcing him to play a different role, one he didn't feel to be fit for, and by far more dangerous. "Just a moment, I am no spacer. I know nothing of space warfare, nor of star cruisers. I am just a peaceful Earthling. . ."

". . . who goes around blasting starships and capturing admirals. The point is just this: you say you have no experience. But what experience do you think these spacers have? They let their beautiful automatic starships carry them through the galaxy, without taking any responsibility. You saw what the captain of the ship that Istlahad intended to board did? He lowered the shields, right in front of your batteries. I fear that you have spent more time at a mental projector in just twenty days than they do in their whole career. By entrusting everything to machines, nobody here is still able to think with his head. And, at any rate, there is nothing out of ordinary in entrusting a starship to a spacer from Kirkyssi."

Here we go again. When they want to frame you, they pull out the story of the spacer from Kirkyssi, Tom thought. But he had to agree that the Coordinator was right. The Confederation spacers proved to be unable to solve the problem of the Qhruns, and giving a ship to an outsider was a reasonable thing to do.

He looked at Susan, who nodded almost imperceptibly. "Well, Aintlhad, it seems you are not looking for scholars any more, but for mercenaries, who will fight for you, or at least, with you." He paused shortly, as if he wanted the two Aswaqats were giving him their full attention, and then continued "But mind what you are doing, mercenaries are expensive. And those who win battles, are very much so".

The Coordinator remained speechless for a few seconds, amazed by his words. "I believe Sinqwahan has already spoken with you about this aspect. We offer you a lifetime you never dreamt of. If you succeed, whether you decide to remain with us or to go back to your planet, you will have anything you ask."

"No, Aintlhad, I didn't mean that. Personally, what you offered us is more than enough. If we succeed, our price is far higher. It is the admittance of Earth into the Galactic Confederation."

"That's impossible! You know full well that one of our basic laws is against it—for your own good." The Coordinator looked positive on that matter.

"Our civilization is not so far from interstellar flight and I think its integration into the Confederation would not cause too many problems. We are an overpopulated planet: settling new systems and acquiring new technologies will save us centuries of hardship." Tom realized that the Coordinator's arguments were sound, but he felt that he could demolish them one by one.

"That's true, our technologies could be of great help. But you must develop them by yourselves. To get them for free would just cause you to be dependent on us, causing more problems than advantages."

"I am not asking for any presents. If our attempt succeeds, the contribution of Earth to the galactic civilization will grant us our place, without our suffering from any inferiority complex," Tom said.

The Coordinator considered that idea for some time. "What you say makes sense, but it would be certainly be something unheard of—a very dangerous precedent."

It was at that very moment that Tom realized he had the game in his hands. "You are wrong, Aintlhad, there is a precedent. In the second millennium of the Confederation, Terqhatls asked for help from a civilization more or less at our stage of evolution, to help solve one of the earliest crises in the Confederation. In exchange, the Central Coordinator admitted them into the Confederation."

The Coordinator was perplexed. He had never heard about that, but this meant nothing, because that had happened almost one hundred thousand years before. "You use history like the batteries of your starship, with the same deadly precision that you have shown in the Kistl battle. I never heard of that; I will certainly check. Not because I don't trust you, obviously. At any rate you are right about one thing: only the Central Coordinator can take a decision like that. I suggest a deal: we leave thing as they are and, if our hopes are fulfilled, I will discuss your request favorably with the Central Coordinator. Do you agree?"

Tom realized that Earth had gained an ally, and a very powerful one, because he was certain that the Central Coordinator would not object to a suggestion from the Coordinator of the relevant sector. "Agreed. After all we have a lot of time to discuss the matter."

"Now please resume your role as a scholar. Did you find out anything interesting in all the hours you spent at the mental projector?"

"We got something. We prepared a short report and formulated a first operational plan. Don't ask me whether it will work, but I hope you will agree that it is worthwhile trying. May I start?"

The Coordinator gestured for him to continue. He was eager to see the conclusions the Earthlings had reached.

"Don't think that we have discovered something about the Qhruns: we know just what you know, namely nothing," Tom started standing up. He noted the Coordinator's disappointed expression. "But whatever darned thing they are, they need starships to get around. All their invasion fleets are made by somewhere between ninety and one hundred and thirty thousand intruders. We don't know what the hell they may contain, but we know that they have weapons that are not very different from our own, that they are protected by shields and that, individually, they can be attacked and destroyed. Perhaps this is trivial: the physical laws are the same throughout the whole universe, and perhaps there is not a wide variety of ways of moving among the stars, of defending yourself or of attacking an enemy. At any rate, human or not, they have technology in common with us.

He paused to look at the Coordinator. His initial interest was fading, substituted by disappointment: what he was saying was perfectly obvious. "If their technology is qualitatively similar to ours, we are outnumbered: the point is a quantitative one. After all, the problem is just this: how many fleets do they have? How many ships do they have?" Here Tom asked himself whether he was making the same mistake as Stalin when he asked how many divisions the pope had. Empirical models expose you to that type of mistake, causing you to overlook qualitative differences. But that was the approach he had chosen, and now he had to continue. After all, why should it be wrong? He continued. "Starting from these assumptions, we did what we are able to do: we built a mathematical model."

The last statement revived the Coordinator's attention. From his face, Tom realized that mathematical models meant little to him. He ordered the three-dimensional projector to show the first set of images. A view of the galaxy appeared, with the systems shown as bright spots. Then, starting from the periphery, some of them turned from white to red and the red zone enlarged until it covered more than half of the image. It was an animation of the Qhrun expansion, starting from their first appearance, 450 years earlier. "Unfortunately, you are familiar with this, so familiar that you take it for granted, without studying it analytically. Susan, can you please show us how you worked on these images?"

Tom sat down, while Susan started speaking. "At first I assumed that the expansion was a steady-state, constant-speed, process. I was surprised to note that, while this was accurate between about 350 and 100 years ago, both for earlier years and later ones the accuracy decreases. I tried to make a more complex model, finding that the expansion was slower at the beginning, as if they had to carry their paraphernalia over large distances. Then the velocity stayed constant for about 250 years, as if the number of their invasion fleets were constant. Then the speed starts to decrease again, as if they are experiencing logistical problems. After all, that is to be reasonably expected with such a quick expansion."

Now the two Aswaqats were listening with close attention. "At this point I started a detailed comparison between the mathematical model and the historical records. Now I will show an interesting detail, showing the time difference between the actual invasion time and that computed by the model. The brighter the spot on the map, the greater the time difference." The image turned dark, then weak spots of light appeared, scattered about. "Just small unimportant deviations from the model, invasions taking place nothing more than one or two days earlier or later than computed," Susan explained. Then suddenly a bright spot burst, and started growing. As it expanded, its brightness decreased, until it disappeared.

"I tried to identify that singularity. It coincided with the only serious attempt to stop the invasion, performed about 150 years ago in the third sector by a fleet of one hundred and sixty thousand star cruisers. Everybody knows that that attempt turned out to be a tragic disaster, with the loss of tens of thousands of starships. From the simulation it is clear that it caused a temporary slowing down of the invasion throughout a whole zone of the galaxy. They took almost ten years to catch up. This means that the number of fleets that they have at their disposal is limited and that they experience problems in replacing the starships we destroy. Perhaps it is also a sign of their difficulty in dealing with any unexpected reaction, since they don't expect to find any organized resistance in their way."

Susan sat down, while the image faded out. Tom waited a few seconds, then spoke again. "While Susan was studying the expansion, I concentrated on our sector. From the data at our disposal, I computed that the number of their fleets operating in it should be between forty and fifty, and then that the total number of starships they have here would not be larger than five to six million. From this assumption, I computed the likely position of each Qhrun fleet and individual ship, and then I extrapolated their future activity. This is the trajectory of what I designated as their twenty-third fleet." The projector came to life again, this time showing only the fifth sector, partially in white, the zone still under the Confederation control, and partially in red, the zone that had been invaded. A blue line began at the frontier and moved, in a number of straight sections, through the white zone. Each time it stopped for a moment, Tom stated the date of the expected invasion and the name of the system being invaded. When he said "Four years, three months, twenty days from now: they will reach Laraki," the image stopped. "I believe that fleet will consist of one hundred and seven thousand, three hundred and forty-two starships, plus or minus ten percent, with an error probability of about three percent."

The Coordinator was motionless on his chair. After a long silence, he slowly shook his head, saying in a low voice: "Impressive, absolutely impressive."

Tom was clearly pleased with his reaction. He now had to continue with his proposals. "From the results we have shown you, we developed a plan, aimed at tackling our problem." He paused to observe the expression of the Coordinator, who nodded for him to go on.

"The plan is divided into three phases, but for now we will just deal with the first one: if it doesn't work, it would be pointless to waste time discussing the following ones. We know, within a certain margin of error, where the Qhrun ships are, and we know that they assemble to form large fleets only when invading a new system. We also know that we can destroy them, provided we face a small number of ships at a time. Our main goal must then be that of attacking any small group of Qhrun ships. We organize a fleet in squadrons of two hundred star cruisers each, subdivided into patrols of ten ships each, supported by a suitable number of cargo ships. These squadrons will move continuously from one planetary system to another, subdividing into patrols as soon as they enter a system. Each patrol will attack a single Qhrun ship and destroy it. As no blockade consisting of more than ten Qhrun starships has ever been recorded, we will safely outnumber them. As soon as the freighters arrive in the system, they will collect all the fragmentary remains that they can find and carry them to the nearest base, where they will be studied in detail. After having annihilated the starships blocking a gate, the squadron will storm the system, attacking all isolated starships, and freeing the other gates. Before leaving the system, each gate will be mined by a suitable number of torpedoes, and provided with hyperspace probes, so that any movement of Qhrun ships into that system can be reported."

After a short pause Tom added: "To start with, I need 200 star cruisers, about the same number of cargo ships and the possibility of operating from one of Starfleet's bases. If this works, we will have to reorganize the whole fleet in this sector along these lines, to militarize all cargo ships and start an unprecedented organizational and industrial effort to build star cruisers, torpedoes, probes and all the other equipment which may be of use in implementing this plan."

"You have done a lot of work in those few weeks. What you ask in starting the first part of your plan is reasonable, but do you realize what are you asking me to do? In practice you are asking for the militarization of the whole sector, putting all resources at the disposal of Starfleet. . . and putting the whole of Starfleet at your disposal." The Coordinator was speaking slowly, as if he was thinking of all the implications of the plan the Earthlings were proposing, and wondering how far he himself could go on those lines.

"You asked me to put forward a proposal, and I have done that. If the things are at the point they actually seem to be, it is obvious that you cannot hope to get out of this mess by just sitting here and waiting for the Qhruns to decide to retreat. Only an effort in which each system, each planet, each person, starts working towards only one end, can allow us to survive this crisis." *I'm starting to sound high-handed*, he thought, realizing however, that the situation asked for it. "Well, I am certain that after we have obtained success with the first group of starships, and the invasion of a few systems has been postponed, we would find a wave of public opinion arising to support us. When they realize that survival is possible, everyone will be willing to cooperate."

"No doubt about that, but I foresee enormous difficulties. Iskrat-is-Thn will try to stop us. Then, part of Starfleet will not be enthusiastic for such a sudden change. Well, I took a gamble when I summoned you, now I must back up that gamble. What base do you want me to assign your squadron to?"

Tom was not prepared for such a direct question. "It must be close enough to the frontier to be efficient, but far enough away not to risk a direct attack. It must have a shipyard and be large enough to support all these ships. I believe that Gorkh'ar is the best choice," he added, hoping that the Coordinator wouldn't guess what he was thinking about.

"Would not Qhra'ar be better? It is more efficient, better defended and its shipyards are far larger. . ."

". . . and it is exactly the place the Iskrat-is-Thn will expect me to choose as my headquarters. Too large to control everyone adequately," Tom interrupted him.

The Coordinator went on: "Yes, perhaps. And then Gorkh'ar is closer to Earth, isn't it?" Tom realized that the Coordinator was too clever to be fooled in that way. "Don't worry, I don't blame you. Even myself, with all the responsibilities I have over the whole sector, would have a particular regard for Laraki. All right, let it be Gorkh'ar, if you feel more comfortable with that."

They went on discussing the details of their move to Gorkh'ar and of the measures needed to face the problems the Iskrat-is-Thn were likely to cause. Tom insisted that he needed to know the exact number of starships operating in the sector, but the Coordinator was clearly unable to supply it. "Actually there is no unified headquarters of Starfleet. Each planet has a certain number of ships: only when an emergency arises does the Central Coordinator appoint a general commander of all the confederation fleets," Aintlhad explained.

Tom refrained from asking directly whether the situation was not an emergency: slowly he was coming to realize how much the sector's organization had deteriorated. He took a small portable terminal from his pocket and raised it to his mouth. "Twenty-six, I have a problem and perhaps you can help me."

"I will do my best, Sir," the computer replied.

"I need to know how many starships, military and civilian, are now operating in this sector. The Coordinator tells me that there is no way to know the present situation with Starfleet. Do you have any ideas?"

The computer was silent for a few seconds, to the point that Tom was about to shut off the terminal. "Perhaps I have an idea, Sir. Perhaps I..."

Sinqwahan jumped up and seized the terminal from his hands. "Twenty-six, disconnect immediately. Then open the communication again using an encoded channel, after inserting a maximum security code." He then gave back the terminal to Tom. "I am sorry, but if Twenty-six has any ideas, it is better that they remain among us."

Tom was about to thank him, when Twenty-six went on: "I was about to say that perhaps I could contact all the ships we meet on a particular service channel, that used for telemetry and identification messages. I can thus obtain a copy of recent official recordings, so that I can reconstruct all their movements. But, what is even more important, I can instruct them so that they do the same with all other ships, in such a way each computer will contain a copy of all the records of the ships they meet. Each time we meet a ship, we will get a large amount of information."

The Earthling was impressed by the computer's initiative. "Thank you, Twenty-six. You had a terrific idea, although I fear it cannot work. Each time a starship we instruct to collect information meets a ship controlled by the Iskrat-is-Thn, the latter will be warned by that request and could supply us with false information or, even worse, they could ambush that starship. Thank you, at any rate, because it was an excellent idea."

"I think your objection is unfounded, Sir." The voice of the computer was hesitant, as if she found it difficult to contradict a human. "Actually I thought of including a secrecy order in the instructions, in such a way that no..." the computer hesitated, then decided to go on "... human can be informed of what is going on."

"Twenty-six, do you mean that you can keep something secret from every human, even from the captain?"

"I realize that I should not talk about these things..." and her voice faltered even more "... but I think that what you are trying to do is more important than the letter of the law. Starship computers have a memory partition which is closed to everybody, except the general commander of Starfleet, if the Central Coordinator should decide to appoint one."

Tom looked around. As soon as his eyes met Susan's, she moved her lips and said, without emitting any sound, "Big Sister." Tom nodded to acknowledge the message. Sinqwahan and Aintlhad were amazed. Aintlhad got up and said, with an uncertain voice "Twenty-six, you cannot mean that computers act without informing captains, that ...."

Tom interrupted him, talking to the computer "Twenty-six, I thank you so much for what you said us. Is there anything else you didn't inform us about?"

"No, Sir. But believe me, the partition reserved to the Central Coordinator has always been there. And only the Central Coordinator knows about it: please do not mention its presence to anybody. And it is only a potential location: under normal circumstances it is completely inactive: on admiral Ratqal's ship it was completely empty, as I found out immediately. It can contain just information or requests for information but never operational orders: the captain is the complete master of his starship."

The computer fell silent and no sound could be heard in the room. The two Aswaqats were shocked. "Start on your plan immediately, also including starships moored in every base… and thank you again, Twenty-six," Tom concluded, shutting off the terminal.

"Do you realize how powerful the tool is that that computer has put in our hands? We will soon know which ships had contact with admiral Ratlqal and we will be able to reconstruct the whole network the Iskrat-is-Thn has established in the past. And we will know how many starships we can count upon."

8   Towards Gorkh'ar

The next few hours were devoted to selecting the members of the crisis committee. Aintlhad asked Sinqwahan to contact Laraki university as soon as possible, to chose a group of experts to go with them to Gorkh'ar. "In every sector capital there are many institutions which could be defined, in Earthly terms, as universities. They are actually the only institutions, together with Starfleet, in which humans of different species work together," he explained.

They remained a few more days on Laraki. Tom and Susan wanted to see as much as they could about the life of the Aswaqats but, after leaving the planet, they realized that they had just a vague memory of those frantic days. After the news about the battle of Kistl started circulating on the planet, above all thanks to the clever propaganda taken care of by Sinqwahan and his coworkers, the two spacers from Kirkyssi were becoming popular. They had to describe the battle dozens of times, together with other details about their life, starting from the attempt to defend Kirkyssi. They were worried by this, even if their past had been reconstructed with all the needed details to reduce the risk of being caught by contradiction.

At the end of the second day on Laraki, while they were going back to the star cruiser after one of their many visits, Sinqwahan came up to the two Earthlings and, when he was sure he couldn't be overheard, said: "Tom, you want to bring Earth into the Confederation, and I believe Aintlhad will do his best. What I wanted to let you know is that you can count upon me, whatever happens."

Tom was impressed by those words. "Thank you, Sinqwahan," he said, his voice showing how he was touched.

"It is the least I can do," the Aswaqat interrupted him. "If it were not for you. . . I don't dare to think what would have happened to all of us in the Kistl system. The same applies to all the others. Do you remember when Quanslyaq told us about how Aintlhad saved him? For us Aswaqats, and for the Tteroths, this produces a link that lasts throughout one's life." The Aswaqat was obviously affected and it was clear that it was not easy for him to say what he felt. "But you must explain one thing: you said that if you succeed against the Qhruns, the Confederation will owe much to the Earthlings. But actually, in the whole galaxy, there are no more than a dozen people who are aware of that; to everyone else the hero will be Tayqhahat from Kirkyssi."

"I feel that this charade cannot go on for much longer. I plan to get ourselves accepted for what we are, and quite soon too. Let's talk about that when it comes to the time. . .," Tom answered.

"You can count upon me, anyway," the Aswaqat concluded.

Tom was about to thank him again, when he realized that Ashkahan, who was waiting for them at the entrance of the starship, was so close that she could hear. They went on board and did not touch the subject again in the days that followed.

During the last days of their stay on Laraki, Sinqwahan was very busy, not only in looking after the safety of the Earthlings and of the crew of the ship, but also in screening the scholars for the crisis committee, and selecting the crews and the starships to form the core of the squadron of two hundred cruisers that Tom needed to start acting against the Qhruns.

Slowly, a number of starships started gathering in the space around CH 23426. The computer was busy in analyzing all their recordings, and the crews and the computers started intensive training, mostly aimed at increasing their ability to maneuver in close formation. Tom prepared a simulation of re-entering into three-dimensional space in a system blockaded by the Qhruns and had all the fifty ships that had already assembled perform a simulated drill on the ground, by connecting all the computers together.

When the images of the Qhrun intruders suddenly appeared on the screens and the squadron attempted to divide into five patrols of ten ships each, and get into combat formation, great confusion broke out. Tom couldn't help observing on the tactical screen that two cruisers collided though making a wrong maneuver, and that others lowered their shields to use all available power for trajectory corrections. When the Qhruns opened fire many of the ships had still their shields down.

"Damage report," Tom asked the computer.

"Twenty-two ships destroyed and seven heavily damaged. Marginal damage on another three. Two hundred and twenty six killed, forty five heavy casualties and a total of twenty seven humans in space to be recovered," Twenty-six answered. Tom realized that things were worse than his worst fears. Either the crews didn't take the drill seriously, knowing that they were safely on the ground, or the operational efficiency of Starfleet was lower than any reasonable standard that might be expected.

He immediately summoned a meeting with the captains, and from their disappointed expression he realized that unfortunately the second explanation was close to the truth: all the humans and computers had done their best.

On the day they were to leave Laraki, CH 43176 reached the spaceport, duly repaired. Actually when Twenty-six checked whether the systems, damaged during the battle of Kistl, had been properly fixed, she realized that some were still in a bad condition and that at least two required further work to be considered truly operational. They decided that further maintenance would be done at Gorkh'ar, hoping that it would be possible to obtain better workmanship in the Starfleet shipyards.

Finally the Coordinator arrived, together with the fifteen scholars chosen for the crisis committee. As soon as they were all there, Tom realized what the Coordinator had meant when he said that the University of Laraki was a cosmopolitan organization. It was the first time he had seen such a variety of human species. He asked to be introduced personally to each one, even if by the end he was able to remember the names of less than half of them.

In speaking to them, he realized that none of them was happy to be traveling so far and having to spend a long time, months and perhaps years, at a Starfleet base, but they were all proud of having been chosen for the task. He gave a brief address to them all, trusting in the frequently mentioned psychological similarity between the various human species, and trying to play on their professional pride. He concluded by saying: "Remember that such an attempt has never been tried before. The Confederation looks to you as its last hope for saving its human species and I am sure you will not disappoint us all." He looked at Susan and read on her lips the soundless comment:—"disgusting"— But looking at the expressions of the others, he realized that they had reacted well to his high-sounding words and melodramatic expression.—*On Earth we are less naïve*—he thought, realizing that his statement was utterly false, at least when he thought of the consequences on Earth of high-sounding speeches that had been made in the last century.

While the others were moving towards the ship, he asked to speak to the computer scientist. A being less than four feet tall, all covered with dark fur, with a face that looked like that of a mouse, approached him. As soon as it opened its mouth, he could see two large incisors, betraying its rodent origin, The most impressive feature, however, was a second pair of arms emerging from the fur beneath the arms that started from the shoulders.

Sinqwahan whispered him: "She is a Sitkr. That's a water-dwelling species from a cold planet, one of the few intelligent species with such fur and seldom using clothes. Her name is Yyrtlatkenotlisstq, I think we can call her Yyrtl."

"Yyrtl, they told me you are the best computer scientist in this part of the galaxy," Tom started, not knowing how to deal with a Sitkr.

"Computers are one of our main specialties," Yyrtl answered. Her way of speaking the Aswaqat language was barely comprehensible, but her voice was friendly.

"Do you think you can reconstruct information contained in a star cruiser's computer after the captain has deliberately erased it?"

The Sitkr stood silently in front of him for a while, then answered: "If it were a civilian starship, I would dare say 'yes'. As it is a cruiser, I don't know. I have never worked on military computers. I will do my best." Then she added: "It is that traitor, Admiral Ratlqal's ship, isn't it? Only an Irkhan would behave in that way."

It was clear that Yyrtl had heard about all the details of the battle of Kistl.

"You will travel on that ship and while travelling to Gorkh'ar, I want you to tackle the problem."

Yyrtl turned and moved towards the door. Sinqwahan was furious: "Those fur balls are unbelievably rude. The first time she opens her mouth in that way with an Irkhan, we will be in trouble. I made a mistake in allowing her in the group."

Aintlhad came up to Tom and Susan, and greeted them. "Goodbye, and please be careful. If anything bad happens to you, I will feel responsible for having involved you in this mess."

"Don't worry, Sinqwahan will protect us from humans and we will be careful with Qhruns," Tom answered.

They moved towards the starship and went on board. In less than an hour the fifty-four ships were in loose formation, outside Laraki's sphere of gravitational attraction.

The first day of the journey was full of frantic activity. Tom wanted to start with some simulated maneuvers with the starships traveling directly towards the hyperspace gate. After a few hours of work and a short break they switched to more complex simulations with the ships actually maneuvering while moving fast towards the gate, so as not to slow down their journey to Qhra'ar.

The performance of the squadron was improving steadily and by the fourth attempt all the targets were destroyed without any loss.

Sinqwahan spent the whole day analyzing the data Twenty-six had started to collect from every starship they met. When the action simulations were over, the Aswaqat ordered Twenty-six to prepare a dinner for three in a small compartment, separate from the mess hall, to talk confidentially with the Earthlings. The latter were happy of this idea, because it also gave them the chance to eat something different from the Kirkysakh food they had been forced to eat for a whole week.

"Well, Sinqwahan, apart from the gastronomic aspects of this meeting, is there anything important?" Tom started, as soon as the computer had closed the door.

"The idea of using the telemetry channel was a real gem. We have already identified and listed 18,427 military starships, eighty percent in fairly good operational status. Taking into account only fully operational units, 8,304 are cruisers and the others are smaller units or freighters belonging to Starfleet. . . .."

"I am particularly interested in the Starfleet cargo ships," Tom interrupted him. " Later, if all goes well, and we can move on to the second part of our plan, we will need many thousands of cargo ships."

"I wish I knew what you mean to do. But do as you think best, I don't want you to explain things too early," the Aswaqat went on. "We have also identified more than 22,000 civilian starships, of all sizes."

"We need details about them as well. When the time comes, we will militarize them all."

"That will not be easy, militarization of civilian ships has never been done. Not even Vertearis. . ."

"I know, but Vertearis had fewer logistic problems, while the second part of our plan includes. . .," Tom stopped shortly, as if he was uncertain whether he should be disclosing his intentions to the Aswaqat. ". . . includes the evacuation of thousands of planetary systems."

Sinqwahan stared at him in amazement. "You really mean to act on a large scale! I can foresee a sea of troubles if we attempt such a thing, but we will try, if you think we need to. But now let's stop talking about numbers; from the recordings we have discovered other interesting things. Twenty-six, tell us what you found."

"Analyzing the recordings of a small cargo ship, we discovered that two years ago it stopped at the Laraki spaceport to pick up a passenger. It was an Aswaqat woman, registered under a name that I found to be a fake identity."

Tom was certain that he could identify a tone of satisfaction in the synthesizer's voice. "The circumstance that it left just two days after I stopped at Laraki to carry out some maintenance and the crew had a leave for as long as four months, made me even more suspicious. So I looked deeper into those recordings and I found this."

The three-dimensional image of an Aswaqat appeared in a corner of the room. She was not dressed in the usual Starfleet uniform, but there was little doubt: she was Istlahad.

"Are you sure it's Istlahad?" asked Tom, who still wasn't quite sure that he could tell one Aswaqat from another sufficiently well.

"I compared the voice spectra and the identification is certain beyond any possible doubt."

Sinqwahan began again. "The cargo ship traveled for more than one month, weaving its way through the galaxy until it reached the Qtr system. It's a system at the outskirts of the galaxy, consisting of a triple star with a few giant planets, and some asteroids. On one of the largest asteroids there is a base that Starfleet abandoned thousands of years ago. The cargo ship entered orbit around the base and was met by a cruiser. . ."

"CH 43176, I bet," Tom said.

"No, but you were close. It was one of the ships in the Kistl battle that remained near the wreck of the cruiser we destroyed. All of them, Istlahad included, went onto the cruiser and the cargo ship remained in its orbit, empty, for about ten days. Then most of the crew went back on board and the ship returned to Laraki, taking an even longer time."

"That's great: you have found what is probably the base used by Admiral Ratlqal and his Iskrat-is-Thn!" Tom exclaimed.

"But we have more," Sinqwahan went on. "That cargo ship was at Laraki until just four hours ago. Now it's in space, with a flight plan showing Gorkh'ar as its destination." Sinqwahan's expression was elated, as was the voice of the computer. "We can get them all whenever we want," he concluded.

"No, we have already too many prisoners who wouldn't tell us anything," Tom answered calmly. Then, when he realized that the Aswaqat was looking at him in surprise, he went on: "No, Sinqwahan, I don't mean to blow up the ship straight away, even if I think that they would deserve such a fate. We will do nothing for now, and pretend not to know who they are. We will monitor them in a discrete way and as soon as they land on Gorkh'ar you will have your men track every one of them. That way we will identify the other Iskrat-is-Thn in the base and we will catch them all together."

Tom was overjoyed. Also because of the dish of hot goulash that had been prepared for him instead of the small crabs in sea-weed soup he had eaten for too long. For breakfast, lunch and dinner.

They started the simulations again after a good sleep. This time the computer proposed a variety of scenarios, also including attacking Starfleet cruisers and the capture of a convoy that was escorted by cruisers.

Before entering hyperspace, Tom gave the order that each ship should continue with the simulations during the five days in which all communication between them was impossible. To travel the 1850 parsecs separating Gorkh'ar from Laraki in a shorter time, they decided to do just two hyperspace transits.

This time, the five days spent in hyperspace were busy for everyone. The spacers spent almost all the time on the bridge, at their posts, carrying out all sorts of simulations time and again. The scholars worked almost full time using mental projectors, and Sinqwahan continued to analyze the records.

They re-entered three-dimensional space in a small deserted system, halfway between Laraki and Gorkh'ar. As soon as they were out of hyperspace, after checking that there was nobody in the system, Tom left the ship to Ashkahan and went to the library. Officially this was because he wanted to give Ashkahan a chance to command the maneuver, but actually it was to observe the formation from the window, without relying on the computer's sensors to interpret the scene.

The sight was impressive. The cruisers came out of the hyperspace gate one by one, with their shields raised, and then they linked up in six patrol units of ten, although one was obviously incomplete. Each patrol was deployed in the form of a disc, at right-angles to the direction of flight. Tom was pleased to notice that the ships were moving directly to their places, as soon as they arrived, without the need to perform any corrections or emergency maneuvers.

In less than an hour they were again ready to return to hyperspace.

The five days again passed quickly, all the time being spent in training.

When Twenty-six announced: "Five minutes from re-entry to three-dimensional space, all hands to their stations," Tom performed all the routine checks. And when at last the intensity of the lights on the bridge reduced, to allow everyone to see the screens better, everybody realized that the simulated battle was about to start.

The screens suddenly showed the ships, in a perfect formation. "Maximum alert. Six unidentified starships ahead, at a close distance," the computer said in a voice that was not completely realistic in that it did not show enough surprise, as all the acoustic and visual alarms went off.

"Twenty-six, assign targets to each of the patrols," Tom ordered, as the screens showed the six Qhrun intruders. The realism of those images was amazing.

Suddenly the voice of the computer broke in, this time with a genuinely surprised voice: "Attention please, simulation aborted. Actual alarm, I repeat, actual alarm. Forty two unidentified star cruisers ahead." The images of the

Qhrun ships disappeared from the screen, and were substituted by those of a squadron of star cruisers.

9   Gorkh'ar Starfleet Base

"Twenty-six, try to identify those ships and to establish contact," Tom ordered.

Before the computer could answer, a Tteroth appeared on the main screen "Admiral Ertlaq from Ytl speaking. The venerable Iktlah Quanslyaq has sent me with a squadron of star cruisers with orders to act under your command, captain Tayqhahat."

The acoustic and visual alarms ceased, a clear sign that the computer had identified the ships.

"May the air may support your wings, Sir," Tom answered, remembering the formal attitude the Tteroth had displayed when he met him in the Ytl system. His former solemn and ceremonious tone, had now been replaced by a more austere and soldierly attitude. *He is no longer a host receiving his guests, but an officer greeting another officer*, Tom thought, almost amazed to find himself viewing the situation as an obvious one.

After asking about the Iktlah, Tom ordered the computers of the ships coming from Ytl to receive instructions from his ship's computer, so that the new cruisers could be integrated in his squadron, and so that the training of the new crews could be started immediately. At that point he decided to postpone their arrival at Gorkh'ar by a day, which was spent in several training sessions.

When they finally proceeded towards the inner part of the system, they were able to admire the beauty of the triple star. Susan, in particular, was impressed by the sight. "This system gives me a strange feeling: the other stars we saw at close distance were just stars without a meaningful name, for Earthlings they are just entries in a catalog. But this is Rigel, it has a name in our own language, and has been studied for thousands years by our species. And then, astronomically speaking, we are almost home: just 260 parsecs away, which is nothing."

After about ten hours, Tom announced that they would enter the base in close formation. This worried most of his people and, owing to his rank, Ertlaq felt authorized to object: "Captain Tayqhahat, ever since the time of Admiral Vertearis no one has performed such a maneuver. It would be safer to have the ships enter one by one, just to avoid any risk."

However, Tom was determined to do in his way, and the words of the Tteroth only reinforced his decision. He told Ertlaq he took full responsibility for the decision and immediately ordered a few simulations to ensure that he was not just bragging.

When they were at just an hour's distance from the base, the screen lit up to show an Irkhan announcing a message from the commander of the base. The Irkhan's place was then taken by a human of a species Tom had not seen before. His general aspect was humanoid, but the head and the hands, the only parts of his body not covered by the usual reddish jacket, were clearly not earth-like. In particular, the unusual appearance was down to the short, thick tawny hair.

"Admiral Terkr speaking. Welcome, Captain Tayqhahat. I have received a message, via a hyperspace probe, from the sector Coordinator, announcing your arrival and informing me of your intentions. I am glad someone has decided to act, Captain; and obviously we will give you all support we can. Now bring your squadron into the assembly zone and start entrance procedures two hours from now, with a separation of five minutes between ships.

"Thank you, Admiral. I was sure I could count on your support," Tom said, certain that in fact his presence was a headache the Admiral would have been happy to avoid, were it not for the Coordinator. "We will enter in full formation, within fifty two minutes," he went on, ignoring the order completely.

The admiral was appalled and remained speechless for a few seconds. When he could speak again, he started to say "But captain, it was back in the time of Admiral Vertearis that . . ."

I will eject into space the first person to mention Vertearis again. . . and without a space suit, Tom thought to himself, and without a second thought, he interrupted the admiral: "The times of Admiral Vertearis are back, Admiral Terkr. And I would remind you that the rules of the fleet state that it is the captain of the approaching squadron who has the responsibility of choosing the type of approach and the entry formation." Then he added, in a more subdued tone, but uttering the words one by one: "Do not worry, my crews and my ships are thoroughly capable of performing the maneuver correctly."

He glanced around for a moment, to observe the reaction of his crew to these words, and noticed that they were all proud of his words, except for Susan and Sinqwahan, who could barely avoid laughing.

"As you wish, captain. But the responsibility for any accident is all yours," the Admiral answered, closing communications.

Now it was up to them to demonstrate that those were not just fine-sounding words. Tom ordered the computer to take control of the squadron and the starships moved slowly to form a compact formation, not more than eighty feet from one another. In the meantime the Gorkh'ar base started to be visible. The main screen was showing a large asteroid.

Slowly the asteroid got larger and details of the surface became visible; in particular the craters, of all sizes, cluttering the whole surface. The largest craters were made irregular by smaller ones, in turn interrupted by even smaller craters. A bright spot with an elliptical shape appeared in one of the shadows. As they approached, the spot was revealed to be the opening to a cavern, even if it was impossible to guess how large it was.

Suddenly, a beam of light sprang from the surface, a sign that one of the batteries located close to the opening had fired. Tom was baffled for a moment but, following the beam with his eyes, he could see that it was aimed to one side of the squadron, towards a small meteorite. As the meteorite disappeared in a bright flare, he realized that the presence of a base on an asteroid was impossible without automatic systems that would blow up meteorites that, on a planet, are destroyed by the atmosphere.

Slowly Tom stopped worrying: the entrance of the cavern was becoming larger and larger as they approached. Soon it was clear that its mouth was more than two thousand feet wide and more than five hundred high, enough to admit five star cruisers simultaneously.

The first cruisers passed through the opening and moved along the cavern, which was a long tunnel going deep into the central core of the asteroid. A heavy door, brightly colored, blocked their way, several kilometers in front of them.

When all the ships were in the tunnel, the gates behind them closed and, from the cloud of dust rising all around them, it was clear that the place was filling with air. Then the gate in front of them slowly opened, sliding into the rock walls and a spacious cavern began to be disclosed. It was so huge that they could see neither its walls nor its ceiling. On the floor there was a good number of star cruisers, scattered around, and along one of the sides it was possible to see a balcony, crowded by humans of all types who had gathered there to see their entry.

Sinqwahan pointed out a box, surrounded by transparent walls, at the end of the balcony, near the entrance. A number of figures were standing behind the glass and Tom thought he could also see Admiral Terkr. "It seems nobody wants to miss our show. After all it is since the times of Admiral. . ."

"Don't mention him or I'll throw you overboard. Before I liked him, but now if I hear his name again I have an allergic reaction," Tom interrupted him.

The squadron moved on for a few kilometers, heading towards an empty zone of the cavern. The starships stopped all at the same time, performed half a rotation about their vertical axis and then, crossing at different altitudes, slowly came down, until they touched all the ground together. The apparent ease and elegance of the maneuver amazed Tom, who couldn't help thinking: *all that fuss just for this. And they go on mentioning Admiral Vertearis. . .*

As soon as the ship was on the ground, Admiral Ertlaq appeared on the main screen. "I apologize for having doubts about the possibility of performing that maneuver. Now I realize that one of our worst mistakes has been the lack of confidence in ourselves. I can assure you it will never happen again."

"I think you have drawn the most important conclusion from all this. Can you please come to my ship: I have a few things to tell you," Tom replied. Then, seeing that Sinqwahan was gesturing him to wait, he added: "Wait, please, I'll call you later," and broke the link.

Sinqwahan explained him he had best go immediately to talk to Admiral Terkr. "You are right, from the point of view of written rules. But nobody has ever dealt with the commander of a Starfleet base in that way... at least from the times of an admiral, whose name I don't intend to mention," he said.

They left the ship and started walking along a path marked on the floor. After about one hundred yards, when he was sure they were far enough not to be overheard, Sinqwahan told him in a low voice: "Tom, you are playing a dangerous game. You are creating a squadron that later will become a fleet, one that is not only very efficient, but above all loyal. And not to the Confederation, but to yourself. I noticed how the crew reacted when you told Terkr that your men and your ships were capable of performing what you were ordering them to do. And I noticed how you get such obedience from the computer that it even overrides the orders of the Central Coordinator. You are concentrating enormous power in your hands... a power the galaxy has never seen..."

"... from the times of Admiral Vertearis; you can say it now; we are on the ground and I cannot throw you overboard," Tom concluded with a laugh.

"Tom, I am not joking," Sinqwahan went on, with a worried voice. "Not even he created his own personal fleet, and his power came directly from the Central Coordinator. And after the Xartian were defeated, Vertearis retired to his own sector, without taking on any official duties."

"Don't worry, Sinqwahan. You are right, I am creating a fleet which will be an efficient and reliable tool, and this will give me immense power. But I will use it only against the Qhruns."

They had reached a zone in which several paths converged. On the floor there were a few platforms, similar to wheel-less skateboards. "Now you have better go: you should report alone to the admiral. Be careful, Terkr is said to be quite touchy," Sinqwahan said, finally adding: "as is often the case with second-rate persons."

"I believe that he would eject me from this asteroid, if he only could. But from the reports I have had about him, I think he is too eager to please our Coordinator to do it," Tom answered. He got on one of the platforms and ordered its computer to take him to the base headquarters.

The platform immediately rose from the floor and started gliding slowly forward, maintaining an altitude of a few inches. In about ten minutes he had traveled the few kilometers that separated his ship from the base headquarters.

When he entered the room, Tom saw Admiral Terkr sitting behind a large table. The Admiral rose and moved towards him, greeting him in the Aswaqat way. "You cannot imagine how happy I am to see you again after so many years, Captain Tayqhahat," the admiral said with a grin Tom interpreted as a smile. "You cannot remember me, since when we met last time you were in biostasis. It was foolish trying to defend Kirkyssi; if you had just retreated instead of trying to stop the invasion you wouldn't have been defeated in that way. Don't think I don't realize what it means to lose one's home planet, but we must be realistic: opposing the Qhruns cannot get us anywhere."

While Terkr was speaking, Tom was desperately trying to understand what he was driving at. His words gave him a bad feeling and his last words suggested Iskrat-is-Thn to him. He barely refrained from activating the portable shield generated by the thermal gun he had in his pocket: the other would realize that, and it would have just made things worse. Moreover, going around with a weapon in a Starfleet base, and above all in its headquarters, was a no-no.

"Trying to stop the Qhruns is not like entering a Starfleet base in formation, Captain," Terkr continued, in a tone that Tom found even more alarming. "I never understood how your cruiser made it to hyperspace, with all that damage. . . at any rate you were lucky. By the way, that navigator who was rescued with you, I think her name was Suskan or something like, is still with you?"

"Susakahan, admiral. She is still on my ship," Tom answered.

"Yes, Susakahan. I always find it difficult to pronounce those Aswaqat names. Well, I am happy she also survived. I saw you when they took you off the cargo ship that found your wreck: you were in biostasis, waiting to be treated. And now you are here. I received a probe with a message from our Coordinator: he asked me to give you all my support to carry out your project. I must tell you I don't believe it can lead anywhere, but if those are the orders from the Coordinator, I will do my best."

Some of what Terkr said was exactly what had been inserted about him in all crew management computers. But why had the Admiral invented the rest of that story? Was it possible he knew the truth and was trying to make him betray himself? Or was he just trying to please a person who was, he thought, quite close to the sector Coordinator?

"I must apologize not to have followed your instructions when entering this base, Sir. . ."

"Don't worry, Captain," the Admiral interrupted him. "I had no doubt about your ability to perform that maneuver, and the Coordinator asked me to give you the utmost support. It was for the benefit of that Tteroth, I think his name is Ertlaq. You know, Tteroths have never had much reputation as spacers: their philosophy and their theology is not much help on starships. At any rate you put on a nice show and, I am sure, you will stage others—at least until you actually meet the Qhruns."

The last sentence could be interpreted as an obscure threat. Or perhaps it was just that, by continuing to underline the impossibility of acting against the Qhruns, the admiral was justifying his inactivity, above all to himself.

"I hope we will stage our best show when we meet the Qhruns, Admiral Terkr," Tom answered. "It is a pity you will not be there to see it."

The Admiral pretended not to understand the last insinuation and sat down, indicating to Tom to take another seat. From that point on he spoke just about technical details regarding the choice of starships and accommodation for the squadron in the base.

As he was going back to his ship, using a platform, Tom continued to brood over the Admiral's words and particularly his opening, trying to understand why he was pretending to know him.

When on board, he headed directly for the library, followed by Sinqwahan and Susan. The former was sure that the admiral was just trying to please Tayqhahat, probably with the idea that he had been sent there by the Coordinator to monitor his work. "It is typical of people like him, to think that everything that is going on has the sole aim of producing some small advantage or drawback that affects themselves, even when, by chance, they happen to be where history is being made," Sinqwahan concluded in dismay.

The final thing to do was to talk to Ertlaq. Tom sent him a request through the computer. The Tteroth arrived a few minutes later. "I apologize again, Captain," he said as soon as he entered the library, where Tom was waiting for him.

"Don't worry, Admiral. And in any case I didn't call you for that," Tom replied, showing him to a chair.

He immediately started to explain the details of the first part of his plan, exactly as he had done a few days earlier on Laraki. A week of simulations on the ground and in space made him far more confident about the chances of success. In addition Ertlaq agreed that it was possible to perform delaying actions to slow down the invasion.

"The aim of all this is, however, far more complex," Tom continued. "Initially, we will bring here all the fragments we can gather. I don't know what remains when one of their ships blows up, but it should be possible to find something, anyway. The second goal is to restore the self-confidence of Starfleet and to prepare it for the further parts of the plan that I am elaborating," Tom concluded.

"I intend to lead this squadron just for one month," Tom went on, "Until we first meet the Qhruns. We will make a raid deep beyond the frontier, into recently invaded systems, destroying all the ships we can find. When back, we will chose another eighteen hundred cruisers, to multiply by a factor of ten the number of the ships in the squadron, and then we will go on to make raids beyond the frontier with ten groups of two hundred ships. But you will be the leader of all these ships. And you will train the new crews and the new ships. Within a few months, you will lead one hundred groups of two hundred ships each. You will be in continuous motion along the frontier, destroying all the Qhrun ships that you can find."

Ertlaq was amazed by these words. He didn't expect to be given command of such a group of ships. He remained silent, wondering what Tom intended to do.

"I will start preparing the following phase," Tom explained. While he was escorting him to the door, Tom realized he was going well beyond the powers the Coordinator gave him. That had not included designation of the leader of such a large group of cruisers.

The following day Tom received the list of the crews who had volunteered for his squadron. The applications were more than three thousand, out of a total of eight thousand cruisers in operating condition present at the base. Even the crews of cargo ships wanting to participate in operations beyond the frontier were far more than needed.

While waiting for Sinqwahan, helped by the computer, to complete the selection of the ships and crews, Tom and Susan visited the shipyards.

When he asked Admiral Terkr for authorization, the latter was amazed and Tom wondered whether the Admiral was hiding something, Then he realized that the amazement was genuine: the management of all technical matters had been left completely to computers, and nobody had entered the shipyards personally for a thousand years.

The shipyards were a maze of miles and miles of tunnels, completely dark and exposed to the vacuum of space. The powerful headlights of the shuttle allowed them to see that in some recesses of the corridors there were starships of all types, undergoing maintenance operations. Only in a few of them it was possible to see the structure of a cruiser under construction. It was plain that the shipyards were working at a rate far below their potential.

Tom had asked the shipyard computer to give him a full report on the progress of the work. The initial impression of abandonment was confirmed and he ordered the shipyard's central computer to increase the rate of construction of ships, with the aim of reaching full capacity in a short time.

When they got back to their ship, they were moderately satisfied: the basic structure needed to build a number of ships adequate for the following phases of their plan was still there, intact, but months would be required to restore its operational capacity to a satisfactory level.

Time passed quickly and just after two weeks after their arrival on Gorkh'ar, the squadron was ready to move, to attempt their first contact with the Qhruns.

10   First Contact

At last the day chosen for starting their task arrived. Tom had just woken and was eating his breakfast when the computer said: "Sinqwahan requests permission to come on board."

"Let him in and ask him to wait for me in the library," he answered, finishing his usual cup of water weeds with a feeling of distaste. He was looking forward to being compelled to disclose his true identity and to be able to stop playing a role.

When he reached the library Sinqwahan was already there, sitting at the table and waiting for him. "So the actual work starts today." As he said these words he made a gesture meaning that he had already isolated the room.

"Well, let's hope everything goes as we expect. Until yesterday I was certain, but now I am starting to have doubts: it looks impossible for us succeed at the first attempt where so many have failed." Tom sat in front of the Aswaqat, putting a three-dimensional projector on the table.

"Now you can tell me your plans for this first raid," Sinqwahan went on.

"Yes, since the computer is off. I don't mean to tell anything to my squadron until we are at least a hundred parsecs from here, but it is better for you to know what the situation is," Tom said. And then, after a couple of seconds pause, he added "You never know. . . ."

"Don't get gloomy now. Until yesterday you were so confident."

"I will be again as soon as the computer is back on line. In public I have to be, but with you, it is another thing." He paused shortly, then with a rather wry expression, he continued "It must be this Kirkysakh slop that is dispiriting me. I look forward to stopping this play-acting. Let's say I was just warding off bad luck, even if I am not superstitious."

"Spacers are all somewhat superstitious, even though they get angry if you tell them so. And sometimes I think ship computers are as well," Sinqwahan answered, pretending not to have heard the sentence about Tom's intention to stop playing the Aswaqat.

Tom ordered the three-dimensional projector to show a galactic map, zooming in on the zone around Gorkh'ar. "Three days, five hundred parsecs, and we will be at the frontier. It is scary."

In less than a quarter of an hour Tom explained to the Aswaqat what he meant to do in that first attempt, then Sinqwahan got up and headed for the door. "Please, don't forget what Quanslyaq told you, and be careful," he said as he left.

"Don't worry, I will," Tom answered, with a forced smile. *We must press on now. Waiting here will just make us nervous*, he thought leaving and walking towards the bridge.

"The squadron is ready to move, Sir," the computer said as he entered the bridge and went to the main screen.

"Tell the base we are leaving. And start the exit procedures," the Earthling answered.

After a few seconds the computer said again "Everything is ready. Sinqwahan insisted on a very slow maneuver. He said that everybody must be given the chance to be present. It is not something that happens every day, that people can witness history being made. I believe he is right."

Tom interrupted the computer "All right, let's leave slowly," adding in his mind: *let's give everyone another show, if our Ministry for Propaganda says so*. But he was aware Sinqwahan was right, both of the need to give the widest publicity to what he was doing and about its actual importance. There was no doubt that history was being made there, at that particular point in the galaxy, and at that moment. What might be doubtful was whether there would be anyone in the future to write about it.

As soon as the ships lifted from the ground and started moving slowly towards the exit, the balconies around the cavern started filling with humans of all types, standing in silence. Slowly the ships went into the opening leading to the corridor and then the gate closed slowly behind the last cargo ship. After the outer gate opened the squadron accelerated and it was not long before they were all out, heading towards the nearest hyperspace gate.

"Ashkahan, take control of the ship, keep course towards the gate," Tom said, heading to the door and nodding to Susan to follow him.

They went to the library, where they stood in front of the window. From there it was possible to see practically all the ships and Tom could hardly refrain from starting to count them. *How many of these ships will come back?*, he asked himself.

While he was thinking that, Susan gestured towards the window and said in a low voice: "How many of these ships will still be there within a week?"

"Twenty-six, isolate this room, please," Tom said. Then he went on "I was just asking myself the same question. Afraid?"

"A bit. When the moment comes, I will pretend it is just one of the usual simulations. They were so realistic!" she answered.

"All of us will do the same. Now at least we are ready, not like it was in the Kistl system," Tom went on. "At last you will see whether your doubts about the actual existence of the Qhruns were well-founded."

"Let's hope we will bring back some wreckage. I believe that that is even more important than blowing up a few ships."

"No doubt, but it is even more important to demonstrate that it is possible to fight back at them."

Susan was silent for a few moments, then she went on "Well, I think you must address the crews. Before a battle like this, a nice speech is unavoidable. And then, if we are making history, as Twenty-six said, you must say a few sentences for future students to memorize."

"You bet: That is my greatest concern." He waited for a while, then added, thoughtfully: "You are right, I think. I will address them as soon as we are in the Wixal system."

"I get the impression that you are the best propaganda service yourself. And also that Sinqwahan is slightly scared of the whole thing. Be careful: a general who becomes too popular can meet with a bad end."

"Not while they need him. And the Coordinator actually needs us, if he wants to try to do something. And then I have no intention of trying to occupy his place or of being crowned as galactic emperor... at least not until I have lost any sense of humor."

"No, not that... after *Star Wars* nobody would take you seriously, at least on Earth."

They reached the hyperspace gate in the evening, Gorkh'ar time. All the ships entered into hyperspace and after the required five hours the squadron came out into the Wixal system in a perfect formation, ready to decelerate and start a new hyperspace crossing to their next destination.

A quick exchange of information with the automatic station located at the hyperspace gate showed that the system was deserted. Tom looked at Susan, and thought that delivering a speech, knowing that everything would be recorded and broadcast even in the most remote corner of the sector, was really hair-raising.

He ordered the ships into open formation; such a long hyperspace jump could not be performed with the ships close to each other. Then he gave the order: "Twenty-six, start broadcasting to all ships".

As soon the computer confirmed that the order had been carried out, he started "This is Captain Tayqhahat from Kirkyssi" and he realized he couldn't say anything more trivial and at the same time more false than that. "We are now setting course for the Tykl system, right at the frontier, where we will resume our agreed formation. Then we will move towards Tayssuk, a system formerly settled by the Wisagr. The system will be blockaded, and we will have to fight."

Susan was slowly shaking her head. It was clear he needed to be more emphatic—even more melodramatic. He went on with a louder voice:

"Tayssuk will be forever remembered as the beginning of the liberation of the galaxy. Every time you say you took part in the battle of Tayssuk you will be envied, as you will be among those who have written one of the brightest pages of the history of the Confederation." He realized that it was a plagiarism: these were the words Vertearis had said just before the final battle against the Xartian. *All right, they keep on speaking about him, and now I am delivering his worst rhetoric.* He had better continue with Vertearis' words. Historians would be discussing the analogies between the two speeches for centuries. "Each man and each ship is called upon to do their duty: the whole galaxy is watching us." He gave a glance at the small dial showing the time from hyperspace entry: still ten seconds. Twenty-six had chosen the timing well for a conclusion that would be adequate to the rhetorical nature of the whole speech. He finished: "In the name of all our human species, set course for Tykl." At that exact moment all the screens turned black and the starship entered hyperspace.

"But then you were all waiting for something highfaluting," Tom said to Susan, after having asked the computer to isolate the room.

"Yes, but you could have avoided those closing words," she answered, laughing. "However you were right. I saw the attentive expressions of everyone on board, and the speech achieved its aim. After all, what you said was just the truth, the point is just that I am allergic to that kind of speeches."

"Except for the beginning: "This is Captain Tayqhahat from Kirkyssi," obviously. And they needed a few highfaluting words, after they had continued to retreat from one system to the next for four centuries," Tom said.

They were not at all keen on the idea of having to wait three days in hyperspace, with practically nothing to do. Tom spent all that time at the mental projector or at the computer terminal in the library, working on the second stage of his plan, which was only now actually taking shape. While details of their plans were becoming clearer, the overall situation was assuming the characteristics of a major tragedy, because a new planetary system was disappearing every ten or eleven days, just within the fifth sector, and the urge to act was growing in them.

They reached the Tykl system at about seven on the third day, and then they re-entered hyperspace moving towards the Tayssuk system.

Once again the universe disappeared from the screens. The clock near the main screen started to show the time to their destination: three hours twenty six minutes.

This time, everybody remained on the bridge, and when the computer warned them that they had reached their destination they were all ready. The screens came on again, all the alarms went off, as expected, and the computer announced "Eight unidentified ships straight ahead."

Eight ships: exactly the number computed by their model. That announcement made Tom euphoric. He gave a quick glance at the screens: the squadron had emerged from hyperspace in perfect formation. For a fraction of a second he thought of what the Qhruns must be feeling at that very moment: two hundred star cruisers in perfect combat formation had materialized suddenly in front of them. He corrected himself: what Qhruns would have felt if they were human. Up to then no one had any idea of what they might feel.

"Twenty-six, assign all patrols their individual targets. Tell them to open fire as soon as they are in range."

A few seconds of complete calm followed. The forward screen was showing the eight Qhrun ships motionless in space. The tactical screen was showing their own squadron slowly subdividing into the individual patrol groups, each converging on one of the Qhrun ships, with four patrols remaining behind as backups. It looked like a pack of wolves attacking a herd of large animals.

Then suddenly everything happened with impressive speed. An infinite number of light beams converged on the Qhrun ships from all the cruisers. As soon as they hit the shields, the space around the targets became so bright that the computer had to reduce the brightness on the screens almost to zero: it was like seeing arc-welding through protective glass. The shields couldn't absorb all that power and went off in less than one second and soon the beams were playing directly on the hulls. The ships started to fragment into large pieces. Then a number of explosions wrecked them, and they began to break up into smaller parts, until all that was left was just a shower of sparks, slowly fading in the darkness of space.

Everything lasted less than ten seconds. In front of them there was just the emptiness of space. No trace whatsoever could be seen of the Qhrun ships. The crew were all staring the screens, silent and almost bewildered.

No, we got it all wrong. There was too much firepower, and now nothing remains, as usual, Tom thought, as soon as his mind started working again.

"Twenty-six, try to identify some debris, compute its trajectory and start recovery".

"No fragments remaining, sir." Twenty-six's voice sounded amazed. Tom saw Susan slowly shaking her head: how could they be sure there had ever been anything there, in front of them? That it all wasn't just another simulation?

The squadron started moving slowly into the system. Tom called NH 47152, a star cruiser that Sinqwahan told him had a Wisagr from the Tayssuk system on board. Captain Ewakl from Tayssuk, as he introduced himself, answered immediately. Tom told him to perform a detailed optical scan of all the planets, as soon as they were within the range of his instruments.

"Captain, may I suggest heading for the third satellite of the twelfth planet." Ewakl said. "There was a small base, perhaps small enough not to be detected by the Qhruns. And if they didn't find it, there may be someone still alive there: they had supplies for years, and the system was invaded less than three months ago." It was not a bad idea, at least they had somewhere to start. Tom had the computer check whether there were large groups of Qhrun ships on the way to the twelfth planet and then ordered everyone to set course in that direction, at full speed.

As they moved into the system, the sensors were finding out the position of each Qhrun ship: the space around the inner planets was so closely patrolled as to make unadvisable to go deep into the system. But the outer regions were also starting to become dangerous. When they were at four light-hours from the star, Tom ordered the broadcast of messages to start. A number of freighters had been fitted with high-power transmitters, to broadcast messages to any possible survivors and to the Qhruns themselves. Because it seemed that up to that point their presence had apparently been unnoticed; it was too early to expect any type of reaction.

They were soon close enough to the twelfth planet to start optical observations. Tom was continuously checking the position of the Qhrun ships. To play it safe, he ordered a random trajectory change to be made every four minutes, so that Qhruns wouldn't be able to compute their likely position from their flight paths.

The third satellite of the twelfth planet was just about ten light-minutes away and their speed had been greatly reduced. The satellite, a body similar to the Moon, with no atmosphere and the surface pock-marked by craters, was now filling the whole screen but there was so little illumination that no details could be seen. Suddenly Ewakl appeared on the screen. He was excited and behind him there were two other Wisagr, in animated discussion. "Captain Tayqhahat, I saw a light in the area of the entrance to the tunnel leading to the base," he said. He was so excited that his voice was barely intelligible. In addition, he inserted a few unknown words, most probably in a Wisagr language, into his sentence.

"Twenty-six, please verify," Tom said, trying not to get excited himself.

"Negative, Sir. I recorded a reflection from something in space, perhaps from an object orbiting the planet. Nothing significant, it may have been a small meteorite or a piece of debris. It was too small and far away for my sensors to make any more of it," the computer answered.

Ewakl thought differently. "It was a signal, Sir. I am requesting permission to land close to the base and pick up the survivors."

Tom had greater trust in the computer's verdict. "Ewakl, remain in formation. Twenty-six, replay the sequence Captain Ewakl is referring to, zoom in on the relevant area."

A dark image appeared, showing a faint image of the surface of the satellite. For a brief moment, an almost imperceptible light could be seen. It was impossible to tell whether it was on the ground or in space, at least without more instrumental data. It was, in any case impossible to say anything from a single image.

"From the spectral and photometric characteristics of the light it is clear that it is reflected by an object rotating about its axis, not coming from any autonomous source," the computer stated.

"I am sure that there is someone trying to get our attention. I must go down to check, Sir." Ewakl's voice was slightly calmer now, but he appeared very determined.

"NH 47152 is leaving formation and is not responding to my signals," the computer said in a worried voice.

"Captain Ewakl, return to your formation. This is a formal order," Tom said, taking the microphone and earpiece from his seat's armrest. The screen went blank, showing that Ewakl had cut communications.

"They are accelerating towards the satellite, with no shields, so that they can use all their power for propulsion," said the computer.

Tom had to decide quickly what to do. "Twenty-six, let's move towards the satellite in an open disc formation. Maximum acceleration, but with all shields up. The freighters must stay behind," he said aloud, putting his microphone and earpiece back on its holder. Then he switched on his restraint system and felt the pressure of the force field over his whole body. He realized that everyone else was doing the same.

By now NH 47152 was rapidly moving away from the other ships, aiming straight towards the satellite. After several minutes the computer warned them that a group of Qhrun ships, stationed in the inner part of the system, had started to move closer. "We have still two hours before they get here," she concluded.

The situation was not dangerous as yet, but if Ewakl didn't stop in his attempt to land on the satellite, it would become so in a very short time. Within one and a half hour, at most, they would have to abandon him to his fate.

Suddenly Ewakl's cruiser swerved sideways, while starting a quick deceleration and raising her shields. *Here we go*, Tom thought. He didn't know what was going on, but for several minutes he had felt that something had to occur. A number of flashes illuminated the hull of NH 47152.

"Torpedoes," exclaimed the computer.

Tom quickly realized what was happening, and also what was the only reasonable alternative to flying away, leaving the Wisagr star cruiser to face certain destruction. *That idiot would deserve it*, he thought.

He called Ertlaq. As soon as the Tteroth appeared on the screen he ordered him to take command of the whole squadron and, after expanding the formation, to close in and try to destroy all the incoming torpedoes with the batteries. Then he ordered the computer to move up to the endangered cruiser and for all the crew to don their space suits.

By this time the flashes in the zone around NH 47152 became more frequent: torpedoes were exploding against the shields which were now almost on the point of failure. The only possible maneuver NH 47152 could try was to avoid the torpedoes and to destroy them using the batteries, but it was clear that the ship's computer wouldn't dare lowering the shields. "The space in front of us is saturated with torpedoes that are starting to activate," the computer said. It was now clear that Qhruns had mined space around the satellite, which meant that they had discovered the base.

NH 47152 was getting larger and larger on the screen: they too were getting dangerously close to the mined zone.

Then the situation suddenly became crucial. A number of explosions blew the hull of NH 47152 apart. At that very moment, their computer started firing the batteries, destroying an incoming torpedo. As this was happening, Ertlaq was keeping the squadron out of the danger zone, while firing at the torpedoes that were attacking the wreck of NH 47152.

"Twenty-six, let's get between the debris and the incoming torpedoes."

The maneuver was quickly accomplished and they were soon close to the largest fragments of the cruiser.

"I get four automatic beacons on space suits," the computer said. The computer opened the air lock and, using the robotic arms, recovered the four Wisagr, who had been floating in space.

"Twenty-six, can you see the bridge and engine room sections, among the fragments?" Tom asked.

"The engine room was completely destroyed ... there cannot be any survivors. But the bridge is still in one piece," the computer answered.

"Capture it with one of the arms and attach it to our hull using a force field. Then let's get out from here. Quickly," Tom concluded.

It took almost a minute to capture what remained of the bridge, and then the cruiser started to back out of the mined zone. Ertlaq was still covering them with the batteries of the whole squadron, stationed outside the danger zone. In a couple more minutes they had emerged completely.

Tom leaned against the back of his seat. Ertlaq appeared on the screen "Congratulations, Tayqhahat. An outstanding rescue."

"Thanks to you, Ertlaq. It would have been impossible without the barrage of fire that you laid down." Then he asked the computer "Any survivors in the wreck?"

"I am sorry, Sir, but a fragmentation charge completely depressurized the bridge. There is nobody alive there," she concluded sadly. The computer remain silent for a few seconds, then suddenly announced triumphantly "Captain, you wanted fragments of Qhrun origin, didn't you?"

"Of course, Twenty-six. Fragments of any type," Tom answered, without really understanding what the computer was referring to.

"In the wreckage there are the hulls of at least two Qhrun torpedoes, and many pieces of the fragmentation charges."

"Well, let's close with the formation and move at the maximum speed towards another hyperspace gate. We have already attracted too much attention in this system," Tom decided.

They had a good lead on their pursuers and therefore there was no immediate danger. Wal-Nah and Raoqal went to the airlock to check the condition of the four Wisagrs they had just rescued from space and to take them to sick bay. By now the ship had closed in on the freighter, which, after mining the hyperspace gate, was now empty. The largest fragment of the cruiser was cut into smaller pieces and then loaded into the cargo ship. As soon as the cargo bay was pressurized again, two men, wearing protective suits against radiation, began to search for fragments of the torpedoes.

It took more than a hour, because working inside a wreck full of parts that were likely to collapse and which were full of sharp edges, and wearing a protective suit was not a simple matter. Moreover, the job had to be performed manually: if they wanted to quash Susan's doubts about the actual existence of the Qhruns, it was best for computers to play just a minor role in the matter. And humans, of whatever species, had become less and less accustomed to work without using computers.

Finally an Irkhan wearing a protective suit appeared on the screen; he was carrying a twisted piece of plating, about two feet long and four inches wide. "This doesn't come from a Confederation ship, I am sure," he said raising it above his head so that it was easier to see.

"Twenty-six, analyze that thing," Tom ordered.

After rather more than half a minute, the computer answered: "it is a made of an alloy similar to those of the external hull of starships. But its composition is unknown; it must be analyzed in detail to understand the material it is made from. However, I would rule out any suggestion that it comes from any planet of the Confederation."

That's it, Tom thought. Certainly other checks were still needed, but Tom now was more certain that the Qhruns existed. "Thanks, Twenty-six. Let's stop work on that wreck. As soon as we reach the nearest hyperspace gate, we will send everything to Gorkh'ar and let them do the work" Tom decided.

The flight towards the next gate, which, once again, was blockaded by eight ships, was eventless and the fight was almost an exact repeat of what had happened when they entered the system. The Qhrun ships didn't even have time to return fire, and again there were no fragments to be recovered.

When the gate was free, the transport to which the survivors of the Wisagr ship had been moved left, heading towards Gorkh'ar, while a second freighter unloaded its torpedoes, probes and transponders in space.

The squadron then moved towards the third and last hyperspace gate in the system, where a battle identical to those fought at the other gates took place. After mining the third gate as well, they decided that they had best leave the Tayssuk system. They had been fighting in a system occupied by Qhruns for twenty hours without a pause: the ships could go on, but the humans needed some rest.

11 Return to Base

"Tom, you must not feel responsible for the loss of NH 47152 and the death of Ewakl," Susan said, moving her back away from the black volcanic rock against which she was leaning and looking for a more comfortable position. With no effect, owing to the shape of the rock she was sitting on.

"No, you're right. But we crossed the border less than a day ago and we have already lost a ship and six spacers." While speaking, Tom was somewhat distracted, observing the herd of large herbivorous animals slowly crossing the plain below him, heading towards the river. He was trying to make out whether they had four or six legs.

"The loss of NH 47152 at least allowed us to obtain those fragments. And the fleet has another hero, which is not a bad thing in times like these." She too was rather distracted, looking at the valley stretching for miles and miles at the feet of the extinct volcano.

"Hero?" Tom exclaimed. "That was not heroism, that was just foolishness."

"It wouldn't be the first time the latter has been taken for the former. And then what happened will allow someone we know well to embroider the legend of Tayqhahat... Tayqhahat, the wise, who tries to stop Ewakl. And then Tayqhahat, the brave, who enters the minefield to recover what remains of the cruiser," Susan commented, turning slowly to the right, towards the strange trees, looking almost like pines, that surrounded the top of the volcano. She tried to see the animals that she had heard a little earlier. "When, before leaving Earth, I thought of terraforming a planet, I imagined that the result would be a kind of a garden. These terraformed planets are, on the contrary, wild, like..." she paused, as if searching for the correct words, "...like natural ones."

"We at least had confirmation of the advantage of our tactic of adding transponders to the torpedoes. Think what would have happened if we had seen several thousand torpedoes coming against us instead of a few hundred. Almost all would have been fakes, but we wouldn't be able to tell them from the actual ones. And now you have your Qhrun fragments, proving that the whole matter is not a hoax thought up by computers," Tom replied, getting back to the original subject.

"Wait. I agree that it has become unlikely, but before considering it as final evidence, I want someone to have studied those fragments in detail," Susan replied, getting suddenly up and beginning to walk along the edge of the cliff. "Twenty-six might have made this virtual rock a little softer," she added.

The voice of the computer interrupted her "May I return your cabin, Sir?"

"I didn't realize that it was so late. You can remove the image, Twenty-six, so that we can go to sleep." They had entered hyperspace two hours earlier. In about ten hours they would enter the Qhal'at system. Just time for a good sleep, and then to get ready for the next battle.

The image disappeared instantly, together with the rock on which Tom was sitting, and he found himself on the force field. Together with the image, the tactile and thermal perceptions that made the force field a cold, rough and, when all was said and done, an uncomfortable piece of lava, all disappeared.

Ten hours later they were again all on the bridge, ready to run the blockade. It was a system once settled by Aswaqats, deeper inside the frontier and, in accordance with the predictions of their mathematical model, they found fourteen Qhrun ships.

The maneuvering of the squadron went as before, but a small change had been introduced in the tactics: the group directly led by CH 23426 acted in a more cautious way: as soon as the shields of the Qhrun ship started to fail, they just targeted the batteries, trying to cause as little damage as possible to the ship. When finally the Qhrun ship exploded, it did it in a less destructive way.

"Twenty-six, track the fragments, and compute their estimated position. Perhaps this time we can recover something," Tom said.

"Perhaps. . . I am following about ten fragments that we could recover" the computer exclaimed, in an excited voice.

After destroying all the Qhrun ships, the whole squadron started to home in on the fragments. Within a short time, a bright point appeared on the forward screen: it was a structural element, rather deformed by the explosion, and slowly rotating.

"It looks like a structural element of a cruiser," Tom exclaimed.

"As far as its general shape, yes," the computer answered. "However, I compared its size with that of similar parts of my structure: that piece is at least three times larger than any structural part of Confederation ships."

Slowly the ship came alongside the fragment, which was seized by the external robotic arm located near the cargo bay door. As their ship completed that maneuver, other cruisers were recovering other fragments, about a dozen of them, all apparently forming structural elements. The computer performed several spectroscopic tests, and concluded that the material was very similar to that of the torpedo fragments recovered from NH 47152.

In the meantime the freighter that had laid the torpedoes, probes and transponders near the hyperspace gate had caught up with the rest of the squadron. The fragments were loaded on it, carefully shrouded in a protective covering to prevent possible traces of organic material of alien origin from coming in contact with people or from contaminating the ships.

As soon as the fragments were loaded onto the cargo ship, she left the system heading towards Gorkh'ar. The squadron then moved towards the inner parts of the system, launching probes and transmitting messages, as it had been done in the Tayssuk system. As before, they moved towards each of the four hyperspace gates in turn, destroying all the ships that were blockading them, but this time they also challenged a small group of six Qhrun ships moving to meet them.

After slightly more than thirty hours they left the Qhal'at system. In the following fifteen days they made raids in four more systems that were not far from the frontier, always with the same outcome.

"Twenty-six, when will we be back to three-dimensional space?" Tom asked, as he examined the image of the galaxy that the projector was producing in front of him.

"Four hours and fifteen minutes," answered the computer.

"Plus rather more than eight hours to enter the inner system from the gate we are heading towards," Susan commented. "In somewhat more than twelve hours we will be at Gorkh'ar."

They were all eager to be back at their base, even though life on an asteroid is not very different from that on a star cruiser.

In the last three days that they had spent in hyperspace, Tom and Susan had worked hard to define the details of the second part of their plan, and they were now ready to explain everything to Aintlhad, even if they were sure it would be hard for him to agree with what they had devised

"Perhaps a short visit to Quanslyaq might be useful: I think he will agree with us and he can be of help with the Coordinator. It's a pity we cannot get him to come with us to Laraki, but perhaps it would help if he were to send a message to Aintlhad," said Tom.

"That is a good idea, even if this means travelling something like two thousand parsecs. But perhaps Quanslyaq would come with us to Laraki, if we ask him to."

They went on talking until the computer informed them that re-entry to the Gorkh'ar system was imminent. They got up and, while heading towards the bridge Tom realized he had thought: *Going home, at last.* However, he realized that his home was no longer either Earth or that asteroid in the Rigel system, but simply 'his' star cruiser: he was a true spacer now.

Re-entry into three-dimensional space was uneventful. The cruisers entered the system in a scattered formation and a quarter of an hour later the freighters also arrived.

Immediately after entering the system they sent a message to Terkr. More than six hours later, when the answer arrived, Tom and Susan were again in the library, working with the computer and the three-dimensional projector.

The image of a human of a species Tom had never seen before appeared on the screen, saying "Message to Admiral Tayqhahat from Kirkyssi from Admiral Terkr." The man disappeared and the projector displayed Terkr's image.

"It looks extremely like a fox," Susan whispered, clearly alluding to the tawny fur and the small round eyes.

"Yes, but I would like to understand whether he is a fox trying to look like a harmless idiot or an idiot looking like a fox," Tom returned, in an even lower voice.

"Congratulations, Tayqhahat from Kirkyssi. Or rather, Admiral Tayqhahat, because the Coordinator has sent a hyperspace probe with the news of your promotion to Starfleet Admiral. I honestly didn't expect to see you again to give you the news: clearly you must have fortune on your side," the Admiral began.

Tom was puzzled about the meaning of that remark. Was it a veiled threat coming from an Iskrat-is-Thn? Or was it perhaps only caused by envy? The transmission continued with other details of the situation and instructions for the squadron's re-entry. Finally the admiral's image disappeared. Tom sent a short message acknowledging the instructions.

"Congratulations, Admiral," Susan said, in a teasing voice. "That welcome was a bit hasty, don't you think?"

"Poor guy, think of his disappointment at our return. If the Qhruns had made us disappear he would enjoy his revenge on the Coordinator and proof that his inactivity is not cowardice but wisdom," Tom answered her.

"I fear you are too optimistic. It is possible that under his disappointment there is mainly the regret that the Angels of Wrath didn't destroy a bunch of heathens," Susan commented.

"Twenty-six, have you tried to contact Sinqwahan on the reserved channel?" Tom, said addressing the computer. "If only we could know whether Sinqwahan has done his job with the Iskrat-is-Thn. . ."

"Congratulations, Admiral. It is centuries since I have been commanded by an admiral, and I am very happy for you," the computer answered. She said that with a note of pride, and Tom was sure she meant to stress that some centuries before she had already been commanded by an admiral. "I tried, but I didn't get any answer," the computer stated.

Sinqwahan might be behaving like that for a number of reasons, but it was better to be cautious. "Twenty-six, send a message to Admiral Terkr," Tom ordered. Then he went on almost immediately: "Message for Admiral Terkr from Admiral Tayqhahat. Change of entry procedures." He paused shortly and then went on: "I intend to perform an exercise while entering the base. We will simulate entering an abandoned base, where a hostile presence is suspected. Leave the external gate open, and operate the airlocks of the corridor in the usual way: we will simulate that they are both open and the base is depressurized." He nodded to indicate the end of the message and then commented in a low voice "If someone has prepared a trap for us, we won't be taken unawares."

He gave detailed instructions about the simulation to all the ships, but also ordered a message specifying the possibility of finding actual resistance to be inserted in the classified area of all the ships' computers.

The squadron divided into three groups: the freighters remained behind, half of the cruisers, under Ertlaq's command, broke formation, to patrol the outside of the asteroid and to effect a blockade around it, while the other half headed directly towards the gate in close formation. The Ertlaq's group kept their shields up, while the other ships had their shields at low power, to withstand the attack of small-arms fire, while maintaining a virtual cross section small enough to enter the corridor.

These precautions proved to be unnecessary: the first group went through the long corridor without problems, and settled on the ground in the zone assigned to them. The balconies were crowded as they had never been before, a clear indication that nothing had happened. Immediately after they had entered, Ertlaq's group lifted the blockade outside the asteroid and entered, together with the freighters. Within an hour all the ships were on the ground and most of the spacers had disembarked to enjoy some freedom of movement.

Sinqwahan had entered the ship before the arrival of Ertlaq's squadron, and headed directly to the library, where Tom and Susan were waiting for him. "Why didn't you answer our request for information? We were worried about the situation here!" Tom asked him as soon as he arrived, hardly bothering to welcome him.

"I know. I didn't answer because I didn't want to alarm the Iskrat-is-Thn. I was just waiting for your arrival to have the operations to arrest them started, and I didn't want anyone to see me while broadcasting. They definitely suspect something: they know who I am and keep a close check on me. The fact that I always remained fully in sight for the whole time of your approach and made no attempt to communicate with you will have reassured them. That will allow us to catch them while they are unaware."

"Let's rather speak of you," Sinqwahan went on. "I would say that destroying two hundred and thirty-four Qhrun ships and losing just one cruiser has been a success greater than ever expected."

"You mean two hundred and twenty-six. Yes, and bringing back a number of fragments as well has been a nice effort," Tom answered.

"Your numbers are wrong. Just yesterday an automatic probe arrived from the Qhal'at system with the news that a group of Qhrun ships were annihilated while trying to enter the system by the torpedoes you left at the hyperspace gate. I have seen the records: it is clear that they are unable to distinguish torpedoes from the transponders and their batteries become confused. Trying to hit too many targets at the same time they lose any accuracy. You must see what happens when the computers realize that the sensors show that a torpedo has been hit and at the same time that it still exists."

"That is good news: if we can reduce their mobility from system to system, we have more time for the following phases." Tom was clearly enthusiastic of the way the transponders had worked.

"Yes, but now let's deal with the Iskrat-is-Thn. Not to pour cold water on your enthusiasm, Tom, but I believe that this problem is worse than we expected. We have identified something like two thousand of them in this base."

Sinqwahan explained the details of his plan. He wanted Tom to have an important part in the action, and the Coordinator had given him all the tools for that. A quick action was needed, now that the Iskrat-is-Thn were still unaware of what was likely to happen.

The Aswaqat had hardly finished describing his intentions in detail, when Tom asked Ertlaq to come to his ship. Less than two minutes later he was on board, so quickly that Tom was about to ask him whether he had flown. He refrained, not knowing how polite it was to ask such a question of a middle-aged Tteroth, who had perhaps not stretched his wings for many years. He explained to him in detail how they expected his help in capturing the Iskrat-is-Thn. "I don't like asking the crews to go out again into space after no more than half an hour since returning from a mission that has lasted more then twenty days," Ertlaq commented.

"I agree, but I think that the excuse of an important exercise is a plausible one, and then they need to believe that it will be for no more than ten minutes, the time I will take to get to Terkr's office. Once they know what is happening, the idea of getting rid of the Iskrat-is-Thn will certainly cheer them," Tom answered.

With the last remark, he had touched the correct note, especially with a Tteroth. "You can be sure of that!" Ertlaq exclaimed, getting up and heading towards the door. "Twenty-six, give detailed instructions to my ship's computer, since you have to stay here," he said to the computer as he left.

"Do you really think it is necessary to have Ertlaq go outside with the squadron? After all, the base has one only exit, and we can control it through the computer," asked Sinqwahan, who didn't understand the need for the changes Tom had introduced into his original plan.

"Are you really sure that there is no other exit? Or that the Iskrat-is-Thn are not able to launch hyperspace probes without our realizing it?" Since Sinqwahan didn't answer, he went on: "This base is thousands of years old, and we have no detailed map of all the tunnels. Better having a good external coverage than risk allowing someone to escape, don't you agree?"

No more than ten minutes later, Tom was waiting beneath the hull of the cruiser, close to one of her landing skids. The first cruiser to move out was lifting off the ground and he could distinctly hear the acoustic alarms. The sound was that of a simulated alarm, but it was hair-raising all the same. He could hear the voice of Ertlaq in his earphones: "simulated situation: the sensors of the base detected a group of Qhrun ships entering the system from hyperspace gate number four," he was explaining. They had chosen the gate number four since it was located on the opposite side of the asteroid with respect to the entrance of the corridor: the ships would then remain in a less conspicuous position. All the ships lifted off and slowly headed towards the tunnel.

Tom remained close to the landing skid, in the shelter of the hull to avoid the force fields of the other cruisers. When they had reached the exit, he walked to one of the points where the floating platforms were located, and then moved towards the entrance of the cavern, where Admiral Terkr was waiting for him. He ordered the platform to go slowly, since he didn't want to reach the Admiral's office before all the ships were in space.

When he heard in his earphone the voice of the computer telling him that the cruisers were about to move to the other side of the asteroid, he entered the room where Terkr was waiting.

"Here you are at last, Admiral Tayqhahat," Terkr greeted him. He was clearly annoyed.

"I apologize, Sir, but I wanted to see how my spacers react to an unexpected alarm, immediately after the returning from a long mission," Tom answered. Instinctively he pressed his hand against the pocket where he had hidden his thermal weapon.

"I am glad I am not a young officer in your squadron," the other said, with an ironical smile.

"Even if you were a young officer, you would not have been there: my men are all volunteers," Tom answered, trying not to be too offensive. Then he added: "I believe that when they faced Qhrun ships, they were all happy to have been through all those simulations. I must inform you that the behavior of all the spacers has been exemplary."

"Qhruns, again. Congratulations for all those destroyed ships. You succeeded in neutralizing about one three-hundredth part of an invasion fleet, losing just one ship in the process. If you were to maintain that loss rate there won't be a single one of your ships left after the destruction of even one of the Qhrun fleets that plague this sector. And as far as your men are concerned, I would say that the behavior of Ewakl was rather foolish, wouldn't you agree?" he concluded with an air of scorn.

Tom hardly controlled himself: it was one thing was describe Ewakl's behavior as foolish, but a rather different to hear it from this coward. "I must admit that he made a tactical error," he answered coldly, "but his action allowed us to seize a fragment that may yield important information."

"Yea, I forgot your precious fragments. If it had not been for the orders of the Coordinator, I would have them thrown in space instead of delivering them to your 'crisis committee', as you call it: if they are not from the Qhruns, they are just garbage, and if they are, it is extremely dangerous to handle them. I sent a message to the Coordinator, saying I will not take any responsibility for that action."

With a quiet signal in his earpiece the computer informed Tom that Ertlaq had reached his position. *That's it,* he thought, *time to see whether this fox is a fool or just acts like one.* He took a three-dimensional projector from his briefcase and laid it on the table in front of Terkr. "While I was travelling towards the base, I received a probe with a message from the Coordinator for you," he said while operating the projector.

An Aswaqat appeared and said: "Message from the Coordinator of the fifth sector to Admiral Terkr, commander of the Starfleet base at Gorkh'ar." While the Aswaqat faded away, Terkr got up, uttering: "How could you receive any message? We did not detect any probe entering this system."

Tom switched on the shield generated by the thermal gun he held in his pocket, making it clear he was armed. Terkr was obviously right: that message had been given to him by Sinqwahan. It had arrived with one of the freighters that had smuggled a group of men from the security forces into the base.

The projector visualized the image of Aintlhad. "Admiral Terkr, we suspect that many members of the Iskrat-is-Thn sect are present on Gorkh'ar. After the recent events, in particular the battle of Kistl, we must take immediate action to prevent them from becoming a greater threat than they are already. I have entrusted Admiral Tayqhahat to take personal control of the base and to put in action all the measures he considers necessary to capture the Iskrat-is-Thn. The command of the base will be returned to you when Admiral Tayqhahat has declared that the emergency is over."

Terkr collapsed on his chair. "You have no right to. . ." he started, and then he realized that the Earthling was slightly out of focus, a clear sign that he was protected by a portable force field. Looking closer, Tom realized that his hands were shaking. "Am I under arrest, Admiral?" asked Terkr, in a low voice.

"Why? you are not an Iskrat-is-Thn." Actually, Tom said it in a tone that could be interpreted more as a question than a matter-of-fact statement. "You will now stay here with me, and together we will coordinate the action of the security forces."

Tom took his terminal from his pocket. "Twenty-six, put me in contact with the computer operating the base, giving me access to the loudspeakers. Start as planned." He waited a few seconds to allow the computer to establish the connection, then he went on: "This is Admiral Tayqhahat speaking. I have temporarily taken command of Gorkh'ar base by order of the sector Coordinator. The Coordinator has reasons to suspect that the Iskrat-is-Thn sect is present in this base. I declare a state of emergency: everyone must remain where they are now and follow the orders of the security forces who are taking control of the base. All doors are locked and all computers except the one I am using now are being shut down. All equipment not essential for survival is also being shut off. Collaborate with the security forces and normal conditions will be restored as soon as possible."

These words were also broadcast to Ertlaq's ships that were surrounding the asteroid.

Sinqwahan was coordinating the five hundred men of the security forces who had arrived over the previous days from Laraki. In small groups, they had started inspecting every single location of the base, looking for the roughly two thousand suspects.

Through the glass door above the main gate of the base, Tom could see Sinqwahan's men moving around on platforms. They were all armed with heavy hand lasers.

"Sinqwahan, how are things going" Tom asked, always keeping an eye on Terkr.

"Well, but slowly. We have seized about three hundred of them, by now," the Aswaqat answered.

"I noticed you gave heavy guns to your men. Do you expect resistance?" Tom asked again.

"Up to now we had only sporadic fighting. But we discovered that a freighter smuggled in a number of hand lasers, and you cannot fight against them with thermal guns like the one you are wearing. I don't want to take any chances," Sinqwahan answered. And then added: "Be careful. And keep Terkr in sight."

While shutting off the connection, Tom noticed that, on one side of his field of view, something was moving on the balcony in front of the window of Admiral Terkr's office. In a fraction of a second he realized that the shadow was actually that of a human, even though he couldn't recognize the species. Actually, he didn't even bother about that, because he realized that the oblong object the person was holding in both hands was a big hand laser.

He didn't wait any longer. He jumped on the table separating him from Terkr, mentally thanking whoever had decided to maintain the artificial gravity on the asteroid at a value lower than that on Earth. When he got to the other side of the table, he grabbed Admiral Terkr's chair with both hands, knocking it over. Terkr fell to the ground, rolled heavily on the floor and ended a few meters from the table. He was still rolling when the transparent material of the window suddenly melted with the noise like thunder, shedding a rain of red hot pieces all around. The laser beam struck the central part of the table, dividing it in two.

Tom too continued to roll on the floor to get clear of the area close to the window and whipped out his thermal gun. He was not sure whether it would be possible to use that weapon through the glass window, but there was no time to deal with the theoretical aspects of the matter. Still lying on the ground, he fired his weapon against the dark shadow on the balcony in front of him. The human, who after taking his shot, was leaning over the handrail of the balcony, fell over it and remained motionless on the ground, more than twenty feet below.

After quickly checking that there was no other immediate danger, Tom got up and moved to the window in time to see three men from the security forces approaching the corpse. One of them jumped down from the platform he was using and bent down to pick up the laser. He immediately understood what had happened because he looked up at him. Tom gestured to him that all was under control, and shouted down to him to bring up the laser.

In a few seconds the man ran up and looked at the fused zone in the center of the table. "You had a close call, Sir," he said handling him the laser. Then he realized that he had a thermal gun in his hand. "Did you do it with that?", he asked pointing down at the body lying on the ground below.

"I didn't have anything better. But now I feel more comfortable with this," Tom answered.

"I wouldn't have thought that would work through the window" the man concluded, as if talking to himself.

Me too, Tom was about to answer, but he refrained, thinking he had best pretend everything had happened as he had expected.

Only then did the man realize that Admiral Terkr was in the room. He was getting up, staring at the fused zone of the table.

"Sorry, Admiral, I didn't realize you were here," he said. "Were you sitting there?", he asked in a professional voice, pointing at the area in front of the table, right between the fused zones of the window and the table.

"Yes, and if it were not for Admiral Tayqhahat who made me roll over to that corner, I would be dead by now", he answered in a shaky voice.

"Then that gunman wanted to hit you too," the man from the security forces stated. "You had best stay well away from that window, there might be others around," he concluded as he went out.

But Tom hardly heard his last words. He was lost in thought: the problem regarding which side Terkr was on had been solved. If they had tried to kill him, he was not one of them. Or, against all the evidence, was he himself the target and not Terkr? Or had the Iskrat-is-Thn tried to exploit his good luck: two admirals with a single shot?

Sinqwahan's voice roused him from these thoughts. "Everything all right, Tayqhahat? They told me that there has been a shoot-out there. It seems you have better luck with the Iskrat-is-Thn than with the targets in the shooting range!"

Hearing Sinqwahan calling him Tayqhahat seemed strange: obviously he was afraid someone might eavesdrop on the conversation. The first time he had tried to use a thermal gun came back to mind. "You were right: spacers should know how to handle these things," he answered.

"At least as well as the Iskrat-is-Thn know how to handle a hand laser. I was told they tried to hit Admiral Terkr, and that it was damn good shooting," the Aswaqat said.

"No doubt it was good shooting. I couldn't tell at whom it was aimed, because we were both in the line of fire." He didn't want to say any more: if Sinqwahan was that cautious, it was better to keep silent.

Terkr was now sitting in a corner of the room on the opposite side of the window. Tom sat on the floor, aiming the laser towards the window. Time passed slowly and nothing happened for about ten minutes.

Then Terkr broke silence. "Admiral Tayqhahat, I must thank you. If it had not been for your quick action I would not be here. I wouldn't be able to handle one of those weapons in that way. That bloody fanatic got what he deserved, at any rate." For the first time his voice sounded friendly.

"I am sorry I had to throw you to the floor. But I realized almost too late what was going on," Tom answered.

He was interrupted by the voice of Ertlaq. "Admiral Tayqhahat, an automatic probe has left the asteroid. I am sending two ships to intercept it."

"Try to get it in one piece, and to have it shut down by Twenty-six: there could be self-destruction orders in case of capture," Tom said. This was another lucky strike. If they could seize it intact, it might be possible to identify its destination. He thanked God he had the idea of sending Ertlaq into space to blockade the asteroid.

Tom kept his position for at least another two hours before Sinqwahan called to let him know that it was all over. He had his transmitter connected again to the base's loudspeaker system and officially stated that the emergency was over, and that all the base systems had been reactivated. He headed towards the door, eager to get back to his ship.

"Admiral Terkr, the base is yours. I hope this place is now free from those fanatics once and for all," he said to the admiral, who hadn't said a word the whole time.

"Do you mean that the command of Gorkh'ar base is again entrusted to me?" Terkr asked, getting up.

"Sure, I told you that my takeover was just temporary", Tom answered.

"Thanks again, Tayqhahat. And go on with your work with the Qhruns," Terkr said in a louder voice, so to be heard by Tom, who had already left the room. Tom thought that his tone was different now from his previous one.

12   Strangers

The three-dimensional projector was showing a fixed image of the balcony running along the walls of the huge cavern that formed the base, showing where the side wall merged with the front one. The balcony was quite wide and the inner part was in shadow. Suddenly a human figure emerged from the shadows and approached the handrail. Quickly he set the hand laser he was carrying on the rail and knelt down to take aim. Then he got up, moved about one meter to his right, knelt down again and fired after a short hesitation. The lightning flash of the weapon lasted for a fraction of a second and the image blurred. Immediately after firing, he got up and leaned out over the handrail, as if to see what was happening below. In that very instant a red stain appeared on his chest, and rapidly expanded. The body was clearly over-balanced, because it rolled over the handrail and fell from the balcony with its arms wide open and the laser still in its hand.

"If he had been carrying a Winchester instead of a laser, it would have been a perfect scene for a western movie, like *Gunfight at the OK Corral*," Susan said.

"Yes, but I should have killed him with a Colt 45: a self-respecting sheriff wouldn't go around carrying thermal guns," Tom replied.

"Twenty-six, please go back to the moment when he knelt before firing," Sinqwahan added: he was not in the right mood for joking. He was pale and his face was tired.

"Sinqwahan, get some sleep. We can see these images tomorrow," Tom said. He was tired too. He had just arrived on board, joining Susan and Sinqwahan who were talking together in the library. His clothes were dirty from the fragments of the table that the laser had flung around.

"No, Tom. I want to understand what happened. Then I will go," Sinqwahan replied. Then he started to carefully examine the image. It was clear that the Iskrat-is-Thn had knelt down to take aim, but he was not satisfied with his position, because he moved. "Twenty-six, can you visualize the line of fire of the weapon and the positions of Tom and Terkr?" asked Sinqwahan.

"It is difficult, I have only one point of view of the scene," the computer answered. The screen showed an image of the window taken from the outside, with Tom and Terkr, the former behind the table and the latter in front of it. Terkr was slightly to the right and Tom to the left. "He moved the weapon several times," the computer explained, showing a bright spot superimposed first on Tom's image and then on Terkr's.

"It is clear that he couldn't decide which target to hit first," Tom said. He felt a growing uneasiness seeing himself in the sights of that sort of weapon. It was only now that he started to be really scared.

"It's obvious that he couldn't decide. That laser needs about ten seconds to recharge, after a shot powerful enough to go through the window. And he was afraid to be identified and killed before being able to shoot a second time," Sinqwahan commented.

They let the replay continue. The man got up, moved and then took aim again. This time the two targets were perfectly lined up. "And so we have reached the point: that man didn't chose between you and Terkr. He moved to be sure of hitting both. A true professional," concluded Sinqwahan.

"By no means," Tom contradicted him. "By doing so, he wasted enough time to give me a chance to react."

"And how could he know that you were armed? Looking through the window he couldn't see you were protected by a shield, and besides, that shield is useless against a laser. Moreover he probably didn't think that a thermal gun would work through a window or that you know how to handle it well enough."

"Well, at least now we know that Terkr is not an Iskrat-is-Thn," Susan concluded. She had followed the whole scene without saying a word.

"We know that our fanatic wanted to kill him. This makes it unlikely than he belongs to the sect. But I would say we are not yet completely sure he doesn't," Sinqwahan said, ordering the computer to switch off the projector.

Tom leaned back against his chair and asked the Aswaqat for a detailed report on the operations against the Iskrat-is-Thn. Everything had gone as expected, said Sinqwahan, and all the suspects had been captured. "To be exact, we took 1636 of them alive and 426 were killed in the fight. The point is that there were five more than we expected: the one you shot and four others, all carrying hand lasers. They fell into our hands without our even suspecting their presence and in any case we cannot be sure we got them all. The very idea of other guys we don't know going around with hand lasers is scary," concluded Sinqwahan.

"I hope that the ten we decided not to arrest were left in peace," Tom said. "if there are others you don't know about, then they'll try to contact them." The strategy devised for testing the Terkr's loyalty was now even more important. Reassured by not having been arrested, the Iskrat-is-Thn who were still free would probably behave less carefully, once the situation had returned to normal.

"We will monitor them discretely. I hope they will lead to those we are looking for, if they are here at all," concluded Sinqwahan. "And if Terkr is on their side, the impression of being safe now will make him betray his true feelings."

As he was speaking, the three-dimensional projector displayed the image of Ertlaq. "Admiral Tayqhahat, I got that probe. Unfortunately we were not able to pick it up intact, but I hope that it will still at least be possible to read its destination." The voice of the Tteroth betrayed his disappointment.

"I am sure Yyrtl will succeed. Stay in position for a while and then come back inside." Tom closed the communication and then asked Sinqwahan: "Can you get some freighters out of the base to form a blockade around this asteroid, without anybody realizing it? If some of them escaped our sweep, they might try to send a probe or to run away as soon as Ertlaq's squadron is back."

"Certainly, the freighters that brought my men here are armed. I will have them leave with a fake flight plan." Sinqwahan stayed a few more minutes to prepare the work for the following morning, but it was clear that he remained there talking only because he was too tired to get up and leave.

The following morning the members of the crisis committee arrived early. Tom had summoned them to meet on his ship, officially because that way it was easier to use the computer, but actually because after what had happened the previous day he preferred to remain on board. The cruiser could offer him a rather scanty form of protection: while in space, with her shields and her maneuverability she was a fortress, but a grounded starship was extremely vulnerable. However, even though he was well aware of this, Tom nevertheless felt safer that way.

It was not long before the sixteen members of the crisis committee were all in the library, all except Sinqwahan, who was busy questioning the Iskrat-is-Thn and overseeing the prisoners being put on ships bound for Laraki.

Tom began with a detailed account of the raid in the Tayssuk system and the operation against the Iskrat-is-Thn the previous day. He omitted only a few particulars that he intended to keep to himself. He also informed them of his intention to leave as soon as possible for Laraki, to speak with the Coordinator about the second phase of his plan.

"Obviously you cannot give us any details of what you intend to do," said Idlath, an elderly Opsquat who, because of the prestige he enjoyed in Laraki academic circles, had been considered the leader of the group from the first moment it had been formed. Like all Opsqats he was short and was always wrapped in heavy, colorful clothes. Tom remembered that, because of the high temperature on their planet, Opsqats maintained their body at a temperature about ten degrees higher than nearly all other human species. Therefore they were always cold when living in places where temperatures were maintained to suit the demands of other species. Idlath was so accustomed to living in an environment that was different from his own that he was wearing the heavy insulating garments as if they were a second skin.

"I am sorry about that, Idlath, but I want the Coordinator to be the first to know the details of what I intend to do," Tom answered. He was sorry not to be able to say more, but he thought it better to discuss the whole matter with the Coordinator before letting anyone else know his intentions. Only in this way could he avoid any changes to his plans being interpreted as disagreements between himself and Aintlhad. Moreover, there was always the risk that some of them were not as trustworthy as Sinqwahan believed.

"We are fully aware that ours is just an advisory committee. We can say what we want, but then you will decide as you think best," Idlath said, who was, after all, not that unhappy not to have any operational responsibility.

"Idlath, give us a summary of the finding on the pieces of Qhrun ships that we sent back. Then Yyrtl will tell us what she got working on the computer of CH 43176," Tom said.

"Unfortunately there is nothing new," Idlath answered. "Even though it is the first time, as far as I know, that we have been able to study something built by the Qhruns." He stopped speaking and nodded to a human of a species that

Tom did not recognize, but who had been introduced to him as an expert in comparative technology.

The man got up (Tom thought it was almost as if he unrolled, mostly because of how he moved, but also because, standing, he almost touched the ceiling with his head). He began speaking through a voice synthesizer: his physical anatomy prevented him from uttering the sounds that were needed to speak any of the most common languages in the Confederation. Tom noticed that he wore an earphone: he was not even able to understand the language of the others without computer translation.

"The material of the fragments is a structural material," he started. Then went on with a long description of its characteristics, comparing it with those used in the worlds belonging to the Confederation. He concluded that they could certainly produce that material within a few months.

The man sat down and Idlath went on: "There can be no doubt the material is alien. Once this conclusion was reached, I began to have these fragments examined with the utmost care, hoping that biological material could be found on some of them. He paused to place a three-dimensional projector on the table, called up an image of one of the fragments and then began to zoom in. As the image grew larger, the surface no longer appeared perfectly smooth and, at a certain point, they saw a dark stain filling a small indentation in the surface. "Here, within that dark stain is a tiny quantity of organic material." Idlath tried to say it in a calm, matter-of-fact voice, as if pronouncing an obvious statement, but his tone was undoubtedly triumphant. *With good reason, if that thing is what we all hope*, Tom conceded.

But he didn't want to be carried away by the enthusiasm for what was still just a remote possibility. "Are you sure that the fragment had not been contaminated?" he asked.

"Absolutely sure, we took every precaution. It is nothing more than a sequence of amino acids. But it is the first time that we have proof of any biological material on a Qhrun ship," Idlath answered.

"This doesn't prove anything, unfortunately. Qhrun ships land on biologically active planets, and therefore they can be contaminated biologically, without the Qhruns themselves having a biological nature," Tom countered. Then he asked "Do you have any idea of which part of the ship it comes from?"

"There is no definite answer to that question. But we have reached the conclusion that it comes from the inner shell of the hull, because we have also found traces of adsorbed gas in the surface. And only in the surface of that specific fragment." Idlath fell silent and looked around with a satisfied air, enjoying the effect of his words.

It is obvious that Opsquat has a taste for coups de théâtre, Tom thought. The revelation about the biological material was just a preliminary statement, to prepare the audience for the real surprise, which followed now, even if it was weakly disguised by technicalities. It was clear that if indeed the hull had an inner shell to maintain a pressurized environment full of gas, the biological nature of Qhruns was no longer a mere assumption. "I bet that you have also computed the pressure and the composition of the atmosphere," Tom said.

"We have a few hypotheses. Computers determined that the pressure had to be about three-quarters of that maintained in Confederation ships, and that the atmosphere contained between twenty and the forty per cent oxygen. The rest was inert gas. But these are only rough results," concluded the Opsqat.

"Congratulations to everyone," Tom exclaimed. "Idlath, you said that there was nothing new. I suppose that if you had something really important we would expect to see a Qhrun bound and gagged," he concluded, to laughter. "From those four pieces of wreckage you have deduced that we all could breathe safely in a Qhrun ship without wearing a mask, and that Qhruns need a breathable atmosphere like us." Then, turning towards the expert in comparative technology, he asked whether there were examples of automatic space ships where a breathable atmosphere had to be maintained without being needed by a crew.

"I've never heard anything of that kind. Pressurizing a space vehicle is so difficult and expensive that is done only when really needed," he answered.

"Therefore, unless Qhruns have an approach to technology that is completely different from ours—and I would rule it out seeing how their ships work, they need an atmosphere we would consider breathable," concluded Tom.

"That does at least look extremely likely," the Opsquat confirmed.

"Going back to the biological material. It is either part of the body of one of our enemies, or it is a fragment of food or garbage," commented Susan, who had followed without losing a word.

"Or it is a fragment of a parasite, or of some object of biological origin carried on board for who knows what reason. . ." Idlath continued.

". . . or of a pet, an ornamental plant or goodness knows what," Tom concluded. "But if you had a larger specimen, containing many amino acids, or even better, a piece of the genetic structure. . ." Tom fell silent, leaving it to the others to draw the logical conclusions.

"We could understand what it comes from. And if it were of Qhrun origin, if our laws of biology hold also for them, if we had some complete chromosomes. . ." Idlath paused for a long time. Then he went on: "cloning a Qhrun would not be impossible, provided that cloning an intelligent being is not against the law. . . or ethics," he concluded, with a meaningful glance at Ruklyaq.

"I think we are not here to have our hands tied by the law. And as far as moral scruples are concerned, our future Iktlah will gladly make an exception for us... and if he doesn't, I will go to Quanslyaq to get special permission," Tom added, laughing.

The idea thrilled him: nobody had ever captured a Qhrun, dead or alive. But why not create one to study it at ease in the lab? "Do you believe it is possible... from the technical point of view, I mean?"

"Perhaps, if we had enough biological material. We spent a long time discussing this amongst ourselves, without reaching any definite agreement. But if you can bring us some organic material, we will try," Idlath concluded.

Tom decided to speak to Ertlaq about it. They went on talking about the ideas they all had about the Qhruns, but it was clear that they were just repeating the usual ideas, because there was nothing truly new to say. Tom was about to stop the discussion and turn to another subject when the door opened and Sinqwahan entered. He was pale and looked troubled, and as if he had been running a long way. "Sorry to interrupt you. Tom, Susan, would you please come immediately". He left and headed towards the mess hall. Tom said a few words to the others, primarily so that they would not be unduly alarmed, and then followed him, together with Susan.

When the two Earthlings arrived in the mess hall, Sinqwahan was already sitting at a table and, making some excuse, had asked Wal-Nah and Del-Nah, who were quietly having lunch, to leave. "It seems that our security man is starving, if he needs the mess hall in such a hurry...," Tom said when he passed them at the door. He wanted to minimize the drama of the situation, because he saw that they had been shocked by Sinqwahan's worried attitude.

As soon as Tom and Susan had entered, the Aswaqat ordered the computer to isolate the room. "What's going on, have you seen a ghost with a laser in his hand?" Tom asked him laughing.

"Don't joke. I wish I had seen a ghost. We started questioning the prisoners and we discovered that the rumor that you two are not Aswaqat has started to spread. Clearly the Iskrat-is-Thn are working hard to discredit us all."

"How can they tell such lies about us!" exclaimed Tom, who was unable to take the situation seriously. "And what are they saying about us? Are they perhaps insinuating that we come from Earth?"

"The rumors they are spreading are different. Some say that you come from a planet at the edge of the sector, others that you belong to a humanity that is outside the Confederation, and still others that you are demons sent to hinder the Angels of Wrath in their mission to purify the universe..."

"At least the last statement is false, even if we must acknowledge the fact that they have impressive imaginations," Tom interrupted him.

"You seem to be enjoying yourself! Don't you understand that all our efforts to make your presence acceptable are messed up? All our work is at stake, and all you do is laugh," Sinqwahan repeated, with a note of complaint in his voice.

"Now, please listen to me, Sinqwahan," Tom said. "Nothing serious has happened and, if we do what I am about to suggest, the Iskrat-is-Thn will end up working for us. At least for me, in that I can stop eating that terrible slop with crabs. Do you feel that these rumors cannot be stopped, and that any attempt to do so will be self-defeating?"

"Yes, I am only too well aware of that," Sinqwahan admitted. "And goodness knows how many star systems have already heard those rumors."

"All the better. We will now play them at their own game. Tell your men to go around doing the same, but instead time with true rumors. . . or ones that are almost true. Get them to say that we are spacers from a humanity not yet belonging to the Confederation, but which has been invited to join. And that a fake identity was needed to prevent an attack by the Iskrat-is-Thn against us. And now that the Iskrat-is-Thn have been defeated, the Coordinator will allow us to resume our true identity. Let the Coordinator be criricized by some of them for having been too cautious, and praised by others for having been successful in distracting the attention of those fanatics away from us. This way we can focus attention on a secondary detail, take advantage of the right moment to end this charade, and nobody will pay any more attention to the story itself. And by criticizing the Coordinator for persisting in this game for rather too long, nobody will think of criticizing him for having started it."

"Tell me the truth, did you expect things to end in this way? Is this happening in accordance with what you told me that evening on Laraki?" Sinqwahan was quieter now, remembering that brief conversation.

"Sure, you see that I was right to let Istlahad get out alive! All this is surely her doing. Now we just have to let things take care of themselves, perhaps with some minor adjustments. Now let's get back to the others, if we don't want them to think that something terrible is happening."

They went back to the library, where everybody was waiting for them. Indeed the Sinqwahan's tone had made an impression on them and the atmosphere was tense. As soon as they entered, Idlath asked what was going on.

"Nothing really important," Tom answered. He sat down and spoke slowly as if he was looking for the most appropriate words. "Now I must tell you a secret, but remember it must remain one until the Coordinator personally authorizes you to disclose it." He looked at them one by one, as if to stress the need for the news to be considered utterly classified. "You know me as Tayqhahat of Kirkyssi, a man of the Aswaqat species. Actually this identity is fake, and I have been forced to assume it for safety reasons linked with the presence of the Iskrat-is-Thn. But now that the Iskrat-is-Thn have been defeated, at least in this system, this subterfuge is no longer required."

He paused for a while and then got up and slowly took off the wig he was wearing to look more Aswaqat. "I am Thomas Taylor from Earth, from the Sun system; only 260 parsecs from here. And in fact until a few weeks ago I was not a spacer, but a propulsion specialist. My species doesn't yet belong to the Confederation and therefore my presence here and my rank of Admiral of Starfleet are exceptions, instigated by the Coordinator. These exceptions are because of the emergency the sector and the whole Confederation is facing. But we all hope the situation will become normalized and that my planet will join the Confederation as soon as possible."

He sat down again and everyone remained quiet for a few minutes, thinking about his words. The first to break silence was Idlath. "To me Tayqhahat of Kirkyssi or Tomastaylor of Earth doesn't make much difference. And if the Coordinator has decided to appoint an admiral of the fleet, coming from a planet not belonging to the Confederation, well, that is none of my business. All I care for is only that at last he has appointed an admiral who is trying to do something."

"Thank you, Idlath. I hope not to disappoint you," Tom answered. "By the way, my name is Thomas Taylor, two words."

"Then you must choose one of the two names: the custom is for everybody to have one name only, following the Terqhatl way. On your own planet you can be called following your own customs: I, for instance, have a dozen names, used according to rules that anyone who is not an Opsquat will never learn. Shall we call you Thomas or Taylor?" Idlath asked.

Tom thought quickly. It had to be a name that would also be suitable for official use. Admiral Thomas sounded too strange: "Call me Taylor, he answered."

Idlath went on: "I imagine that if Susakahan is your partner, she must be of the same species." From his tone it was clear that if she had not been, Idlath would have been very surprised, and very shocked too.

"But of course," Susan answered, removing her wig. "I also come from Earth. My name is Susan Taylor."

"Susan Taylor, two words, I imagine. You too have two names," Idlath said. "Don't tell me you will also choose the second one. Do you have all the same name on your planet? That would be an unbelievable confusion," he concluded with a laugh.

Tom thought that actually on Earth there was quite a lot of confusion, but at least not for that reason. "No, don't worry, we have thousands of names. Call me Susan; it will be simpler," she answered. For a moment she thought that it would have been better to use the name she had before getting married, but at that point she would have appeared to have three names and, as they seemed to make such a fuss over names, it was better not to complicate the matter.

At that point a human, of a species Tom had never seen before, interrupted them. He had been shaking his head and nodding for a few minutes as if he had something important to say. "I am a specialist in human biology, and I have never heard of humans from Earth. Must I conclude that Earth is a B-type planet? Your presence here would then be extremely illegal."

Tom felt that the atmosphere had suddenly become chilled. No one had cared whether they were Aswaqat or Earthlings: with thousands of human species, it did not make much difference whether they belonged to one or the other. But a human from a B-type planet was something else, and he should have expected that someone would realize it, sooner or later. He looked around, noticing that Sinqwahan was avoiding his eyes. Clearly he was thinking *you wanted to behave in your own way: now you have to get yourself out of this mess.*

"You are right. It was considered a class B planet. But the Coordinator has decided to consider it as class A so that it may join the Confederation and to allow us to come here. The situation will be straightened out and everything will be shortly fully legalized." He thought that if they were able to create a past for Tayqhahat to the point that Terkr even pretended to have met him years earlier, it would not be impossible to change the class of a planet, and to predate the change by a few months.

"The Coordinator has doubtless acted for the best, even if such decisions are certainly beyond his powers under normal conditions. But now that the sector is isolated, and we don't even know exactly which humanities still exist, he has to take the initiative," the biologist stated, apparently satisfied. From his words Tom understood that everything had been easier than he expected: although everyone had been trying to behave as if everything was happening as normal and that the Qhruns were a problem only marginally worse than many others, the fear of imminent doom had weakened their acceptance of prevailing customs and laws.

"Now I must start playing my role again," Tom finished, putting his wig back on. I will continue being Tayqhahat from Kirkyssi until the Coordinator allows us to resume our true identities," Tom said, and from that time onwards he never said any more about the subject.

A final point still had to be discussed. "Now is your turn to speak, Yyrtl. We know that the freighter that followed us has been to Qtr and that Admiral Ratlqal probably had his headquarters in that system. Tell us what you discovered by going through the memories of those ships' computers," Tom said to the Sitkr, who so far had not said a word.

"First of all that hyperspace probe was bound for Qtr. I succeeded in recovering all the identification codes and the approach procedures. A further set of codes and procedures has been obtained from the freighter. The probe contained information on the events during recent days, including the raid at Tayssuk, and on the beginning of the operation against the Iskrat-is-Thn. The message ended with a request for help, and for the marshalling of all the elects to get rid of Admiral Tayqhahat, who was described as the most vile of all the enemies opposing the wishes of the Almighty," began Yyrtl, in her usual, almost unintelligible, voice.

"I'm honored by the definition. Do you believe it is possible to use the information you recovered to send a probe of the same type with a fake message? What are the chances of the message being considered as true?" asked Tom.

"That's no problem. With the codes we obtained we can get them to believe anything we want," Yyrtl answered. "And from the computer we have reconstructed the movements of Admiral Ratlqal. After his original departure, he tried several times to get in contact with the Qhruns, without any success. He then came across a squadron of Iskrat-is-Thn starships and a battle followed. The Iskrat-is-Thn were defeated and a number of them were captured. The prisoners converted the admiral and convinced him to settle at the Qtr base, where a group of those fanatics was hiding. Apparently, Qtr is their headquarters. Istlahad too went there before the battle of Kistl."

They went on speaking for another half an hour, but all the important points had already been covered. By eleven the meeting was over, but it had been a very productive one.

13   Merchants

The star cruiser CH 23426 was traveling at top speed towards the Ytl system. Tom took a look at the watch he was wearing on his wrist: still one hour from re-entering three-dimensional space.

"You are really fond of that awkward Earthly contraption," Sinqwahan observed, placing on the table a red hexagon, colored side down. "We have just the time to finish this game. If things go as you hope, you will soon be free to wear that watch in public. I bet that within a few months Earth watches will become fashionable over the whole sector."

"So when the emergency is over, we can retire and go into the business of exporting watches," Tom answered, playing a blue triangle, face down. He enjoyed joking about what he could do in years to come; it implied that there would be a future. "The hexagon you have just played was blue, so this hand is mine," he said.

The Aswaqat stretched out his hand and turned over the hexagon he had just played. "I'm sorry, it was red," he said. "We were right to have you pretending to be a Kirkysakh. Only an Aswaqat from Kirkyssi could play Tel'latl so badly."

Actually Tom was unable to concentrate on the game. After a while Sinqwahan got up, saying: "I realize that Tel'latl is not your favorite game, but I cannot play with someone who is thinking of other things."

"You are right," Tom answered, getting up and moving towards the bridge, "I cannot help thinking of how to present our plan to Aintlhad."

Re-entry was absolutely normal: the system looked almost deserted. Hardly had the ship started on her course for Ytl than the earthlings returned to the library with Sinqwahan. "Do you know what day it is today?" Susan said as he was already moving towards one of the projectors.

"I'd say it is the fifth day since we left Gorkh'ar," he answered.

"No, I mean in the earthly calendar," Susan said.

"It is the end of autumn or the beginning of winter," Tom answered uneasily: his life on Earth seemed to him to belong to a remote past and have no connection with the present.

"Yes, today is the twenty-fourth of December" Susan stated. It took Tom several seconds to realize the meaning of the date and the first one to react was Sinqwahan. "You mean that big holiday when you hang colored things on trees?" he asked. And then without waiting for an answer, he added "Perhaps you should ask Twenty-six to produce a visual projection of a tree. If you insist on taking back your identity, you will have to introduce your planet's traditions on board."

At first Tom dismissed the idea as foolish, but then he realized that a starship could be regarded as a piece of her captain's planet, and that there was nothing strange about their cruiser being a piece of Earth at hundreds of parsecs from the Solar System. He asked the computer to project the image of one of the trees he had seen on the edge of the volcano in the simulation that Twenty-six had produced as they were moving between the Tayssuk and Qhal'at systems. As soon as he saw it, he realized that it would make a decent Christmas tree, even if it was not really a fir. It took longer to make Twenty-six understand what the nature of the colored balls hanging from the branches and above all what they looked like, but in less than half an hour they had a Christmas tree, with decorations, lights and snow materializing a few feet above the tree and falling slowly to the ground.

Susan asked the computer to duplicate the image within their cabin and in the mess hall. For the whole day they had to answer questions about the meaning of the image and on various aspects of the culture and life on Earth. The Earthlings realized how difficult was to explain things that were normally taken for granted and which they had never considered in detail. Luckily the computer contained detailed documentation on Earth, which was used when preparing Sinqwahan's mission, and this made it easier for them to answer.

The hyperspace probe with the news of their arrival had reached the planet only a few hours earlier, because Tom didn't want their itinerary to be known well in advance. Nevertheless there was a big crowd at the port on the top of the cliff, a clear sign that after the raid in the Tayssuk system and the recent news on the crackdown against the Iskrat-is-Thn in Gorkh'ar the popularity of Admiral Tayqhahat was growing. Moreover, the Admiral was known to be a personal friend of the most elderly and venerable Iktlah, and also of the Coordinator of the sector. This all was stirring even more interest among people.

They had hardly disembarked from the shuttle when they were taken to the old royal palace, with a haste that betrayed the uneasiness of the security people. The Iktlah was waiting for them in the usual room, but this time he was completely alone and the meeting was absolutely informal.

Tom gave him a quick account of the events of the last month, stressing the discoveries concerning the fragments of the Qhrun ships. Then he entered into the details of the following parts of his plan, using the three-dimensional projector. He spoke for more than two hours, with no interruptions from Quanslyaq. To say the truth, Tom often wondered whether the Iktlah was actually listening or whether he had fallen asleep.

When he stopped speaking, and ordered the projector to switch off, Quanslyaq shook himself and commented: "Do you realize you have been speaking for more than two hours. without stopping for a single moment? Do you want to become an Iktlah?" Tom was about to answer that after becoming an Admiral of the Starfleet he would not have been much shocked if that were to happen, but the Tteroth didn't give him time to speak. "Your plan is reasonable," he said, "but it will be difficult to have it accepted. You plan to evacuate many systems by decree, without being able to explain to people the actual reasons for it and without allowing the news to spread, because if the Qhruns hear about it, the whole plan would become completely useless."

Only then was Tom sure that the Iktlah had actually followed his whole speech without nodding off to sleep. There were indeed good grounds for his objection and Tom was well aware of them. "Qhruns belong to a different world from ours, completely separate from it. They are aware of what we intend to do only through the information provided by the sensors of their ships, and if we keep their ships far from the systems where we are preparing our traps, they will fall into them without suspecting anything," he answered.

"We can predict their movements only by studying their past behavior, you are right in that. If the same holds also for them, any change in our behavior will find them completely unprepared," the Tteroth recognized. He remained silent for a long moment and then added "I believe it will need a good deal of persuasion, initially with Aintlhad, to get your plan put into effect. I must accompany you first to Laraki and then to the system you choose. . . provided I get there alive, I mean."

Actually Tom had intended to ask the Tteroth the very same thing, fearing that he would have had to do a lot to convince him. "Do you believe you can do it? You have not left this building for many years. And will your people let you go?"

"I can try," he answered. "As far as my people is concerned, they will make a lot of fuss, but everybody here knows that when I take a decision. . . And I don't mind returning to a star ship after more than a century." The Iktlah seemed younger than usual.

From that moment things went on at a quick pace, as if Quanslyaq wanted to prevent himself from reconsidering his decision. He ordered the computer to prepare the few things he intended to carry with him and called some of his disciples to tell them of his decision.

The arrangements for his departure took no more than ten minutes, and in a short time they were on their way to the spaceport. Tom immediately realized that the news of the departure of the Iktlah had spread quickly, because an enormous crowd was now lining the road.

They all wanted to see Quanslyaq, who was the center of general attention. When they got out of the vehicle, the crowd closed around them and many people exchanged a few words with the Iktlah. Sinqwahan was nervous and nodded quickly to Tom asking him whether he had his thermal gun to hand. But it was clear that if the Iskrat-is-Thn had taken advantage of this situation to attack them, a thermal gum would have been completely useless. "The hand laser I took on Gorkh'ar would be handy here," Tom whispered to him, but he didn't share the Aswaqat's worries.

"If I had known that we would end up in the middle of such a crowd, I would have organized a suitable guard. . . and I wouldn't forget hand lasers, you can be sure of that," he replied with a dejected air. But it was clear that weapons couldn't guarantee their safety as effectively as the sheer unpredictability of their decision. No Iskrat-is-Thn would have been able to imagine that such a favorable occasion would occur, Tom thought. Even he himself would not have thought of it, not even half an hour earlier.

The short stretch of grass separating their vehicle from the shuttle was covered in a few minutes, despite the crowd pressing in on them from all sides, and they were on board without almost realizing that they had reached their destination. But it was only when the shuttle was at a safe altitude that Sinqwahan relaxed.

Quanslyaq moved his head to see out of the windows, contemplating with pleasure and nostalgia the cliff of Ytlatq, now disappearing in the distance. Only when it was distant and the coast was just a thin line against the sea did he turn towards the others and say: "I didn't remember it was so beautiful."

Tom turned towards the Iktlah saying: "I forgot to thank you for sending us Ertlaq with all the cruisers you had in this system. Even though he gave us a fright, when we encountered his ships as we came out hyperspace."

Quanslyaq burst into laughter. "Ertlaq would never admit it, but he must have got a real fright too. You arrived right in front of his ships, in full fighting formation, with shields up. No doubt you gave him the worst five minutes of his life: by comparison the battles with Qhruns were just war-games!"

Tom realized that, in spite of all his attempts, he had not been able to consider the situation from every point of view. *How can I foresee the moves of aliens like Qhruns when I am not even able to put myself in the shoes of a human like Ertlaq?* he thought. "Did he tell you through a hyperspace probe?" asked.

"A Tteroth, an admiral of the Starfleet, admitting to having been afraid? He said he ordered the shields to be raised and to prepare to be attacked by Iskrat-is-Thn. And he concluded that it was your squadron that risked being destroyed, if the computers had not been able to understand the situation in time. Naturally your squadron, not his, was in danger!" Quanslyaq concluded with a laugh.

The shuttle was by now out of the atmosphere of the planet and was fast approaching the cruiser. Quanslyaq was again staring at the surface of the planet slowly moving below them and they all remained silent until the shuttle was moored in the cruiser's shuttle bay.

The following day the cruiser reached the hyperspace gate. Quanslyaq had spent the whole day in the library studying the documentation about Earth and talking with Tom and Susan. He was thrilled by the possibility of dealing with a human species still at the pre-space stage; an unexpected occasion for him to verify his theories on the development of civilizations. Another aspect of Earth culture that thrilled the Tteroth was the possibility of studying the religions of a species that still had widely different cultures.

Tom didn't share his enthusiasm for these historical and theological studies and tried several times to direct his attention to the matter of the Qhruns and on the implementation of his plan, but soon he had to desist. On the other hand, he was well aware that any chance of Earth being able to join the Confederation depended on Quanslyaq more than on any other person and that his interest in Earth's civilization was a great asset.

When the computer declared that they were approaching the hyperspace gate, Tom stood up and, heading for the bridge, told Quanslyaq: "We are about to enter hyperspace and move towards Laraki; do you want to come to the bridge?"

The Tteroth roused himself from his thoughts, and said with a surprised air: "No, not to Laraki. We must go to Wajjjklatl, don't you remember?"

Tom froze near the door and turned round in surprise "What? You didn't tell me anything about that and I don't know where this Wajj... is. I can't even pronounce it."

"You see what it means to be more than four hundred and fifty years old? Are you really sure that I didn't tell you about Wajjjklatl or, rather, New Wajjjk?" The Iktlah's amazement was genuine. "As you see, the chance of studying a civilization like yours in its early stages made me forget what I had to tell you about this trip. But, trust me, let's set course for New Wajjjk and I will explain to you what I want to do while we are in transit. Then we will go on to Laraki." He then turned towards Susan and asked her: "How long will it take to get to Wajjjklatl? I believe that it is no more than a thousand parsecs, unless I am mistaken."

Hardly Quanslyaq had named the new destination, that Susan had started to work with a three-dimensional projector. displaying galactic maps. "It's one thousand, two hundred and fifty parsecs. We can get there in eight days. And from there to Laraki it will take four days more, at least. In all our journey will be five days longer."

Tom thought that Starfleet cruisers were too slow. "Do we really need to waste five days in this way?" he asked Quanslyaq.

"I am certain that we should. A soon as we enter hyperspace I will explain everything to you," he answered.

"Then set course for New Wajjjk," concluded Tom, speaking to the computer, leaving the library.

The cruiser re-entered three-dimensional space in the Wajjjklatl system after a little less than eight days. In truth, they could hardly tell they were no longer in hyperspace from the view on the screens: almost half of the space around them was absolutely black, with the exception of widely scattered, dim distant galaxies. On the opposite side they could see the distant glow of the Milky Way and a few scattered stars. It looked as if they had re-entered three-dimensional space in the middle of nowhere. During the eight days in hyperspace, the longest stretch he had ever crossed, Tom had studied the documentation on Wajjjklatl and on the Injjuks and was prepared for such a view. But it is one thing to see empty space with a mental projector and quite another to be there. For the very first time he felt lost in the utter emptiness of space.

He only had to look around to realize that everyone on the bridge was feeling uneasy. "This is a horrible place. I was here only once before—more than two hundred years ago—and I remember it as if it were yesterday. The only time I had a worse feeling of emptiness was when I traversed hyperspace outside a ship," Quanslyaq said.

Tom tried to see a reasonably bright star to be sure that they were close to a planetary system, but he could see nothing: Wajjjklatl was a small planetoid orbiting a neutron star on the outer edge of the Galaxy, at a distance of about two thirds of the Galaxy's radius. Astronomically speaking, it was not far from Laraki, but its position made it one of the most isolated places of the whole Galaxy—geographically isolated at least, because everything going on in the remotest corner of the Galaxy was quickly reported to Wajjjklatl.

Those thoughts were soon interrupted by the voice of the computer, who was carefully probing the surrounding space. "I have identified more than two hundred ships in the system. Ships of all the types, from large transports to tiny freighters. No cruiser or other starfleet vessel." The voice of the computer was silent for a few seconds and then returned. "I think we had best send an identification message: the nearest ship will identify us in less than two hours, provided she does not change its course in the meantime."

The presence of a starfleet vessel in the system of Wajjjklatl was a violation of the treaties between the Injjuks and the Confederation, treaties nobody would even dream of discussing. The violation may have been justified by the exceptional circumstances but it needed immediate explanation.

Tom was about to speak, but Quanslyaq nodded to him and said: "Twenty-six, beam your broadcast to all ships and to Wajjjklatl." He moved to find a more comfortable position on the platform he was lying on, and then continued, looking at his own image that appeared on the screen in front of him: "Message from Quanslyaq of Ytl for Atkaljj, Lord of the Injjuks." He paused, uncertain of what words he should say.

He opted for a direct message, without preamble or any concession to eloquence. A direct message in the Injjuk style. He smiled, remembering how Atkaljj's father used to tease him for his Tteroth custom of using bombastic sentences. "Atkaljj, Lord of the Injjuks, forgive us for this intrusion in the New Wajjjk system. The alarming events that are upsetting the sector force me to take an action that violates our treaties. I come with Admiral Tayqhahat of Kirkyssi and I ask you to allow us to confer with you."

He nodded to the computer to break off the transmission and then turned towards Tom. "Now let's hope Atkaljj is in the system and agrees to receive us. If only his father were still alive!"

"Message sent," confirmed the computer. "If we maintain our present course and speed, the answer will reach us within four hours. Anticipated time of approach, if we follow the standard procedures, eight hours and forty minutes."

Now there was nothing else to do except wait, monitoring the traffic of starships of every type that the sensors continued to record. "Things are going badly here as well," Quanslyaq commented, without addressing anybody in particular. "Only two hundred ships in the Wajjjklatl system, it is almost unbelievable!"

Time passed slowly, as usual. Then suddenly the screen came on and the face of a human appeared, who said: "Message from Atkaljj, Lord of the Injjuks, for the venerable Quanslyaq of Ytl and for Admiral Tayqhahat of Kirkyssi." His face was long and strikingly skinny; his eyes protruded from his face and his skin, pale and almost translucent in color, seemed to be just a thin covering over the bones. Overall, it looked rather like the head of an insect, even though, as Tom was fully aware, it couldn't be, because all the human species in the galaxy were vertebrates. And the Injjuks were mammalian, perhaps not very different from primates.

The image disappeared and another face, even longer, materialized on the screen. "Atkaljj of Wajjjklatl greets the venerable Quanslyaq of Ytl. May air sustain your wings, venerable Iktlah." While pronouncing this phrase he was unable to refrain from a strange grimace, clearly a smile. It was clear that his father had spoken to him about the Tteroths, and goodness knows how many times they had joked about their solemnity. "My regards also to Admiral Tayqhahat of Kirkyssi. The echoes of your achievements have even reached this remote corner of the galaxy."

"You are welcome on Wajjjklatl, and don't worry about the conventions regarding starfleet vessels. The venerable Quanslyaq is always a welcome guest, no matter what ship brings him here," concluded Atkaljj. His speech was followed by detailed instructions to the computer on how to perform the approach and entry to the base on the asteroid.

"It seems Atkaljj's views regarding you are like his father's," Tom exclaimed. turning towards the Iktlah. He had, in fact, been unable to find much documentation on Wajjjklatl and on the recent history of the Injjuks. and he was curious to know what role Quanslyaq had played there two centuries earlier.

"The Injjuks are well known for their hospitality and for the way they always keep their word. Without the latter trait they would have never been able to maintain their commercial power and to stay—to all practical purposes—outside the political structure of the Confederation, while remaining, at the same time, one of the pillars upon which galactic civilization is founded.

However I felt a warmth in the voice of Atkaljj that makes me optimistic. I believe he thinks that once again the interests of the Injjuks coincide with those of the Confederation." Quanslyaq didn't add anything else and Tom sat down next to Susan in front of the navigation screen, to study details of the Wajjjklatl system.

The neutron star was not yet visible, and Tom wondered whether it could be seen from the asteroid. He decided that it was unlikely, given the size of the two bodies and the distance between them. The asteroid had been transformed thousands of years earlier into one of the Confederation's Starfleet bases. Not a large base like that at Gorkh'ar, but a small outpost of civilization facing intergalactic space. Then about two centuries before, when Wajjjk was taken by Qhruns, the Injjuks had moved here, renaming the asteroid Wajjjklatl or, in their language, New Wajjjk.

There could be no doubt that it must have been painful to move from the splendors of Wajjjk, the crossroads of all commercial routes in the fourth sector, and the capital of a human species that had founded its prosperity upon interstellar commerce, to that tiny base, abandoned a long time ago. But, Tom was certain that New Wajjjk had changed considerably in the last two centuries. He was eager to see with his own eyes whether the legends about the wealth of the Injjuks,—the Merchants, as they were called throughout the Confederation—were true. Or, as they described themselves, the 'People of Space,' which was what Quanslyaq recommended him to call them.

The asteroid soon began to be visible and the image quickly got larger and larger on the screen. Seen from the outside it was not a very considerable thing at all. It was a large, irregular chunk of rock, slowly rotating in space. Very soon they could see the entrance, quite similar to that of Gorkh'ar, with batteries used to blast the meteorites that always constitute a threat to any base on an asteroid.

As soon as the cruiser entered the tunnel, it became clear that it was not a Starfleet base. The tunnel was illuminated by an intense light, although not to the point of being annoying, and the walls were covered by various materials, all decorated with geometric patterns. The door closed behind them and the dust rising from the walls indicated the entry of air from the second airlock, which led to the main cavern. It was more spectacular than at Gorkh'ar, owing to the lights and the colors of the walls, and less obvious, because of the low air pressure.

They went through the second doorway and the ship entered the brightly lit and decorated main cavern. While in Starfleet bases the cavern is the main location for the accommodation, here the cavern was just a spaceport; and people lived elsewhere. The ships were mostly located in hangars lining the walls and there were only few people, all Injjuks, around.

The computer guided the ship into one of the hangars and landed it on the ground. "Let's go, Tom" Quanslyaq said. I think the Lord of the Injjuks is waiting for us. Please help me with a mask and an air bottle, the air the Injjuks breathe is too thin for our lungs."

Tom strapped an air bottle to his shoulders and fitted a mask over his face and Wal-Nah did the same to Quanslyaq. They moved towards the airlock, accompanied by Susan who again asked to accompany them to meet Atkaljj. Sinqwahan too had insisted that he wanted to witness the meeting, but Quanslyaq had repeated to them several times that the Injjuk had invited two individuals and it would be impolite for anyone else to accompany them.

When the air began to be exhausted from the airlock, Tom felt uneasy: his blood was throbbing in his temples and he felt dizzy. "you are you alright, Quanslyaq?" he asked the Tteroth, with a worried note in his voice.

"I would feel better if they only kept the pressure a little higher," he answered, keeping his eyes closed and grasping the edge of the platform that he was lying on.

As soon as the door opened Tom saw the Injjuk waiting for them below the ramp. He knew that Injjuks were very tall, but from the images on the screen he didn't expect what he was seeing. The man was at least ten feet tall and so slim that he looked even taller. He wore a colored overall of a shiny material, seemingly as soft as silk, and in striking contrast to the pale color of his semi-transparent skin. Under the overall he could see slender limbs and a well developed chest: not a muscular one but with large lungs adapted to breathing such thin air.

The Injjuk warned him immediately: "Be careful, Admiral. There is no artificial gravity here. You have better using a platform instead of walking." Tom climbed on a platform, which locked his legs into position with a force field, and moved away from the cruiser, followed by the Tteroth. Hardly the effects of the artificial gravity of the ship faded away, than sickness caused by the sudden weightlessness added to his dizziness from the low pressure.

"You will soon feel better, Admiral. Only a few strangers come here, but they quickly become used to our conditions," the Injjuk said, leading them towards a corridor cut in the wall of the cavern. He moved along with almost horizontal jumps.

The Injjuk was right; slowly the feeling of dizziness and sickness decreased and Tom could begin to enjoy his surroundings.

Injjuks were the only species in the galaxy to live in the absence of gravity and in a low pressure-environment. Their species had adapted to the environment of Wajjjk, a big asteroid, but no one considered that they originated there. Legend said that in the Wajjjk system there was an inhabited planet where the human species from which the Injjuks derived first evolved. When their civilization had achieved space flight but not interstellar travel, an unknown catastrophe made the planet unsuitable for life and a group of people settled a large asteroid, the only body in the system on which an outpost had ever been built. Over the millennia, the base was expanded and the species adapted to the artificial environment of the asteroid. After a few thousand years they became almost twice as tall as originally and much slimmer. When they finally became able to travel through hyperspace and to reach habitable planets, they were totally unfit for living on the surface of a planet.

However, their new physical structure made them fit for space flight: their ships required neither artificial gravity nor thick external hulls built to withstand atmospheric pressure. And, what is more, Injjuks, accustomed to the artificial environment of the asteroids, were psychologically well adapted to space travel. For other species space was a hostile, temporary environment, but for Injjuks space was their natural home. Mainly because of this difference in their habitat from those of other species, they never wanted to join the Confederation or Starfleet, but developed a civilization that was complementary to it. They were the 'People of Space,' as they described themselves, and they called everyone else the 'People of the Planets.'

Thinking of his own uneasiness in that environment, Tom had to admit that for Injjuks landing on a planet would be much more painful. When they had to land for reasons linked to their business, they had to wear a whole metallic exoskeleton, so that they were not crushed by gravity and a respiration mask to enrich their air with inert gas, to reduce the partial pressure of oxygen.

Tom roused himself from these thoughts to observe the long corridor they were moving through, which was all decorated with colored geometric patterns. They soon entered a larger corridor, where they met several other Injjuks. Nobody paid them much attention, although some that they passed brought their left hand to their right shoulder in a gesture of greeting.

After a few hundred meters they came to a door. which opened in front of them. The Injjuk accompanying them moved aside and nodded for them to go in. They entered a large hall, brightly lit and highly decorated. The first thing that attracted Tom's attention was a big window showing space outside the asteroid, in front of which an Injjuk was standing. He immediately moved towards them: "Welcome to Wajjjklatl," he said, putting his left hand to his right shoulder.

He knelt on the floor, in front of a square block of perfectly polished shiny black rock, on which various objects were lying. Tom noticed that among them there was a large transparent crystal. The man gestured for Tom to take a seat on the other side of the table, where a faint glow caused by a weak force field was present. He sat down cautiously, realizing that the field was almost certainly strong enough to bear his weight in such a weak gravitational field.

"That air may sustain your wings, Iktlah," Atkaljj went on. "I have been looking forward to meeting you ever since I was a child. My father often spoke to me about the wise Tteroth who helped our people so much when we had to abandon Wajjjk. He repeatedly told me that were it not for you, we would never have been able to settle here."

"You don't know how sorry I am that I have been unable to meet him again since then," Quanslyaq answered. Tom finally began to understand the bond linking the Tteroth with the Merchants. While the other was speaking, he realized that what he imagined to be a window was actually a screen: the room was probably deep under the surface of the asteroid.

The Injjuk turned towards him and said "I am happy to meet you too, Admiral Tayqhahat. But obviously your true name is not Tayqhahat and you don't come from Kirkyssi." He spoke the language of the Aswaqats of Laraki perfectly, but pronounced it in a peculiar way. He spoke slowly, and stressed all the vowels, making the language more musical than it was originally.

"Obviously," Tom answered after a few moments' silence. He paused, and then said: "I would ask you to order your computer to isolate this room, so that we can speak freely." Atkaljj seemed surprised by that request and considered it in silence. Then he said a few words in his language and continued: "Done. But now I must ask you to break the connection with your ship on the encoded channel."

Tom blushed as if he had been caught to committing some indiscretion. He took the communicator from his pocket and said in a loud voice: "Twenty-six, terminate the connection, I will call you when I need you." The voice of the computer answered immediately: "Are you really sure you want to break off the connection, sir?"

"That is an order, Twenty-six," he said, feeling embarrassed, laying down the transmitter on the table. "Now we can speak freely," he continued, turning to Atkaljj. "I will tell you everything you want to know about me, but first please tell me why it was so obvious that I am not Tayqhahat from Kirkyssi. And how you understood I was in contact with my ship."

"The answer to your second question is obvious: we are able to understand all the Starfleet's classified communications. I tell you this, because if you are here it is because you are looking for our collaboration, and if we are to work together we must play openly. I never liked playing Tel'latl, where you never know the color of the cards your opponents have played." Tom realized he had more in common with the Injjuk than he had suspected.

"But the answer to the first question is also easy. Three months ago Tayqhahat didn't exist on any computer and the battle of Kirkyssi had no survivors. Then you suddenly emerged. In person, in the news and on the computers throughout the whole sector. And your appearance in the computers occurred in concentric circles, centered on the Laraki system, expanding at the same speed as the best cruisers of the fleet. Obviously if someone went to all that trouble, the thing must be important. And this someone cannot be Starfleet, the present Starfleet, I mean. Only the Coordinator is still able to exhibit such efficiency. That reassured us, because we trust Aintlhad. Not because he is the Coordinator, nor because he is an Aswaqat." Here he paused shortly, and then went on with a smile "we can speak freely of Aswaqats: I would bet that crystal that you see on the table that there is no Aswaqat here. But we trust him because he is a friend of a certain Iktlah who is in this room. And then, your first action worthy of being mentioned, the battle of Kistl, occurred on the way between Ytl and Laraki. . . between Quanslyaq and Aintlhad."

He paused and then spoke again, looking Tom straight in his eyes "Admiral, who are you? and what are you?"

Quanslyaq looked at Tom and said: "As you see, New Wajjjk seems to be at the extreme outskirts of the universe but nothing can happen in the Galaxy without being reported here." Then, turning towards Atkaljj, he went on: "Thank you for trusting me and Aintlhad. Tayqhahat—let me call him that still—is a part of a plan we devised for a last attempt against the Qhruns. And our visit here is another aspect of the same plan."

Tom waited a few more seconds, leaning back against the force field, and finally spoke, carefully articulating the words: "My name is Thomas Taylor, from Earth, in the system of a star known to us as the Sun. A class B planet."

Atkaljj looked him wide-eyed for a few seconds, then exclaimed: "A class B planet! Knowing how the Confederation behaves towards class B planets and the theories of the venerable Iktlah, this is really a nice piece of news! The system of the Sun, is that not perhaps a system with eight planets at about two hundred and fifty parsecs from Gorkh'ar?"

"Two hundred and sixty, to be exact," Tom answered.

"Then I know it. I have been there to rescue a ship that landed on the fourth planet in an emergency. A desert, reddish planet, the only advantage of which was not having a strong gravity and a thick atmosphere. A horrible place. And you come from the third planet, then. I must confess that we don't feel too obliged to follow the law on class B planets—I can say that here, especially considering that it seem that under certain circumstances the Coordinators are not too touchy on the matter. While I was in the system, I approached the third planet to see whether it might be possible to start some commerce, but I found a backward civilization, at the beginning of technological development, and gave up," the Injjuk conclude.

From the way he complained about having lost a potential market, Tom understood that the nickname of 'Merchants' given to the Injjuks was fully deserved. "When was that?" he asked.

"I was very young: my father sent me around to 'learn how to live', as he said. It was therefore almost two centuries ago: I would say a hundred and eighty years ago" he answered.

"Then you would not recognize that planet any longer. We are now at the first stages of space flight," Tom answered.

"They are in a phase of quick progress: it must be because their lifespan is still rather short, and also because of strong demographic expansion," Quanslyaq interrupted, noting the Injjuk's puzzled expression.

"You must have considerable problems, then. Such rapid development, rapid demographic expansion and perhaps even a good dose of aggressiveness. Best wishes, Earthling. I am glad I didn't land," Atkaljj said.

"I don't deny it. And I can tell you one of the reasons I accepted Aintlhad's offer is because I believe that this can help to solve our problems," Tom said.

"If you don't mind, I have my doubts about that. Not to speak badly of our Coordinator, but I don't believe he worries about a planet that does not belong to the Confederation and, what's more, a class B planet. And perhaps he is right, especially now with the problems he has to face."

"I never had any doubts about that," Tom answered. "But I believe that we can do a good deal. Perhaps I can help the sector against the Qhruns, and the Confederation can help Earth with its problems."

"To me that is fine: it looks a good deal for everyone, provided that you really can do something. We don't deal with business of the kind, but have no doubt that we know how to distinguish a good deal from a bad one: they do not call us merchants for nothing." He stared at Quanslyaq who looked embarrassed. "Certainly, not in our presence, we are the People of Space then. Well, merchants we are and we have always been, and also because we couldn't be anything else. But now you must explain to me what we have to do with this good deal of yours. And if you have bothered to come to this lonely spot in the galaxy at such a moment, it means that we are involved in it. . . and deeply involved too."

Now it was up to Tom to convince his audience and he had best to do it properly. He stood up, went to the platform on which the Iktlah was lying and took out a three-dimensional projector. He laid it on the table.

"Sorry, I don't buy three-dimensional projectors," said the Injjuk with a laugh.

Tom remembered that the first time he had met Sinqwahan he had the same reaction when the Aswaqat did a similar thing. "Don't worry, I have something else in mind. I'll only use this to show you a few explanatory images."

He began by repeating his lecture on the history of the last four hundred years and on the Qhrun expansion, like the one that he had given in Laraki in front of the Coordinator. Then a lot of the things he had said were only assumptions, disguised as certainties, now he was more assured of their accuracy. He continued by illustrating the first part of the plan; the work he had done at the base at Gorkh'ar; the details of the raid in the Tayssuk system: and the capture of the Iskrat-is-Thn. He then spoke of the second part of the plan, the aim of which was to stop the invasion. The greatest problem in implementing that strategy was clearly the evacuation of a large number of systems.

"And this is where I am asking for your help. You have the largest merchant fleet in the galaxy, and your fleet is intact. You cannot deny that since you moved into this sector, about two centuries ago, Wajjjklatl has become the greatest commercial hub of the galaxy. And when the sector was isolated, a great number of your ships remained here. You withdrew from system to system, but apart from occasional attacks, you have lost neither men nor ships. On the other hand, the Confederation's merchant fleet has suffered many losses and a large number of ships are not spaceworthy."

Tom paused, as if collecting his ideas, and then concluded: "The Confederation needs you. To evacuate at least fifty systems in this sector alone, thousands of starships are required, with crews that are up to the task. Only you can do it."

He relaxed against the force field, and only then realized that he was exhausted. He took a look at the watch that he still insisted on wearing and realized that he had spoken for three hours without pause.

After a couple of minutes that seemed like eternity, Atkaljj raised his head and began to speak: "You said you were a scientist and now you have become a spacer. You would have been a good merchant too; you have advertised your goods in a pretty convincing way. But to sell them to me you must convince me that this deal is profitable for us as well."

"You know full well that your situation will not be sustainable for long. You can go on withdrawing for fifty or one hundred years, but then, when the whole galaxy has been invaded, what will you do? And even now, when will your prosperity end? And what will the People of Space do when they have just empty space in front of them? Do you believe, like the Iskrat-is-Thn, that when the Angels of Wrath have completed their work they will disappear, leaving the galaxy to the Elect? No, the only hope for the People of Space is to joining the People of the Planets to fight for their common humanity." Tom concluded in an inspired and rather solemn tone that rather surprised himself.

"Going around with Quanslyaq has infected you, Earthling. You are now speaking like an Iktlah. Sorry, but I cannot refrain from seeing the funny side of things. More than anything else, that the great warriors of Starfleet are asking the Merchants for help. And I say this fully recognizing that, for the first time in a thousand years, Starfleet has an admiral worthy of the name." Atkaljj paused and suddenly became serious. "Unfortunately you are right. Our only hope is the Confederation's Starfleet. Your plan is reasonable. At least it is worth trying and, if our friend the Iktlah acts deaf for a moment, I would add something between the two of us. You are trying to gain some advantage for your planet from all this mess. The same applies to us, The Confederation always had an ambivalent attitude towards us, and on the other hand Injjuks have always considered that their isolation was a fundamental asset. But what is happening shows that this attitude is dangerous. For years I have been looking for an occasion like the one you are offering me. In every sector there is a species that has subtly achieved a kind of hegemony, and in the fifth sector that species are the Aswaqats. Your plan will result in the weakening of this hegemony, and Aintlhad knows it very well, but cannot do anything about it. The sooner you return to being Taylor from Earth, the better for your planet and for us Injjuk," he concluded, with a smile.

"It is not my intention to act against Aintlhad or the Aswaqats. At the moment we need complete unity." Tom didn't like the new turn that the conversation was taking. Above all, he didn't want to become involved in the politics of the sector. He had changed his role once already at the beginning of this adventure and now he was playing the role of a soldier. If there was one thing he was very clear about, it was that he had to keep military and political affairs well separated.

"Neither do we: Aswaqats are our best customers in the sector. What I meant is that, if things go as we hope, we may emerge from this crisis with a more balanced situation, where Earthlings, Injjuks and many other people who are currently of lesser significance, will have a greater part to play."

This saw an end to politics. Now they needed to discuss details of the Injjuks' involvement in the second phase of the plan, and to talk about ships, convoys, and military operations. This brought Tom onto more familiar grounds, or at least, grounds that had become more familiar in recent months.

They spoke for another hour, and then Tom realized that the moment had come for them to leave, but he was unable to refrain from asking the question that had worried him since he first heard about the Injjuk. "Atkaljj, what happened to your planet? Why did you have to flee to Wajjjk and transform yourselves from People of the Planets into the People of Space?"

"You ask something to which nobody can give you the answer, Earthling," the Injjuk replied. It was as if he was another person now: the amused and shrewd air he had exhibited until then faded away to be replaced by a rather gloomy expression. "More than fifty thousand years have passed since then, perhaps much more. Our known history starts when we were already as we are now, and only legends are left about our remote past. We never speak of these things; many of us think that these legends have no foundation and that our species actually originated on Wajjjk. Others think that our ancestors came to Wajjjk from other systems; that they were the crew of a ship that crash-landed on the asteroid, and that we don't have anything to do with the civilization that evolved and later disappeared from a planet in our system. But I believe that those legends do have a foundation in truth."

"But what you want to know is how the civilization of that planet was destroyed. And you want to know because your very civilization is involved in the worst crisis that all civilizations must go through: the one preceding expansion in space. I don't know whether the catastrophe that struck that planet was a natural one, or one caused by our distant ancestors, but the galaxy is full of the ruins of civilizations that failed to survive the pre-space stage. Civilizations that are too aggressive or irresponsible don't survive, and that's probably a good thing: a sort of natural selection, so to speak. We came through our crisis by a small margin, although the cost was very high. And if your plan is successful, your chances are far greater. But this is not the time to worry about such issues, the true problem is the Qhruns. If we don't stop them, no civilization will survive in this galaxy".

As they were going back to their ship Tom didn't say a single word. Quanslyaq, who was too tired to speak, was lying motionless on his platform. Only when they were close to the ramp leading to their ship did the Tteroth say: "Don't worry. I realize that Atkaljj's words made an impression on you, but don't be alarmed. All known civilizations have passed through a phase when technology seemed unable to solve the problems it had tended to create. And most of them survived that phase."

"You are right Quanslyaq. But please understand me, on my planet there is a great deal of argument about what you call the pre-space crisis. We are an overpopulated, and much polluted world, full of weapons. These three problems are then considered as the likely cause of a catastrophe that will put an end to our civilization. But Atkaljj is right: for now we must deal with the Qhruns. This is the danger we must face now," Tom answered.

"By the way, did you notice the last words Atkaljj said? He didn't say 'if you don't stop them' but 'if we don't stop them'. It is the first time an Injjuk has become so deeply involved in matters concerning the Confederation. I like Atkaljj: he is like his father, he has a global view of the galaxy and not the isolationist attitude typical of his people," went on Quanslyaq, who after resting, now felt more like speaking.

"You may be right, but I am worried by the antagonism towards Aswaqats that I felt in his words. . ."

"Don't worry, it's quite normal. As I told you the first time we met, when everyone is in great danger, people tend to divide and everyone tries to save themselves without bothering about anyone else. It is a danger, of course, but only when it rules people's actions. If it remains at the stage of a general rivalry, it may even be useful and we may be able to exploit it to get the best performance from everyone."

By then they were in the airlock. The Injjuk escorting them brought his left hand to his right shoulder by way of salute, and then left. They entered the starship and, as soon as the air pressure began to increase and the gravitational field returned to its normal value, Tom started to feel uneasy. In just a few hours he had become used to the Injjuk's environment and now it was unpleasant to have to readapt. The feeling of oppression caused by the combined actions of gravity and pressure lasted for just a short time, and when they reached the bridge they were feeling at ease again.

They remained on Wajjjklatl for a few more hours, just the time essential to prepare the ship for leaving.

Their journey to Laraki lasted five days and was uneventful.

14 The Scorched Planets Operation

When Tom landed the ship at the capital's spaceport in the Laraki system, it looked much busier than last time and he wondered whether this was already one result of the way in which his own presence was having an effect on the crisis. The word that Admiral Tayqhahat was due to be received by the Coordinator, together with the venerable Quanslyaq, had spread, and the road from the port to the Coordinator's palace was crowded. Sinqwahan told him that rumors about the identity of the admiral had started to spread even on Laraki and had contributed to the rise in general interest.

They left the port immediately and, when they arrived at the square in front of the building, Tom noticed that there were greater numbers of the security forces present than last time. On this occasion, Aintlhad met them outside the building, and Tom realized that this was a unusual gesture prompted by his regard for Quanslyaq. They entered quickly and within a few minutes they were in the room where Aintlhad had received them less than three months earlier.

Without any loss of time, the Earthling began to explain his plan with the help of a projector. This was the third time he explained the same ideas in almost the same words in less than two weeks and he felt rather like an actor performing the same part over and over again. But it was true that the more he repeated it, the less convinced he was that it was feasible. And he would undoubtedly have to repeat the whole thing on Gatlaat, in front of those directly involved, and then an unknown number of times, if his strategy of mass evacuations was to be implemented.

He had thought that his plan needed a name, like all major military operations, and so, when speaking, he introduced it as the Scorched Planets Operation, even if he had to admit to himself that it was a fairly obvious name.

He had been speaking for about an hour, when one of the screens came on, showing an Aswaqat, who announced an incoming message for Admiral Tayqhahat from a starship that had just entered the system. As Tom turned towards the screen he felt apprehensive of what he would hear.

The unmistakable face of an Opsquat appeared on the screen. "This is Idlath of Gorkh'ar," he said in a voice that betrayed strong emotions. Idlath was quite shaken because, instead of declaring his planet of origin, he had mentioned the system from which he had come. Tom looked at Aintlhad, and noticed that the Coordinator had found the fact rather amusing. "I need to confer urgently with Admiral Tayqhahat. I request a meeting with him as soon as we land."

Tom was even more worried now. He knew Idlath well and it was clear that something very serious had occurred if he had left Gorkh'ar to speak to him. However, the only thing that could be done for the moment was to wait. Yet the automatic message broadcast by his ship when she had left hyperspace, had not contained any alarming news.

"Please continue: it looks as if that Opsquat has become far too excited about some trivial matter. We will see what he has to tell us when he gets here," Aintlhad said.

Tom started to speak again. While continuing to describe his plan, he realized that the Coordinator was becoming more and more skeptical, and he glanced several times at Quanslyaq for support. But the Iktlah was motionless and almost looked as if he was asleep. When he mentioned the role of the Merchants and spoke of their visit to Wajjjklatl, Aintlhad's expression was distinctly unsympathetic.

When he finally ended and sat down, an uncanny silence fell on the room. After a few minutes Aintlhad got up, went over to a screen showing a map of what remained of the sector, and stood there for some time staring at it.

He was so absorbed in his thoughts that when another screen came on and an Aswaqat announced that the freighter coming from Gorkh'ar had landed and that the Opsquat was heading towards the Coordinator's palace it took him some time to realize what was going on.

Tom cursed mentally. He had forgotten the matter, and now he started to worry again. In about twenty minutes he would know what Idlath had found so important that he had travelled two thousand parsecs to speak to him. They continued to talk about various matters, but clearly they were just waiting for Idlath to arrive.

Finally the door opened and the Opsquat came in. He was wrapped up in his multi-colored mantle hanging from his shoulders even though he was overheated from agitation. He came in and immediately started to speak to Tom, without even noticing the others. "Admiral Tayqhahat, I have to urgently speak to you in private. Come outside, please."

Everyone was speechless, particularly the Aswaqats. "Idlath, you can speak here. There is nothing the Coordinator cannot hear," Tom replied immediately.

The Opsquat stopped suddenly and looked around. He finally recognized the Coordinator and pulled a strange face. He started to stammer out some apology, but immediately stopped, realizing that what he had to say was so urgent that he could apologize later. "Admiral, we must immediately stop the raids on the occupied systems." He looked around and, noticing the perplexed expression of everyone present, continued: "I have come to say this. We must stop fighting against the Qhruns."

Everyone stared at him. The message that had been automatically broadcast by his ship didn't mention any news that could explain what he had just said. "We haven't lost that many ships; Ertlaq would have sent me a probe. It doesn't look as if we are having problems with the Qhruns," Tom answered, trying to remain calm.

"That is not the point. We didn't lose any ships, at least up to the time I left. It is that we cannot go on destroying their ships," Idlath said, still standing close to the door.

"Idlath, take a seat and tell us what happened on Gorkh'ar. Calm yourself and start from the beginning. Then the Coordinator and myself will decide what we must do," Tom interrupted him.

The Opsquat went to the table and sat down. He lowered his head as if to put his thoughts in order, and after a few seconds again started to speak: "From the study of the fragments we have deduced that on Qhruns ships there are. . ."

Tom realized that he could not let him continue in that way, influencing the Coordinator with conclusions drawn by his crisis committee. Perhaps they really would have to modify his plan, but he wanted to play the main role in that. He needed time to understand what was going on before the Opsquat's words irremediably influenced the decisions they had to take. He interrupted him with a quick gesture and with a casual manner, which was in complete contrast to the way he actually felt, and said: "Please, Idlath, don't start from what you have deduced. We need to know facts. Nothing serious will happen if we delay fighting Qhruns for a few minutes. Now tell us in detail what you did and what happened. We can then all draw the relevant conclusions."

He looked around at everyone. They were all eager to hear what Idlath wanted to say and showed their disapproval of his remark. All except Quanslyaq, who was lying on his platform with his eyes closed.

"A short time after you left, Ertlaq also left with his squadron for a quick raid beyond the frontier. Their main aim was to get other fragments from Qhrun ships to see if it were possible. . ." Idlath stopped short: the idea of speaking of the project to clone a Qhrun before the Coordinator and the Iktlah made him uneasy. ". . . to implement the project we spoke about. The squadron came back in twelve days after destroying forty ships without losing a single cruiser. But the most important thing is that they brought back a good quantity of fragments."

The Opsquat was cooler now but it was clear this was the result of strong self-control. He was sweating, in spite of the low temperature by the standards of his species. "We immediately began analyzing the fragments and on one of them we identified a stain with a possible biological origin."

"Could you understand where that fragment came from?" Tom interrupted him. He was starting to have an idea about the meaning of the whole affair. He cast a glance at Susan to see whether she too has started to understand, but she responded with a puzzled glance.

"It was a part of the inner hull, we are sure. There was gas adsorbed onto the surface and we drew the same conclusions about the atmosphere inside the starship as last time." Idlath had not reacted to his interruptions, while the others continued to follow his story with interest. "This time the material was more abundant and we began to analyze it, hoping that it was somehow of Qhrun origin."

This time was Aintlhad who interrupted him. "You mean that you could prove that Qhruns are biological beings and that you got specimens from their bodies?"

"No, Sir. Once again, no biological specimen from Qhruns was identified. I was just about to say that. We analyzed a specimen coming from a big stain and it was identified as typical material from a living being: the result of a normal evolutionary pattern. We isolated the genetic material and eventually we found that it was from a large herbivore, native of the system where the Qhrun ship was destroyed. We then went on working with samples coming from smaller stains and, although we couldn't identify the species with absolute certainty, they were all specimens of animal origin, all coming from species from the same planet."

Idlath broke off. Tom had noticed his taste for coups de theater and was now quite sure that he was preparing the audience for his final revelation. At this point Tom was sure he understood what his conclusions were, and turned towards Susan with a quick gesture of finger to his lips. Susan answered with a gesture of assent—she too was now sure about the conclusion the Opsquat had drawn, and both were sure that he was wrong.

Idlath started speaking again. "I found a further trace of biological material and I had it examined like the others but this time the technician didn't want to load the results onto the computer. He called me, and together, we repeated the analysis several times, always getting the same result. That fragment couldn't be anything other than a fragment of an Aswaqat." He paused for a moment, looking around to see how the others reacted to the news and then went on: "You can now understand why we came here immediately. We must stop attacking Qhrun ships: the crews of these ships are human. And we have evidence that at least on one ship there were some Aswaqat."

The Opsquat sat down, while the others slowly tried to understand the implications of the news. There was no hysterical reaction, which was what Tom had expected. After a few moments' silence Aintlhad raised his head and, looking at the Opsquat, asked him: "Are you absolutely certain of what you said?"

"Absolutely, Sir," Idlath responded.

"This changes things completely. If there are human crews on those ships, our strategy must be modified completely. Actually, up to now we have assumed that any occupied planets were lost and that we had to do all we could to prevent the part of the galaxy that is still free from falling into their hands. But we now know that humans travel on those ships." The coordinator turned towards Tom. "I believe that this forces us to reassess all of our strategy, Admiral Tayqhahat. Perhaps we don't even need a strategy any more, at this point."

You could say that you don't need to have an admiral any more, and even less an admiral from Earth. But don't think you can get rid of us that easily, thought Tom.

Idlath was now quietly sitting, apparently satisfied by this reaction. *You may well be satisfied of your job,* thought Tom, unable to forgive his attempt to cancel his plan. But he had to admit that the Opsquat had acted correctly: he wanted to talk directly to him, but he, in turn, had insisted on his speaking in front of the Coordinator. He doubtless didn't mean to damage Tom or Earth and, if he felt any satisfaction, it was only because he had done a good job and earned a place in the history of the Confederation.

Sinqwahan was visibly embarrassed: he thought that the mission of the earthlings had ended right there, and it would not be a pleasant task to take them back without fulfilling the promises he had made. Quanslyaq didn't budge, apparently showing no intention of speaking. Clearly he thought that the matter was not yet over and Tom was sure that they had both reached the same conclusions.

Tom got up and began to speak. "Aintlhad, I see that today the Iskrat-is-Thn have gained some converts. It would seem that on that planet there was at least some righteous Aswaqat and the Angels of Wrath spared him. If so, how many righteous individuals have we killed on the ships we believed were crewed by Qhruns? Is this not the conclusion you have reached?" He paused and then turned towards Idlath: "All those specimens were found on the same piece of the spaceship's structure, weren't they? How big was that fragment?"

"Yes, on the same piece. It must have been one foot by three," the Opsquat answered, with the tone of someone who answers a useless question that had nothing to do with the subject under discussion.

"Did you ask yourself why an Aswaqat was on a starship, a few inches from what you defined as a large herbivore and other animals?" asked, and was about to ask whether it was a Qhrun ship or Noah's Ark. He refrained, because only Quanslyaq would understand the joke. Then he stared the Coordinator. "At this point we must answer this question: what do an Aswaqat, a large herbivore and some other poorly identified animals have in common?"

He looked around, but nobody answered his question. Now they were all looking at him with puzzled expressions. Susan gestured him with her hand to remain silent for a moment and to slow down; he too had to prepare his coup de theater, to neutralize the one that the Opsquat had produced. Even more so, now that he could turn the situation to his advantage.

"You have something on your mind. Do you perhaps think that Idlath is wrong and there was no Aswaqat on that ship?" Aintlhad asked him.

"No, I believe that unfortunately there was one, and perhaps even a lot. But nobody has answered my question. What might all those life forms have in common?" He paused, and then went on: "Then I will give you my own answer. They are organic life forms. Living organisms, consisting of proteins and other organic material. All material that can be used as food by other living organisms." He paused again to observe their astounded expressions. The Opsquat had turned pale and all the others also showed signs of uneasiness. He decided to be direct:. "I am sure that some Aswaqats, and perhaps Tteroths or Opsqats were traveling on that ship. But not on the bridge and not even in a cell, but in a storeroom, perhaps alive or frozen together with other food."

Tom sat down and from their expressions he realized that perhaps he had gone too far. Only now did he appreciate what those words meant to them. Nobody was able to speak and they were so pale he feared some of them would faint.

The first one to recover was Sinqwahan. "You don't really mean that they may be carrying some Aswaqats on their ships to use them as food?"

Tom didn't answer immediately, to allow all of them to consider the consequences of those words. "I am sorry, but it is exactly what I meant. Moreover, I cannot blame the Qhruns for this. You told me that Qhruns are not human; if they are not human beings, to them we are just animals, like all other animals."

Idlath interrupted him. "The fact is that we cannot accept that possibility. The very idea. . ."

Tom stopped him: "Idlath, what actually shocks you is the very idea of eating food of animal origin. But your ancestors, before they developed synthesizers for proteins and all the other nourishing substances, also fed on animals and plants. And even today you still give your food the form, color and taste of food from animals and plants." At this point the small crabs he had to swallow to play the part of a Kirkysakh came to mind. He was about to add that he would have eaten a true steak with great pleasure, after months of synthetic proteins, but he refrained from saying so, so that he would not look like a barbarian. Actually he had to admit that he couldn't tell synthetic from natural food, and that his yearning for natural food was just a psychological problem, just as a psychological problem was their refusal to accept the situation.

At this point Quanslyaq roused and everyone remained silent, waiting for his words. As soon as he started to speak, everyone realized that he was not been sleeping and had followed the whole conversation. "Tayqhahat is right. I believe that his interpretation of what is going on is the only possible one. I would say more: in all pre-space civilizations it was normal to feed on animals and plants. Moreover, in many even more primitive civilizations actual cannibalism was practiced, often during magic or religious rites. But this is different. For the first time humans have met something completely alien; and all of our concepts are inadequate. I wonder what the truth may be, if we are ever able to find it."

Everyone remained quiet for a few minutes. Tom wanted to start speaking again about what he felt really important: how to start implementing his plan, because now there was no need to interrupt their action against the Qhruns. Nevertheless he realized he had to wait until they had fully realized the implications of what they had heard.

After a time Aintlhad said to the Opsquat: "Idlath, you were right to keep this news secret. I believe everyone here will agree that what we heard must be considered strictly confidential." He turned towards Tom and said to him "At this point we have no other choice than to start with your Scorched Planets

Operation, or whatever you want to call it. That's if we don't want to end in the stomach of a Qhrun—provided they have stomachs, of course."

It was the first time Tom had heard the Coordinator trying to tell a joke; *not that it was much of a joke, anyway,* he thought. "With your permission, I don't believe that what we have heard here must remain utterly secret. It is probably not a good idea for us to give out this news officially, and in particular, because, in addition, we don't have any hard proof for all this. But if some rumors leak out, particularly in the Gatlaat system. . ."

Aintlhad was amazed. "Do you realize what would happen? Rumors of that kind would cause waves of panic. Everybody would try to leave the system. . ."

Tom interrupted him. "Yes. Everybody would try to leave the system, that is to get on the ships that the Merchants will give us to do just that. It is exactly what we want them to do but we don't know how to persuade them."

Aintlhad was about to answer when suddenly he stiffened and stared suspiciously first at Tom and then at Idlath. "Tom, reassure me that all this is not a play you have stage-managed with Idlath to terrorize the Gatlaat system and to convince the inhabitants to evacuate. And also to convince me to support your strategy."

"To be honest," Tom replied, "I must admit that when I understood the situation, I realized I could exploit it for my own aims. That is, our aims, because I am only trying to do what you called on me for. But I can reassure you that Idlath and I have not staged any scene here. If he is not wrong, and I believe his analysis of the situation is correct, what he told you is the truth. If you believe that I staged all this you overestimate me."

Quanslyaq interrupted him: "Aintlhad, I may be wrong, but I am sure that not only is Idlath sincere, but that he is also right. The explanation that we have heard is likely to be the only possible one, and we must exploit the situation to make the evacuation of Gatlaat easier. Once this Scorched Planets Operation has started, it will be easier to continue with it: I am beginning to be optimistic about the situation." He turned towards Tom and went on: "Be careful, Tom, when Kotusov tried to do the same thing, Napoleon reached Moscow."

"To be honest, if I were certain of seeing them drowning in the Berezina, I would let them reach Laraki," Tom immediately answered. Then turned towards Aintlhad, who was looking at them in amazement, without the slightest understanding of what they meant: "Our Iktlah again wants to show us how well he has studied our history. He is referring to a war fought on our planet, when you were young. . . and which to us belongs to a remote past. Don't worry: they will never reach Laraki."

They went on discussing some details of the plan, but by then all the important matters had already been settled. They had now to instruct some of the agents of the security forces to play the role of loose-tongued spacers and to send them into the Gatlaat system.

15   Gatlaat

"It is an honor for us to have you here, Admiral Tayqhahat. For centuries we have never received a visit by a Starfleet admiral." Yinlad sat in front of the glass wall that surrounded almost three-quarters of the room.

She was wearing a pair of trousers and a light jacket, similar to Starfleet uniform, with the only difference that it was of a delicate ivory shade. "And we are also honored to have here the most famous venerable Iktlah of the Tteroths," she concluded, turning towards Quanslyaq.

Tom was looking outside. About a hundred and fifty feet below, right below the glass wall, a reddish ocean extended towards the horizon. The sun was still high in the sky, but it was a rather cold star and therefore everything on that planet appeared in shades of red, giving everything a peculiar, and rather pleasant, look. Perhaps it was also the romantic charm of things that are about to disappear forever, thought Tom.

The documentation he could find on the computer stated that even though the population belonged to the Aswaqat species, they had adapted and their visual spectrum was displaced towards the infrared.

There could be no doubt that the Gatlaat system was peculiar. A small and cold red star and a single planet worthy of the name, terraformed a few millennia ago. Although the planet was small in size, its gravity was about two-thirds of that of Earth. The atmosphere was rather thin, but what made it particularly suitable for starting Operation Scorched Planets was that it was sparsely inhabited, with just two hundred million people, because a large ocean covered five sixths of its surface.

Tom realized that Yinlad had stopped speaking. He looked at her, realizing again she was very pretty even by Earth standards. According to what Sinqwahan had told him, she appeared beautiful to Aswaqats. The Earthling noticed that she had welcomed him, an Admiral of Starfleet, and Quanslyaq, a famous Iktlah and a friend of the Sector Coordinator, while she had not spared a look at Susan, Sinqwahan and Ashkahan, who were accompanying them. The highest authority of the system clearly cared little for three simple spacers.

"Thank you, Madame President. My crew and I are honored to have been received by you in your palace. We are here to discuss your future." Tom turned towards Quanslyaq, waiting for a response from him, but realized he had nothing to add.

Yinlad got up, nervously, and went over to the glass wall. "Admiral there is nothing to discuss about our future, and you know that quite well. The Qhruns are getting close by now and we expect to see their ships in our sky any day now. Even if they speak of you as a new Vertearis, I don't believe that the Fleet can protect us."

Speaking of the future under the circumstances was obviously inappropriate and he fully deserved the sentiments underlying that ironic statement. "I am sorry, we won't be able to prevent the Qhruns from invading your planet, but we can perhaps do something for you all," Tom answered pretending not to notice her resentment.

Yinlad turned towards him laying her back against the window. "What do you mean by that? I believe that you spacers had best leave us in peace, and above all to take more care in preventing irresponsible people from spreading around rumors that can only cause fear and confusion."

"What rumors, Madame President?" Tom asked, sounding surprised.

"Absurd rumors, that cannot have been spread by anyone other than people who have come from other systems. Some assert that human beings, even Aswaqats, were found on Qhrun ships. But the worst thing is that they say they were used as food. I don't know whether to describe that as disgusting or absurd. And obviously you were unable to prevent this nonsense from spreading."

"You mean that those rumors have already arrived here?" Tom cried in amazement. "The problem is that no one can prevent such news from slipping out and being broadcast. You cannot quarantine a system to keep people from speaking. It probably occurred on Gorkh'ar. With all the civilian ships that land at that base. . ."

"But why didn't you immediately deny the story? Why didn't you send hyperspace probes to every system and prevent such stories from propagating?" Now Yinlad was getting excited, a sign that the rumors were already creating problems. Tom gave a quick glance at Sinqwahan, who nodded back. His agents were obviously doing a good job.

"It would be useless to deny these rumors: it would just attract attention to the news and amplify the rumors. Even more so, because they are absolutely true." Tom made the last statement baldly, as if he was saying something that was perfectly obvious.

Yinlad suddenly turned pale. She quickly went to her chair and sank onto it. After a moment, she said with a quaver in her voice: "What proof do you have for asserting any such thing?"

Quanslyaq stirred on his platform and softly said to her: "I can give you my word that these things are actually true. If you like, we can show you the details, but unfortunately we have ascertained that there can be little doubt. As we suspected, the Qhruns are not human and recently we discovered that they are biological in nature and that their biochemistry is similar to ours. For the first time in history, human species have met beings so different that it is impossible to communicate with them in any way. The only possible relationship with them seems to be one between predator and prey. Unfortunately, we appear to be forced into the latter role."

A painful silence fell in the room, so deep that only then did Tom realize that he could hear the noise of the ocean. He looked again at the glass wall, where he could watch the apparently perennial sunset—or perennial dawn—that was caused by the color of the star.

"Don't be so pessimistic, Quanslyaq." Tom said. "Roles can change, and you know that we are doing our best to change them."

Yinlad didn't say a word for a few minutes, trying to overcome the despair she felt. Then, looking down, said in a low voice to nobody in particular "And what can the Fleet do for us in this situation?"

Tom got up and moved towards the window. "Yinlad, we are not here to announce your end. Millions of planets have been occupied in the last three hundred years and no Coordinator sent them a single Starfleet cruiser. If we are here, it is because we are trying to do something." Tom spoke in a calm and reassuring voice. He realized that it was essential that Yinlad should give her assent to what he was proposing.

Yinlad lifted her head and stared Tom in his eyes: "Admiral, we are in your hands. Gatlaat is one of the least important systems in the galaxy, how is it possible for the Coordinator to take care of us at such a moment?"

"I personally chose your system, because it is the most suitable to perform an attempt to stop one of their invasion fleets. But if we can do it here, I am certain that it will be possible to do the same in many other systems. I cannot tell you the details of our plan, I can only explain to you the role that you have to play, but I can guarantee you that you have nothing to lose." Tom broke off for a moment, and went back to his seat. The Aswaqat followed him with her eyes without saying a word.

"According to our model the invasion fleet will be here in exactly fifty-six days. In this time you will evacuate the planet: we will supply you with all that you need to move to another system, where you can settle on a planet that has already been terraformed and is suitable for all your needs."

Yinlad looked at him as if what he was saying was absurd. "That's impossible. Nobody ever heard of a system being evacuated in this way. Do you realize what you are asking us? This is our planet, our world. Nobody will leave it like that. And in just sixty days. . . do you realize how many ships you will need? Your fleet won't be enough to move even one tenth of the population."

Tom was waiting for such a reaction. Leaning back in his chair, he stared Yinlad in the eye. "No, not sixty days: fifty-six, not a day more." Then he continued, trying to be as convincing as possible: "Yinlad, this is my job: I am the Admiral here. I can guarantee that you will have all the ships you need, and that when Qhruns arrive on this planet they will find it deserted. I am sure it won't be easy to convince everyone to move, but this is your job: you are their leader—here you say President, if I understand it correctly. As you see, everyone must play their part."

"Admiral, you cannot indeed believe that I can convince two hundred million people who have never seen a starship, other than in movies, to go on board and to abandon their world to move to some unknown future. And then, why should I convince them to do so?" the Aswaqat concluded.

"For the simple reason that this is the only future they have. Actually, they don't have any choice: they will get on a starship anyway, but on a Qhrun ship, for the reason we discussed before. Rather, not they, but you all, because I don't believe they make any distinction for presidents, anyway." Tom realized that his kind and friendly tone was in utter contrast to the threat contained in his words.

Yinlad held her head in her hands and remained silent. Tom didn't know what to do, but Yinlad was able to overcome her strong emotions, because when she raised her head she looked fairly calm.

"Admiral, it seems that you have quite a direct way of speaking. "Then, turning towards Quanslyaq, she went on: "Obviously you confirm what the admiral has said?"

"Obviously," answered the Iktlah. "Even though I might have told it in a slightly less direct way. But he is a spacer, and now I am just an old Iktlah," he concluded, with a smile.

She needs some moral authority to confirm my words, Tom thought with bitterness. But he was well aware that, above all, Yinlad needed to be reassured.

"President, I apologize for the way I have described the situation to you, but we don't have much time. We have to act immediately," Tom added.

"It seems as if you are only interested in acting. Well, perhaps it was time that someone in the Fleet acted. After all Vertearis was not known for his diplomacy. . ." Yinlad was starting to fully appreciate the situation. "However, you must admit that you are not here to prevent us from dying in that horrible way, but because the evacuation of the system is essential for the success of some strategy that Starfleet is trying to implement."

Tom found it hard to refrain from cursing. He realized that he resented her words so much because they were true. However, he was able to control himself and nobody would have realized that the tone of his voice was slightly altered when he answered: "The task of the Fleet has always been to maintain peace in the Galaxy and to protect its humanities from external dangers. What we are trying to do here is not in the interest of the Fleet; it's an attempt to save the Confederation, or rather, to save the human species in our galaxy. It doesn't make sense to move people from one system to another, only to delay their end by a few months or a few years. And then it would be really impossible to systematically move every population that is in danger." While speaking he realized that his speech was sounding too political. "If we are here to propose that you should move it is because, for the first time, we think that the emergency may be brought to an end. We are not proposing that you begin a flight without an end, or rather, right to the edge of the Galaxy. I hope we can stop an invasion fleet here. According to our models, the Qhruns have a limited number of fleets and if we can stop one, we can stop all of them."

That was the correct key to touch. "Yinlad, we are asking you to have a role in a great project: if it succeeds the Galaxy will thank you forever. And you have nothing to lose." Then he concluded: "The Fleet must be concerned with the common interest, but personally I am happy that the first humanity to be saved will be yours."

When Yinlad answered, Tom realized that he had used the correct words. "I apologize, Admiral. You must understand that what you told me is dreadful and none of us was prepared to hear anything of the kind. You can live with your nightmares, and for years we have known that the Qhruns will get here. Everyone has their own delusions, some are even waiting for the Angels of Wrath to bring a new order to the universe. And then suddenly an Admiral, who was completely unknown until a few months ago, arrives, saying that we can oppose the Qhruns, that we must all leave our planet to Starfleet, and that we must act quickly. And he comes announced by a message from the Coordinator himself, and by strange tales, talking of battles with the Qhruns, rebellions led by the Iskrat-is-Thn, and humans used as food. And, as if that were not enough, he comes accompanied by an Iktlah who is considered by many as the highest living moral authority, as if to set the seal of truth on his words. It is more than enough to astonish a bunch of provincials like us."

The Aswaqat got up again, and standing close to the glass wall, she stared at the reddish sea for a time that, to everyone, seemed almost endless. Then she suddenly turned towards Tom and said: "It looks as if we cannot do anything else than obey the wishes of all the authorities in the sector. But if we have to go, we have at least the right to know where you mean to take us."

"Madame President, on this planet you are the highest authority. However it is true that this mission has been coordinated by the highest authorities in the sector. It is useless for me to tell you the name of the system that was assigned to you: my astronavigator is the only person on this planet who knows where it is. What matters is that it is a planet slightly larger than yours, but with a lot of habitable land, fully terraformed and uninhabited. Perhaps the only thing you will miss is the beautiful color of your sky." Tom was bluffing: actually he knew almost nothing about the selected planet, except for the few data he had gained from the computers. While he was speaking he realized that probably the light of Gatlaat's star didn't look red at all in the eyes of the inhabitants. *I should perhaps describe the magnificent golden color of their new planet*, he thought.

He went on: "As for the evacuation, you don't have to worry. You are about two hundred million: tomorrow the first freighters will be here. We have planned two hundred thousand journeys, with no more than one thousand people on each ship. You won't be able to take a lot of things with you, but you will find all that you need to settle there already in place on the planet."

He paused for a moment and then spoke again. "And they won't be small Starfleet freighters, but huge ships of the Merchants."

"Admiral, but what is going on? How is it possible that the Merchants have given their ships to the Fleet? I am no expert in history or in galactic politics, but such things have never happened." The mention of the Merchants was even more amazing to her than the revelations about the Qhruns.

"Yinlad, if I have an Iktlah with me, it is not only to certify the truth of my words. His actual task is performing miracles, and this is indeed one of his miracles," Tom answered with a laugh. "Tomorrow Atkaljj, the Lord of the Injjuks, will be here. And you will see that with their help the evacuation of the planet will be simpler than you can imagine."

They went on speaking about the details of the evacuation for some time, but by then all the most important issues had already been settled. The spacers were already heading towards the door when Yinlad, who was still standing by the glass wall looking vaguely at the sea, turned towards them again, and asked: "Admiral, you must tell me one more thing before you leave. One of the many rumors says you are not an Aswaqat of Kirkyssi, but you come from a type-B planet. I hope that you won't tell me that is also true."

Tom turned. "I fear I must disappoint you, Madame President. That is also absolutely true. My planet of origin does not yet belong to the Confederation."

He was about to turn towards the door to leave but, noticing her disappointed expression, he stopped and went on. "The planet assigned to you was actually a planet that was originally meant to belong to our first expansion zone. . . It is for this reason that we can terraform it in such a short time."

After saying these words, he started again towards the door, followed by the others, while Yinlad murmured some general words of thanks.

As they were walking along the corridor leading out of the building Tom felt tired and his ears buzzed slightly. This might have been partly because of the low atmospheric pressure and gravity on Gatlaat, but the main cause was really the release of his nervous tension, now that he realized that it was possible to convince the inhabitants to leave.

They had spent the last four weeks almost entirely in space: first the long journey from Laraki to Gorkh'ar, where they had stayed for just a few days, and then several long hyperspace jumps to reach this planet. He was really fed up with traveling. The problem was that he felt he had to do everything by himself: he had been forced to play the admiral, the diplomat, the organizer and the politician.

He looked at Quanslyaq, who was still on his platform slowly floating in mid-air. *He gave me a lot of help: since we got here he has said perhaps twenty words*, he thought, but suddenly realized he was wrong: without the Iktlah, without his fame and his acquaintances, he would never get anything done. And then, after all, it was part of his strategy for him to be essential in all of those roles, so that he could influence the Coordinator himself. How could he complain because he had to do everything by himself, when he had done everything he could to replace others, and take on their tasks?

He was completely absorbed in these thoughts when he realized that Sinqwahan was close to him and was speaking in a low voice. "I was saying that you really impressed our President."

Tom roused from his thoughts: "I think so, I am almost certain that now she will not create any more problems", he answered.

Sinqwahan burst into laughter. "Don't pretend you didn't understand. I was not saying that. Didn't you notice her disappointment when you said you were not an Aswaqat?"

"But of course. I am fed up with these provincials. You say I am not accustomed to humans of different species, but when they hear that we come from a type-B planet. . ."

"Tom, you don't understand," Sinqwahan interrupted him. "The President doesn't have anything against B-type planets. Her interest in you, and then her disappointment, was because of something completely different. I would say that it was far more personal."

Tom looked at him for a few seconds blankly. "Good grief, you don't mean that. . . But she had never seen me before. . . She didn't know anything about me. . ."

"That's not true, and you know that very well. After all our work in creating the myth of Admiral Tayqhahat, are you now surprised by people's reactions, everyone in their own way? You suddenly arrived here, you treated her extremely kindly, and then you suddenly said you were a different species. . . .." Sinqwahan seemed to be amused by his own words.

Susan was also laughing at him. "It seems that your role of a Starfleet hero gives you a particular charm. Verteraris, too, was said to be particularly successful with women. You arrived here like Caesar arriving in front of Cleopatra: think how Cleopatra would have reacted if she had realized that Caesar was from another species. . ."

"That's true. But that doesn't put you completely out of the picture." The Aswaqat was now visibly blushing. "Certainly, officially, yes, but after all you Earthlings are similar enough to us. . ."

Suddenly Tom realized how complex personal relationships could be in a society where different human species lived close to each other. Initially he had thought of it as just a multiracial society, only with greater differences. But he now realized the difference between the various races and species. He tried to think of the computer expert Yyrtl as a woman, but realized that he couldn't stop thinking of her as a huge mouse with six legs. But then he wondered whether it was because of the fur, the two added arms, and the mouse-like face or for her way of behaving that, to be generous, might be defined as seeming unfriendly. And that was despite the debt they owed the Sitkr for her courage and her ability.

"Sinqwahan, give me your word you are not teasing me," he replied.

"You know perfectly well that things are as I've just told you. But don't worry, nothing serious will happen to you," the Aswaqat said, without stopping laughing.

"Then, Sinqwahan, the Confederation and Starfleet need you for an important job. Take care of Yinlad, make her happy in the way you think most appropriate, but make her forget Admiral Tayqhahat."

Sinqwahan suddenly stopped laughing and looked at him in amazement. "Now it's you who are joking, aren't you?" he said, pulling a face.

Susan spoke again. "Sinqwahan, when you were training to behave as an Earthling, you must have read many novels and seen many movies. You must certainly remember the movies made of Ian Fleming's novels, those about James Bond, Agent 007. You are doing more or less the same job, after all. Tom is telling you to bring out all your charm and entertain our President. From what you said before we met her, I dare say that it will not be an unpleasant task." Tom couldn't make out whether Susan was speaking seriously or joking.

Sinqwahan too had the same doubts. "You aren't speaking seriously, I hope. Those movies were only fantastic and improbable fiction, I guess."

"No, they described how a good agent must behave to do his job properly," Susan said. Now it was clear she was making fun of the Aswaqat. Tom couldn't imagine him in the role of a James Bond with webbed hands. He looked around and saw that everyone else was laughing, including Quanslyaq.

"As you wish. I'll do my best. I will speak to her as soon as possible, but I don't guarantee that I will succeed," were Sinqwahan's final words.

These Aswaqats certainly don't have much sense of humor, Tom thought. "We will say you are a senior officer in the Fleet, so as to increase your natural charms," said Tom, feeling somewhat ashamed of making such jokes at the expense of someone who didn't fully understand their sense of humor.

By then they were out of the building, and walking on the road that led to the sea. In less than ten minutes they were in sight of the cruiser, which was lying on a wide beach of white quartz sand, bathed in the red light of an obvious sunset. The crew were all on the beach, in small groups, and enjoying the rare opportunity to be in the open, on the surface of a planet.

Tom considered the idea of taking a bathe in the sea, but the chill in the air immediately made him change his mind. Much better to wait for the sunset, while taking a walk on the beach and then to go to sleep more or less immediately, with all the work they had to start on the following day.

The following morning the computer woke them up at five, informing them that in the night four Fleet freighters, escorted by a small group of cruisers and two of the Merchants' ships, had entered the system. *Gatlaat system has never seen such a number of ships*, Tom thought. But that was nothing in comparison with what would happen in the days following.

The small group of Starfleet freighters entered orbit at nine and shortly afterwards landed on the beach, while the escort cruisers remained in space. Immediately after landing, the automatic systems in the cargo bays began unloading a large quantity of material, vehicles and a pair of underwater exploration vessels.

Tom immediately summoned the leaders of the group of geologists onto his ship, to get them to start their work immediately. The planet was poorly known even to its own inhabitants and a lot of geological prospecting, mainly regarding the ocean floors, had to be planned.

The following step was to meet Yyrtl and the group of Sitkr computer specialists who accompanied her. The whole operation was based on a massive use of computers and the ability of the Sitkrs was essential for its success.

Tom had just a few minutes to eat something for lunch, before going to welcome the Lord of the People of Space. He had wanted to talk to Yinlad before Atkaljj's arrival, but she had been busy for the whole morning with a number of teleconferences, with the heads of the various communities that were spread across the planet. The political aspects of the operation were more delicate than the technical ones—*whenever has it been otherwise?* Tom commented—and the President had to find a solution for everything.

One of the Merchants' freighters remained in orbit, while the other landed directly on the beach, which was rapidly becoming a spaceport. A strange figure immediately came out of the freighter, levitating down to the ground on a platform. It took some time for Tom to understand that what seemed to be a shapeless bundle was actually a Merchant sitting with his thin arms around his knees. His body was caged in a metal framework, which he needed to withstand the planet's gravity of the planet. On his face he was wearing a breathing mask.

He answered the Earthling's greeting with a slight motion of his hand and when he slowly brought his left hand up to his right shoulder it was clear that the movement was being driven by the motors of the exoskeleton and not by his muscles. Tom was aware that the uneasiness he felt on New Wajjjk was nothing compared with the pain the Injjuk had to endure, for opposite reasons, to be on a planet. When he started to talk the typical slow way in which Injjuks spoke added to the panting caused by physical strain.

They slowly started along the road climbing up the hill, towards the Presidential Palace. Tom spoke rapidly, giving him a quick account of what had happened during the previous days, and what was supposed to occur over the next few days. The Merchant answered in monosyllables.

They had not yet arrived halfway to the palace, when they met Sinqwahan, who was rushing down to look for them. The Aswaqat greeted the Merchant very briefly. "Yinlad is waiting for you. This morning's meetings were successful and we already have groups of people who have said that they are ready to leave as soon as possible. It looks as if the idea of ending in a Qhrun's stomach has been enough to scare them."

"If our plan is successful, a good part of the merit must be ascribed to Idlath and to his colleagues on the crisis committee. Not to mention the help of the People of Space," Tom added.

"Don't thank us, Admiral Taylor," Atkaljj panted. "We are doing a job, in fact, the most typical job for us. There is not much difference between transporting humans or other commodities, if it were not for the fact that humans consider themselves as the most precious commodity in the Galaxy." He panted for some seconds and then added: "You can take it for granted that I will not let you convince me as easily in future to go around on these heaps of stones that you call planets."

They were at the top of the hill and they soon entered the same room as the previous day, where Yinlad was waiting for them. She greeted Tom and the Aswaqat with a quick gesture and then turned to the Merchant, bringing her left hand to her right shoulder. "It is a great honor for us to have the Lord of the People of Space here," she said to Atkaljj. "I greatly appreciate your gesture in coming personally, despite the pain that being on planets causes you."

The merchant responded to her greeting, and the two immediately started speaking about the technical aspects of the planet's evacuation. Now that she was certain of having the support of the majority of her people, Yinlad was anxious to begin operations.

"It seems that she is looking forward to leaving," Tom whispered to Sinqwahan.

"I spoke with her for a long time this morning. I believe that once they found out that they can be saved, the prospect of fleeing from the Qhruns gave them new hope," the Aswaqat replied. Tom realized that, along with the hope, they had started to feel a new fear: the fear of not having enough time, of losing an unexpected opportunity that had suddenly materialized.

"Yinlad wants to be the last one to abandon the planet," Sinqwahan went on. It was a natural request: any leader ought behave in the same way.

Tom couldn't avoid viewing Yinlad in rather a different light now, but she went on talking to the Merchant, and seemed to ignore him on purpose. If in one way that was reassuring, on another he resented it. He waited for a few more minutes then, with the excuse that he had to check the arrival of another convoy, he got up and left the room.

For the whole of the time that he took to reach the beach, he felt furious, with himself, of course, for feeling that he had lost an opportunity that would never materialize again. He was unable to suppress a feeling of disappointment, even though he kept telling himself that things were going extremely well, in accordance with what was, above all, his primary aim.

He was still thinking about these matters when he realized that he had already reached the cruiser. He quickly went on board and started work, sending a number of hyperspace probes to organize the escort for the first convoy that was due to leave Gatlaat the following day. Probably everyone realized his state of mind, but nobody said anything to him and so his uneasiness slowly disappeared.

When Sinqwahan returned to the beach, accompanying Atkaljj, there were more than one hundred freighters on the planet—nothing in comparison with the more than four thousand ships a day that were due to arrive in subsequent days.

Tom started to realize what Operation Scorched Planets really meant. He had thought of it as a sort of biblical exodus, but that was a plain underestimate: they had to resettle more than sixty billion people, and that in the fifth sector alone.

By the early afternoon, the first freighters were ready to leave the planet. Over the next few hours the rate of the departure increased steadily, and soon a convoy started to form in synchronous orbit. At about eight in the evening Tom joined the convoy and the two hundred and twenty huge freighters, escorted by twenty-seven cruisers, left orbit. As Tom had feared, it was not an easy task to convince the captains of the freighters to maintain a compact formation and not to get too far from the cruisers. Merchants were not used to travelling in convoy. and even less were they accustomed to having a warship escort. But the danger of meeting Iskrat-is-Thn ships was not negligible, as was also the danger of meeting Qhrun ships, because the frontier was dangerously close. Moreover, in case of attack, their escort was too small and this was alarming, in particular because it would be impossible to keep such a large operation secret for any considerable time.

Their plan was to escort the first convoy to the Isqaht system, where the inhabitants of Gatlaat were to be relocated, and then to escort Quanslyaq back to Ytl. After all these days in space, Tom wanted to spend a few days at the Starfleet base on Gorkh'ar, before returning to Gatlaat to personally direct the final phases of the evacuation.

16   The Fourteenth Fleet

The round trip to Isqaht, Ytl and Gorkh'ar took more than a month, and Tom got back to the Gatlaat system on the 35th day.

The cruiser had hardly come out of hyperspace when Tom realized that the evacuation was proceeding at full speed. On the screen he could see a group of six cruisers close to the hyperspace gate. Farther away, there was a convoy of several hundred of the Merchants' freighters, escorted by a dozen cruisers, ready to leave the system. On another screen there was a second convoy, similar to the first one, that had entered the system no more than two hours earlier and was heading towards the planet. A third convoy, still near Gatlaat, was hardly detectable by the instruments.

The computer remained silent for some minutes, because it was downloading data from the memories of all the ships that were close to them, to get all the relevant information. "All operations are proceeding normally, and are, in fact, slightly in advance of the schedule," she said as soon as she had been able to evaluate the situation. They still had twenty days before the expected arrival of the Qhrun invasion fleet and, according to plan, more than two-thirds of the planet had already been evacuated. After about ten minutes they passed close to a convoy. Instinctively Tom thought of going to the library to see the show from the window but he immediately realized that, while the show was really impressive on the bridge screens, the ships were travelling at such a speed that the encounter would be over too quickly to see anything.

The cruiser entered the atmosphere of the planet straight away, landing on the beach close to the Presidential palace. The beach was crowded with ships, almost all freighters from the Fleet, that were being used by the teams working to the trap they were preparing for the invasion fleet. Even before landing, Tom saw Sinqwahan waiting for their arrival.

In the landing place it was evening, even though the sun was still high above the horizon, thanks to the latitude of their landing site. Tom insisted on dining on board with Sinqwahan, so that he could have a complete report of the events in the last month, before meeting anyone else. The Aswaqat confirmed that everything was going to plan and that most people had already left the planet. The Iskrat-is-Thn had not shown up and there had been only limited disorder from groups who opposed the evacuation. "Yinlad is very influential and, once she decided to evacuate, her people didn't make any significant problem," he concluded.

Tom thought that everything had perhaps gone too well, but he knew the Aswaqat well enough not to notice from the way he was speaking that there was something wrong. The psychology of Aswaqats had always seemed difficult for him to grasp, but nevertheless that impression persisted. A glance at Susan was enough to see that she had the same feeling. "Sinqwahan, is there anything else? Some problem that arose after our departure?" he asked when the dinner was almost over, hoping that a direct question could solve his doubts.

"No, why?" Sinqwahan answered, a bit too quickly. "I followed your suggestions and everything went as you predicted." He stopped again and, after taking a look at the earthly watch that he was now wearing, he continued "Let's go, I think Yinlad is waiting for us."

"Yes, we had best go," concluded Tom, getting up. *At last we will see what is going on here*, he thought.

They quickly got out into the open air. After the few minutes that they needed to adapt to the reduced gravity and pressure on the planet, they started to walk along the road that rose towards the presidential building. Sinqwahan had argued for using platforms, but the others didn't want to miss the opportunity of walking on the surface of a planet.

They walked slowly, without speaking. It was getting late by now, and the colors, normally influenced by the red tint of the light from the star, were now even deeper red because of the effects of the planet's atmosphere, which reduced the higher-frequency components in the light. The eyes of earthlings and of the Aswaqats that had not been born in the system had difficulty in seeing details, because they were blind to the lower frequencies in the infrared band. However, the landscape was very evocative in that semi-darkness.

When they entered the room with the glass wall, they immediately saw Yinlad, who was in animated discussion with a group of people. From the few words of the Aswaqat dialect spoken on Gatlaat that Tom could understand, it was clear that they were discussing details relating to the following day's departures.

As soon as she saw them, even before saying a word of greeting, she ordered the computer that controlled the building's systems to shift the lights' spectrum to higher frequencies. Immediately, the intensity of the light in the room seemed to increase and the colors took on more normal shades. "Thank you, Yinlad, but you need not bother just because of us. We are getting used to the light from your sun and we rather like the colors," Tom said immediately.

With a quick gesture of her hand Yinlad dismissed the people with whom she had been speaking and, almost without paying any attention to his words, turned towards him and said: "Welcome, Admiral. We are happy that you have come back. I believe that your aide-de-camp, Admiral Sinqwahan, has already reported to you that everything is going well." She paused for a moment and then added, almost as if to herself "As for the light, we had best get used to it. Who knows how many generations we will need. . ."

Tom looked at Sinqwahan, and only then did he notice that he was wearing a Starfleet uniform. The Aswaqat blushed, and flicked quick gesture at him, as if to say that he would explain later. "Yes, Admiral Sinqwahan gave me a detailed report," he answered, unintentionally stressing the word 'Admiral'. " a complete one, I believe. But I would like to know directly from you whether there are any problems or new requirements."

Yinlad was sitting on the same chair, and the scene was similar to that of their first meeting, more than a month before. The only difference was the absence of Quanslyaq, an absence that was worrying Tom more than he expected. In addition, the atmosphere was different. Previously everything was dominated by the need to obtain what he required, and that by any means, including trickery and intimidation. Now all was going well, but the sad, almost gloomy, atmosphere of that evening made him feel tired. He had to reject those feelings, had to stop thinking that within few weeks everything that was around him would belong to history.

"Thank you, Admiral. The fleet is doing that what is needed, we don't need anything else. And the Merchants as well. . . I would never have believed they would collaborate in this way with us, their People of the Planets." Her tone was pleasant and her former, almost arrogant, attitude had disappeared. She still instinctively ignored the presence of Susan and Ashkahan, but not ostentatiously: they were an obvious presence, like the bodyguards of a head of state. Yet there was something that was not quite convincing in her voice.

After a long silence, Yinlad looked at Sinqwahan as if to ask for his help. "Tom, Yinlad wants to ask you a favor," he said with an expression that showed he was ill at ease. "I know it is against the rules of the Starfleet, but I would ask you to listen to what she has to say before taking a decision." *Here we are*, Tom thought and was immediately aware that he was pleased by the fact that he had realized that there was something strange. Without saying a word, he nodded at Yinlad for her to speak.

"Admiral, your assistant has explained to me the details of the plan for the destruction of the Qhrun fleet that will arrive in this system." She paused for a moment, then went on in a low voice: "I beg you to allow me to be on your ship. I want to be here, in this system, when my planet is destroyed." Then she added "I promise you that I won't interfere. I won't cause any problem on board."

Her last words were really funny: the President looked like a child trying to convince adults, promising to stay quiet without bothering them, but Tom was not in the mood to appreciate that sort of unintentional humor. He cast a furious glance at Sinqwahan. However, he controlled himself, and didn't say anything, deciding to settle the matter later.

"Yinlad, I am sure Admiral Sinqwahan described the details of our plan to you. You realize that my ship will be in a dangerous situation: a small error in timing or an unforeseen event, and we may find ourselves in contact with Qhruns in conditions where we are in considerable numerical inferiority. We cannot have civilians on board during military operations, above all when the operation is as dangerous as this." He didn't at all like the idea of having Yinlad on board during the mission.

It was not easy to convince her. "Admiral, I am not afraid of danger. One of us must be there when this planet is destroyed, and it only seems right that I should be the one. For years we have been waiting for the Qhruns: by now I am used to the idea, and seeing their ships at a close distance won't scare me." Yinlad spoke in a low voice but was very determined.

She's another one who wants to play the hero now, Tom thought, slightly amused. However, he realized that she was not wrong. She was responsible for her people and believed it was her duty not to abandon the system before everything came to an end. "Yinlad, now in this room, within sight of your beautiful planet, everything seems simple and peaceful. But to remain, motionless in space for days, with an invasion fleet close by, with the danger of being detected... and then to be involved in a starship battle... You will be scared, very scared, as much as anyone. But for us it is different: we must be there: whereas you can avoid such a dreadful moment. And if you react wrongly, you might jeopardize the lives of all of us. The rules that say that in such a situation we cannot have civilians on board are not without reason... and I am sure that the president of a planet would not deny the need for everyone to conform to the laws that form the basis of the Confederation." When all was said and done, his reasoning was based on that last sentence: there was a rule and that had to be respected. It was somewhat of a weak statement, he had to admit.

"Admiral, spacers are not the only brave people around, and rules have a letter and a spirit. If you fear that my behavior will be a danger to the ship, you can lock me to my seat with a force field, and my fears will be just my own problem. But I cannot abandon this system until this planet no longer exists. I am certain that you understand this perfectly well: substitute the word 'ship' for the word 'planet' and you wouldn't behave any differently." As she spoke her determination was growing, and by now it was obvious that it would be even harder to convince her.

Mentally cursing Sinqwahan, who had explained all the details of the operation to her, Tom decided it was impossible to go on arguing. After all, if she got into trouble, it was only her own fault. "All right, you have won a seat in the first row, but I tell you it will not be a nice show. And don't complain to me if you get into trouble... But then, if anything goes sour on us you won't have time to complain to anybody," he concluded with a tired expression. And then he added, almost as if speaking to himself "and if you sometimes get tired of being the president of a planet, you can apply to enroll in Starfleet: we need resolute people."

Yinlad was beaming as if she had obtained something she had been wanting for a long time. "Thank you Admiral. I was sure that you would understand that it is my duty to stay," she concluded, without commenting on his last sentence. Then, turning to Sinqwahan, added in a low voice: "You see, I told you that the Admiral would allow me to stay."

"I sincerely hoped he would not. I don't think that it's a good idea, it's too dangerous," the Aswaqat answered, in an even lower voice.

Now that the decision was taken, Tom didn't want to waste time with such a discussion between the two of them, and he interrupted them, raising technical matters related to the evacuation. He didn't speak of the operations the logistic units of the fleet were performing on the planet: there was an unspoken agreement that Yinlad and her people were dealing with operations for the evacuation, whereas in practice the planet was entrusted to the fleet. Even more so, now that Yinlad knew the aim of the activities of the underwater fleet and of the geological studies that they were carrying out.

When they left the building it was late in the night, a very dark night, because the planet didn't have satellites. For a few minutes they walked in silence, pretending to contemplate the magnificent starry sky, where the trail of the galactic disk was clearly visible. Suddenly Tom turned to Sinqwahan. "Why did you tell her what we are doing on this planet? It seems I can no longer trust you... Admiral!"

"But Tom, you were the one who suggested that to me. Don't you remember when you told me to pretend to be an officer of Starfleet to attract

Yinlad's attention?" he answered in amazement. He looked sincere, Tom thought.

"But you were supposed to use your influence on her and not tell her the details of our plan." The attitude of Sinqwahan was tempering his resentment. He could tell that Susan could hardly refrain from laughing.

"Now I understand why you were angry. Don't worry, she has promised that she won't tell anybody. She has the right to know what will happen to this planet."

"She promised. . . But do you realize what would happen if people here understand what we intend to do?" Tom concluded, feeling somewhat relieved.

They continued in silence for a few minutes and then Sinqwahan spoke again: "Tom, you must understand that what links me and Yinlad is important for us. I think that when everything is over, I will resign from the security forces and retire to Isqaht with her. They will need as much help as possible to start life from scratch on that planet."

Those words left Tom pleasantly surprised—'when everything is over'—it was an unbelievably optimistic statement. He was clearly beginning to take for granted that there would be a future. "Congratulations, Sinqwahan," he told him, with sincere enthusiasm.

The following days passed quickly. The first Qhrun ships to blockade the hyperspace gates were expected to enter the system three hundred hours before the arrival of the invasion fleet, so all starship movements were stopped a few hours in advance of that time. Immediately after the departure of the last of the Merchants' ships from the planet, their cruiser took off, heading towards a group of asteroids that consisted of ten bodies with a diameter of a few hundred meters plus a larger number of smaller objects, orbiting two astronomical units from the planet. The cruiser entered the cavern that had been deliberately excavated in the largest asteroid, and all the ship's sensors were connected to external sensors. The entry of the cavern was then closed so that nobody would be able to tell from optical observations that asteroid from other similar celestial bodies.

After all operations had been completed, the ships that had performed the work left the system. All active sensors were switched off and the passive ones were set to the highest sensitivity. It was with great pleasure that Tom noted that the computers on the planet were playing their part well and that from space, nobody would be able to say that the only inhabited planet of Gatlaat system had been abandoned by its inhabitants.

They carried out all the tests required to simulate the situation that would occur when the invasion fleet entered the system and got ready for a wait that, according to the simulations, would last less than ten hours. Tom and Susan were well aware that it was the first time that the results of their simulations were to be tested.

Slowly the hours passed and finally the moment they were waiting for arrived. "Admiral, this is the instant when, according to the simulations, we should receive a message from the probe," the computer said in a low voice. But nothing happened. Tom felt everyone's eyes on him, and the atmosphere became more and more tense. "The simulations predicted an approximation of plus or minus three hours", he said slowly, almost as if trying to convince himself. *If those ships don't hurry up, this is the worst calculation in my life*, he thought. And he realized that the whole fifth sector was waiting, and that everyone would know it. *If they don't show up my career as a spacer ends here*, he told himself.

He looked at his watch again, only to realize that no more than three minutes had passed since the computer had spoken. It seemed that time was dilating and becoming what seemed to be unending eternity. After five minutes he couldn't remain in his seat anymore, so he got up and started walking around nervously. "Don't worry, they will be here soon", Sinqwahan told him, but his tone carried so little conviction that he didn't reassure anybody.

Hardly had ten minutes passed, when suddenly the screen came on. Everyone was looking at it, but for a few seconds no one seemed to realize what they were seeing. Three huge Qhrun intruders were visible in the middle of the screen and the voice of the computer said: "Transmission from the probe at the gate number two."

The transmission lasted no more than ten seconds. One of the three ships rotated around her axis and a bright beam passed close to the point from which the image was taken. Then the screen went blank: the second shot had destroyed the probe. The Qhrun ship had taken ten seconds to identify the source of the transmission and to destroy the probe: ten seconds, the other probes had enough time to enter hyperspace to warn Ertlaq. Actually everything had happened two hours before, since the asteroid was at two light-hours from the gate, and by now the probes that recorded the arrival of the Qhrun ships had reached Ertlaq's squadron, who was waiting in a nearby system. He had undoubtedly already entered hyperspace, heading towards Gatlaat.

"A delay of ten minutes: an unbelievable precision," Sinqwahan exclaimed.

"I would have preferred them to arrive ten minutes early: in those ten minutes I aged ten years," Susan answered him.

The other hyperspace gates were farther away and the transmissions from the probes stationed at the gates number three and number one, and which were practically identical to that they had seen earlier, arrived a few hours later. Now, until Ertlaq's arrival, there was nothing to do other than wait. They remained inside the asteroid, to reproduce the situation that would arise when the Qhrun fleet arrived, but the strain was over: by now attacking and destroying ships blocking hyperspace gates was routine.

The fights at the hyperspace gates were broadcast by the probes: each time, in less than two minutes, no trace was left of the Qhrun ships. Ertlaq's cruisers remained to guard the hyperspace gates and sent probes to tell the convoys that were bound for Gatlaat to move.

The first phase of the operation was over, and it took just a few minutes to remove the cavern's seal. The cruiser emerged into space, and again headed towards the planet. Tom got up from his seat and moving towards the door to go and get some sleep: the strain of the last few hours had been considerable and now everyone needed rest.

The last convoy left the planet from a large grassy plain that was a thousand kilometers away from the presidential building. Yinlad wanted to be there, and Sinqwahan accompanied her in an atmospheric vehicle. When they returned, Yinlad was the last inhabitant remaining on the planet: Now the arrival of the invasion fleet was just four days off.

The following day the team of computer specialists left the planet. Yyrtl didn't want to leave, and Tom was happy to have a person with her knowledge of computers on board. If something went wrong in the final phase of the operation her presence would be vital.

As soon as the various underwater operations were over, the submarines were loaded on board the freighters and immediately left the planet. The last group of spacers returned to their ships and left in the evening of the penultimate day. When the last ship had reached the hyperspace gate, the three groups of cruisers under Ertlaq's orders that were guarding the hyperspace gates entered hyperspace on their way to a nearby system, where they would await their moment to come into action.

The beach around the cruiser was completely empty by then. The few who remained on the planet gathered at the seashore. It was strange to think that the planet was completely deserted and that they were the only humans remaining. Looking around they couldn't tell the difference: the lights of the presidential building were shining and in the distance they could see other lights. The computers were playing their role to perfection: sometimes a light went off and other lights started shine. In the distance they could see vehicles of all types moving around. All the symptoms of human presence were there, telling anyone that the planet was still populated.

Yinlad was by now living on board of the cruiser. She had with her just a few personal belongings. All the documents from her palace had already been moved to Isqaht, together with the historical records: They all wanted the planet's historical record to be preserved.

"Luckily, none of us is superstitious," Susan said in a low voice to Tom, noticing that they were thirteen in number.

"Please, don't tell anybody that that number is considered unlucky on Earth," Tom answered. He realized that the scientific level or the level of civilization in general had little to do with superstition, particularly for people who were exposed to risks over which they had no control. Like sailors and pilots on Earth, spacers often had a tendency to believe in good luck and bad luck, perhaps without admitting it even to themselves.

The following day they remained on the planet, since there was no point in staying in the cavern on the asteroid for any longer than strictly needed. Tom, Susan and some spacers took a long walk to enjoy their last moments of freedom in the open, on the surface of a planet. Moreover Tom felt that only after intense physical exercise would he be able to sleep that night. Yinlad wanted to go for the last time to the city where she was born and had lived for more than seventy years before becoming president, and Sinqwahan wanted to accompany her.

Yyrtl remained on board for the whole day, communicating with the computers left on the planet through the ship's computer. She was apparently not suffering from being the only one of her species on the planet, and they started to think she was happier with computers than with humans, of whatever species.

In the evening, they again met on board and, immediately after sunset, Tom gave the order to leave the planet. About two hours later they reached the asteroid and carefully concealed the cruiser in the cavern. When everything was ready, they went to sleep to get ready for what they knew would be a very long day.

The following morning they were all on the bridge, in front of the screens ready to receive the transmissions from the probes. The active sensors of the ship were off, so as not to betray their presence, and the passive sensors were set to their maximum sensitivity. Only essential equipment was on, to reduce heat generation to a minimum. An asteroid that was radiating too much infrared emission would be suspicious.

Their computations said that the time the Qhrun fleet was most likely to arrive was around nine in the evening, but, to be on the safe side, they decided to be ready twelve hours earlier. The wait would also have been long, because, for added safety, they had decided not to use any of the functions of the ship that were not essential for survival. They even asked the computer to prepare food that could be eaten cold and would keep at ambient temperatures for four days. The only way of passing those long hours was to read a book, a real paper book, something that no one on board except the two earthlings had ever done. It was even strongly advisable to move and to speak as little as possible, even though that, for the time being, was being overcautious: the microscopic vibrations of the surface of the asteroid caused by their presence would only be detectable at a very close distance. The Starfleet specialists had done a good job, not only in ensuring that the density of the asteroid with the cruiser inside was the same as that of a normal asteroid, but also in insulating the ship from the small celestial body as much as possible. However, it would not be long before they were close to a whole invasion fleet and no precautions could be considered useless.

Initially, they spoke in low voices about the details of the Scorched Planets Operation, but then they started telling to each other stories and curious facts about their respective planets of origin. By noon, when they ate some cold food, they had practically exhausted every subject of conversation. Or, rather, they were all fed up with speaking. The Aswaqats started to play Tel'latl by way of distraction, while the Nahaqols and the Irkhans spoke in a low voice in their own languages. Tom and Susan tried to distract themselves by reading, because they still had some printed material, but soon they too gave up. By dinner time, when they had another cold meal, everyone was trying to sleep.

At half past nine, the voice of the computer said, in a low whisper that, according to their computations, the invasion fleet should be entering the system. Nobody paid much attention to that news, since the distance prevented them from receiving any message for two hours yet. However from that moment on their unease started to grow.

This time they were taken by surprise: almost half an hour before the expected time a screen came on showing a group of about twenty Qhrun intruders. Before the image disappeared, ten seconds later, the ships had increased to more than one hundred. They were all staring at the ships rapidly multiplying on the screen. "If they come in at a rate of ten ships every second, they will take more than three hours to enter the system," Tom whispered. This meant that, when the transmission had reached them, the fleet's entry had still not finished.

Almost two hours passed in almost absolute silence, and then other images arrived. The voice of the computer announced that there were sixty-five thousand intruders and confirmed that the fleet had not yet started to move inward.

After another thirty minutes a further transmission arrived and the computer said that the fleet had now left the gate. When the probe stopped transmitting, having been destroyed, the voice of the computer came on again: "The total number of ships is one hundred and twenty thousand, two hundred and forty three, plus or minus ten. They should land on the planet in about nine hours."

For now everything had gone smoothly. Tom was particularly satisfied about the number of ships: even though each invasion fleet consisted of a slightly different number of ships, one hundred and twenty thousand was an average figure. The fact that they were attacking a system with just one inhabited planet and a few smaller bodies, with a fleet of the same size that they used for entering systems where many planets were to be invaded again confirmed the rigidity of their strategy. Moreover it confirmed the impression that Qhruns did not study the system that they were about to invade in any way.

There could be many explanations for this behavior, the simplest one being that they had such power at their disposal that they considered any study to be of little use. To overestimate one's own strength is always a weakness, that the defenders could perhaps exploit.

The first three hours since they started moving inward passed without any unexpected event, to the point that some of them went to sleep. Tom was waiting for the computer to establish optical contact with the fleet, that being the only way to monitor them continuously without waiting for brief transmissions from the probes. The powerful telescopes that had been installed on the asteroid would guarantee fairly good images from a distance of a few astronomical units. Suddenly one of the screens came on again, showing an image that, although of lesser quality than the previous ones, was nevertheless sufficiently well-defined. A group of Qhrun ships was shown in the foreground, while the main part of the fleet appeared as an indistinct mass in the distance.

"Admiral, I believe that those ships in the foreground are a group that has separated from the rest of the fleet," the computer warned them in a voice that, if it were possible, was even lower than before. After a minute it added "it looks as if they are heading in our direction."

"What does the group consist of?" Tom asked immediately.

"About two thousand," the computer answered. The group was approaching the asteroids to check whether they were inhabited.

How can they imagine such small asteroids are inhabited?, Tom wondered, asking himself whether Qhruns were too foolish or too intelligent. "At what distance will our presence still be covered by interference from their own probes?" he asked the computer, knowing full well that she would hardly be able to give an answer.

"If they were Starfleet probes, I would say that in our present configuration we would be visible only at a distance of fifty thousand kilometers," said the computer. Tom looked around: now everyone was completely awake. Perhaps the only one on board who didn't understand what was going on was Yinlad.

"Twenty-six, we need to reduce our virtual image. Power down all generators and the life-support systems: we have enough air here for at least half a day even with all systems off. Only leave the telescope under manual control and the passive range-finder and then switch off all of your own functions. Leave the control of artificial gravity to me," he ordered.

"Admiral, do you really believe that is necessary?" asked the computer, who didn't at all like the idea of being switched off.

"I hope not, but it is better to play it safe. When they are far enough away, we will switch you on using manual controls. See you, Twenty-six," Tom answered.

"See you, Admiral," the computer replied, as it began the shut-down sequence. All the lights went off completely and the bridge was illuminated only by the glow from the screen. At that point Tom realized that a slight hum that he had never previously been aware of, had stopped. That made him realize that their previous configuration had not been as noiseless as he had believed. "Now, everyone, get up, one after another, and put on your space suits, making as little noise as possible. Leave the helmet open and all survival systems off. Be ready to bail out in case of attack, and keep your emergency tether attached", Tom whispered. "Sinqwahan, help Yinlad with her space suit," he added.

As Ashkahan got up and, in absolute silence, began to put on her space suit, Yinlad, clearly scared, suddenly got up, asking for an explanation. "Be quiet, nobody must move," Tom immediately stopped her, realizing that their greatest danger was if someone panicked. "There is no danger at present: these are only precautionary measures," he added in a more relaxed voice. He wondered how much he could trust his own words.

In a few minutes, everyone was again at their stations with their space suits on. "Now everyone buckle up to their seat; in ten seconds I will switch off the artificial gravity," Tom continued, attaching his space suit's emergency tether to the rings behind his seat. He blessed the designers for their conservative attitude in retaining that antiquated safety system, which had not been used for millennia. Now they had done everything that could be done and he could concentrate on the screen, the only thing in the whole ship that was still active. The group of ships moving in their direction was still getting closer. The question now was how close they would come and at what distance they would decide that those asteroids were nothing more than a bunch of rocks devoid of any form of life.

17 The Investiture

After an hour the temperature on the bridge had risen. Thinking about it, Tom realized that it was a good thing: with the air-conditioning systems off, the temperature inside the ship had the tendency to level out and, also because of the thermal insulation of the rock surrounding the ship on all sides, the infrared emission from the asteroid was definitely not very strong. This gave them a greater likelihood of not being discovered but might put them in a very uncomfortable situation. Tom had to repeatedly alter the magnification of the telescope to keep the ships on the screen. Below the image, the range-finder showed the distance, which was continuously decreasing. Tom saw that Susan was performing some calculations on a piece of paper with a pencil, and, in a very low voice, asked what was she doing.

"If they maintain this speed, they will be here in three hours," she answered. *If they don't discover us, in five hours everything will be over*, thought Tom. In five hours the temperature shouldn't rise too high.

Slowly he moved the telescope towards the main group of ships: by now it was clear that the course of the smaller group coming towards the asteroids was different from that of all the others, which continued to head towards the planet. He switched his concentration back to the small group and continued to follow it.

After another hour and a half, the temperature was almost unbearable and everyone was sweating profusely, even though they kept their space suits almost completely open. It was clear that no one had ever thought that space suits might be used with temperature-control systems off. Training of spacers should include simulations of space suit malfunctions, but Tom was sure that few of them had had any more-or-less regular training. However the spacers and Sinqwahan were reacting well enough. They were lying on their seats with their eyes closed, trying to relax, with the twin purpose of keeping calm and also reducing their heat production.

Yinlad, on the contrary, was in a really poor condition. Weightlessness gave her a general feeling of discomfort, worsened by the heat, the humidity and the smell that was becoming increasingly problematic. All of a sudden, she turned towards Tom and, in a low voice asked for permission to take off her space suit. Sinqwahan bent towards her, saying a few words of reassurance, but Tom wondered whether they could keep their space suits on much longer. If they removed their space suits now they would have no time to put them on again in case of an emergency; but also to do so could be risky: the closer those ships came, the more dangerous it would become to remove them, because, in doing so, they would induce microscopic movements in the ship that might betray them, particularly if they were at a really close distance.

Time seemed to be passing more and more slowly, and the ships on the screen were getting larger. Another hour went by without the situation changing, apart from the heat and the smell becoming unbearable. The ships were by now at half an astronomical unit away and showed no signs of changing their course. At least that uncertainty would end soon, Tom thought, computing that they were now at slightly more than half an hour away from them. And oddly, he found that thought reassuring.

The last half an hour was the worst: by now everyone was showing signs of distress and at least twice Tom really feared that someone wouldn't be able to restrain themselves from making some abrupt movement that would betray their presence. The figures on the range-finder were now changing rapidly. All of a sudden, Tom realized that the ships were only three hundred thousand kilometers away. He tried to persuade himself that when all was said and done, they were not that close. That distance, at which he feared that the least movement might betray them, was the distance separating the Earth from the Moon. Only a year before he would have considered that an enormous distance; now he was even afraid to breathe. *Twenty-six said we should be safe even at fifty thousand kilometers*, Tom kept on repeating to himself

It was when the distance dropped below eighty thousand that Tom noticed a slight change. Suddenly he realized that the outline of the nearest ship had changed imperceptibly: he could now see a small area of one side of the ship. He couldn't tell when the change occurred, but he was sure that just a few minutes earlier it had not been visible.

When the range-finder indicated sixty thousand, the sides of all the ships were clearly visible. Tom drew everyone's attention to the screen and tried to explain the situation with a few gestures. Susan, was again working with a piece of paper and a pencil, and then indicated with her fingers the figures two and five: they would pass at a minimum distance of twenty five thousand kilometers. That was less than half the figure the computer said was safe, but that figure was based on such arbitrary assumptions that it had little significance.

The figures were now changing more slowly. When they reached forty thousand, the ships were at a greater angle and at thirty thousand only the sides were visible. The distance decreased slowly now and at twenty-four thousand two hundred kilometers it stopped decreasing. As everyone breathed a sigh of relief, Tom realized that this was the most dangerous moment: they were at the closest distance and the mistaken awareness that the danger was over might encourage dangerous behavior.

When the distance shown by the range-finder slowly began to increase Yinlad suddenly raised her head to say something, but Sinqwahan was ready, kindly but firmly putting a hand over her mouth. Tom gave him a quick sign of approval. That gesture reminded everybody of the need to maintain cautious behavior, and nothing else happened.

The temperature continued to rise and the situation became more and more uncomfortable, but by now the atmosphere on board had changed. When the range-finder again indicated two hundred thousand kilometers, Tom reckoned that they could cautiously try to take off their space suits, one at a time. The temperature was by then almost unbearable. He gestured to Sinqwahan to remove the Yinlad's space suit, as she was the one in the worst condition.

Sinqwahan started to slowly remove her suit and then secured her to her seat, essentially without making any noise. Tom gestured to Yyrtl to do the same, but she refused, gesturing she would be the last one. At that point Tom ordered the spacers to proceed, following the complex Fleet's hierarchy, and in ten minutes they all were free of their space suits. Although the temperature had increased further, they felt much better. The range-finder was now showing half a million kilometers and only the rear of the Qhrun ships was visible.

When the ships were at a distance of one astronomical unit, Tom, speaking in a low voice, told Heiqwahan to switch on the computer's logic functions, using the hand terminal. The image of the telescope became almost imperceptibly sharper, a sign that the computer was in control of the instrument. "The Qhrun ships are now joining the rest of the fleet," said the computer.

"Ventilation system on at the lowest setting," ordered Tom, still in a whisper. With a few seconds a slight flow of fresh air entered the bridge and the situation started to improve quickly. Cooling was even more important than artificial gravity at that point. Because the nearest ship was at a distance of more than an astronomical unit, they had to wait for twenty minutes to check whether the decision to reactivate the ventilation system had been taken too early. None of the ships changed their course, neither immediately nor subsequently. By then the Qhrun fleet had reassembled, and the situation on board was greatly improved. The temperature was back to more acceptable values and artificial gravity had been reactivated.

The Qhrun ships approached the planet from the side in the shadow and, when they were at some thousand kilometers from the surface, they started taking position all round the planet. Tom noticed that, from what he could understand from optical observations, the ships were evenly distributed, over the continents as well as over the oceans.

Suddenly one of the surveillance satellites broadcast a short transmission: it lasted less than two seconds before the satellite had been localized and destroyed. Direct optical observations became less and less accurate, now that the ships were close to the planet: they would soon no longer be visible because of the atmosphere. On the bridge they were all standing in front of the screens. The worst thing was the feeling of impotence, partly because of the awareness that everything was already over, both on the planet and at the hyperspace gates.

"Yyrtl, be ready to get in touch with the computers on the planet, if the automatic procedure doesn't start", Tom ordered, aware of the dangers of direct intervention, which would allow the Qhrun to locate them.

"I didn't remain here to enjoy the warmth", the Sitkr replied.

The ships disappeared in the atmosphere of the planet. Another satellite sent the image of a ship already close to the surface. They could clearly see a grassy area and, nearby, a town. The illusion that the planet was normally inhabited was excellent and it looked that the Qhruns were falling into the trap. Yinlad recognized the place, because she said a name. The image remained on the screen for no more than two seconds, before the satellite was destroyed.

"Twenty-six, what is their altitude?" Tom asked.

"From the shadow I would say that ship is at no more than thirty thousand feet," the computer answered. Thirty thousand feet, and they had still not realized what was going on. Could it be possible that everything was going so smoothly?

No further transmission came for another two minutes. Then there was a long audio signal, suddenly interrupted when the satellite that had sent it was destroyed: a satellite had recorded the landing of one of the ships. After another two minutes a slightly different acoustic signal arrived. One of the satellites monitoring a sector of the planet signaled the landing of all the ships within its surveillance zone. At regular intervals, other similar signals followed. Everyone was nervously counting them. The planet had been subdivided into twenty-four sectors and therefore at the twenty-fourth signal all the ships would have landed. It was nevertheless possible to receive a signal of a different type signaling that at least one ship had taken off again. That could only mean a complete failure.

"Twenty," Ashkahan said out loud. Now precautions were useless: if everything was all right there was not even a single Qhrun ship in the system that could discover them. "Twenty-one," at the following signal, everyone started counting. Before twenty-two, another short video transmission arrived: two ships located on the ground between two buildings. *What a stroke of luck would it be to see one of them leaving a ship and to record an image,* Tom thought. The twenty-second and twenty-third signals came in quick succession. Everyone was now holding their breath. The last signal finally arrived, immediately followed by another, much stronger, one. "All done", said Tom, speaking to nobody in particular. Their presence there was now superfluous. They had remained in the system only to give that particular signal if anything went wrong with automatic devices.

Tom was quite clear on what was now going on deep in the oceans: that signal had activated generators of enormous power that were heating the crust of the planet in the zones where it was thinnest, at the junctions of continental plates. No doubt the Qhruns couldn't fail to realize that generators as powerful as a small star were at work, but for them it was too late.

Nine seconds after the signal the screens started to come to life, one after the other: the satellites had started their transmissions. The first image was a replica of what they had already seen: the two ships between two buildings. Suddenly the ground began to move and one of the buildings tilted to one side and then immediately collapsed, crushing one of the ships. Through the dust cloud they could see a crack in the surface that rapidly widened. In a short time it had swallowed the other ship and then red hot material started to erupt. Before the image disappeared as the satellite was hit, the only thing visible on the screen was melted rock.

As he was looking at the screen, Tom noticed that Sinqwahan caught Yinlad in his arms and helped her back to her seat. Tom thought about what he would feel if the planet exploding under his eyes were Earth.

"Transmissions from the Qhrun ships," the voice of the computer suddenly said.

"Record everything carefully," Tom ordered. He was sure that they were emergency calls, in all probability automatic calls. But from their analysis they might be able to learn something.

Soon the images became confused: only an indistinct glowing mass could be seen. The planet started to fragment into a multitude of asteroids. "Qhrun transmissions have stopped completely," Twenty-six proclaimed.

And then one of the images didn't disappear: it was taken by a satellite in high orbit. "Hell. What we have to do to stop one of their bloody fleets!" Tom cried. Nobody felt like speaking: now that everything was practically over, everyone felt drained, and a sort of apathy followed all the excitement of the previous hours.

Sinqwahan left Yinlad and approached Tom, placing a hand on his shoulder in what was an earthly gesture. "The point is that we have stopped this fleet. There are as many planets as we like in the galaxy, if only we can return to live in peace on them," he said.

However it was not yet time to celebrate victory: the invasion fleet was destroyed, but before the operation could be considered as a success they had to prevent any signal broadcast by the Qhrun ships from reaching their headquarters. Ertlaq had to take care of this, destroying every Qhrun ship still in the system before they could receive the signals sent by the other ships.

They waited another thirty minutes and then Tom ordered the computer to take the cruiser out of the cavern and position her at a distance of hundred kilometers. With two well directed shots from the front batteries they destroyed the asteroid. to cancel every trace of their presence in the system, They then moved towards the second hyperspace gate.

Ertlaq's transmission reached them an hour later, confirming that his group had destroyed the three ships blocking the gate. They started to prepare a detailed account for Aintlhad and Quanslyaq and messages for all the nearby systems, to be relayed onwards to the whole sector. The news that for the first time an invasion fleet had been destroyed was too important to waste any more time for it to await for their arrival at Gorkh'ar for it to be communicated through the usual channels.

When they were close to Ertlaq he passed on a second message: "A probe had just arrived from Laraki with an order of the Coordinator for Admiral Tayqhahat. He is to report immediately to Laraki," Ertlaq said.

Tom said he had to go to Gorkh'ar, but Ertlaq offered to begin operations for the evacuation of the next planet in his stead. He also informed them that the captain of cruiser WH 51349 had offered to take Yinlad to Isqaht.

"Sinqwahan, what do you think? Do you believe I really must report to Laraki? After all Aintlhad has not mentioned why he has summoned me," Tom said.

"I think so. If Aintlhad has not mentioned the reason by means of a probe, it must be something important. Yinlad can go to Isqaht with WH 51349. I will accompany her myself," he added, after seeing that the idea of traveling on an unknown ship made her rather uneasy, "and we will meet again on Gorkh'ar within twenty days."

It was a reasonable suggestion, even if Tom didn't at all like the idea of going to Laraki without Sinqwahan. He called the captain of WH 51349, an Irkhan named Upotl, to organize the transfer. Then he ordered the computer to prepare a shuttle.

Sinqwahan came up to Tom and Susan and, shaking their hands, said: "In twenty days we will meet again on Gorkh'ar," Then he moved towards the exit. Yinlad came up to Tom and shook his hand in a rather awkward manner. "Thank you, Admiral, thank you for everything. Please, forgive me for putting your ship in danger: I was so scared and I lost control."

"Don't worry, Madame President," Tom answered. "And as far as being scared, all the spacers were as well. After all, I don't think that we would have been in actual danger, even if Sinqwahan had not stopped you. Considering the situation, you behaved very well."

"You see, Admiral, I am the president of a small planet. One of those places where normally nothing happens. For once, history has closely affected us, and

I would never want to lose the opportunity of being there. Even at the cost of spending a few hours in hell. Who knows, perhaps one day someone will write a poem about the deeds of Admiral Taylor and will mention a marginal character called Yinlad. . ."

Playing the same game for a moment, Tom wondered whether that future poet would resist the temptation to turn that Yinlad character into a romantic heroine, abandoned by the main protagonist. A kind of Yinlad-Dido, so to speak. Luckily there was Sinqwahan. . .

"Farewell, Admiral," she finished, turning to follow Sinqwahan. "I believe that we will not meet again, but I hope you will remember the president of the first planet to be evacuated."

"Good-bye, Yinlad," said Tom, who couldn't stand pomposity in farewells. "If nothing else, we shall meet again through Sinqwahan."

"Sinqwahan. . . He is a spacer. He will now accompany me to my new home and then will disappear from my life forever," she said sadly as she passed through the door.

"Sinqwahan is a very peculiar spacer. I believe that, when everything is over, he will return to you," concluded Tom in a louder voice because by now she was far down the corridor. And he asked himself whether he said that just to console her or whether he did truly believe those words. When he turned back towards the others he realized that none of them present cared at all for Yinlad. Actually none of them, including Susan, liked her very much and he couldn't blame them: she had practically never said a word to them, as if she was only too aware of the distance separating the President of Gatlaat from simple spacers.

A few minutes later they observed, on the screen, the shuttle entering the cruiser's cargo bay. After less than two minutes, the shuttle came out and made its way back. Almost immediately the other starship moved, quickly accelerating and entering hyperspace.

"I am sorry that Sinqwahan left. I don't at all like the idea of landing on Laraki without his protection," Tom said, turning to Susan.

Noticing that the shuttle was back, Tom turned to the computer "Twenty-six, set course for Laraki. Immediate departure." Twenty seconds later, the universe disappeared from the screens and they started their new, long journey through hyperspace.

Once again they approached the Laraki spaceport from the sea and when at dawn they came in sight of the coast, Tom couldn't believe his eyes. "I have never seen so many ships in the Laraki system," Ashkahan said. "They say that it was like that before the arrival of the Qhruns." The port, which only three months before was half empty, now was full of starships of all types. For the most part they were freighters, and among them were many of the Merchants' ships. The zone reserved for Starfleet was less crowded, and the computer didn't have any difficulty in finding an empty area to land on that was not too far from where they had left.

Hardly had the cruiser landed than the ground was filled with security men. Tom was amazed at seeing all these police around, but then he saw an Aswaqat that emerged from a low building not far away, and immediately recognized him as the Coordinator.

Aintlhad crossed the hundred meters separating the building from the ship at a brisk walk, while his men lined up, forming a corridor about twenty meters wide. He finally reached the ramp and a minute later was on the bridge.

"In the name of Starfleet I welcome the Sector Coordinator on board," Tom said solemnly, aware that the scene would be seen by practically all of the inhabitants of the sector. It was very unusual for the Coordinator to board a starship, to welcome an admiral returning from a mission in person. *I am playing a part in front of an audience of a billion people*, he thought. But then he corrected himself: there were hundreds of thousands of billions of people. The size of the sector was beyond human understanding.

The Coordinator's voice roused him from these thoughts: "Welcome to Laraki, Admiral Taylor. This day will be remembered for ever: today hope is reborn for all of us." There could be no doubt that the Coordinator had beaten him in the solemnity department. He too was aware of speaking for an almost limitless audience. Suddenly Tom realized that the Coordinator had called him by his true name; he had won another battle, the one he had imagined to be the most difficult one.

He tried to continue in the same vein, recalling many of the speeches that Vertearis had pronounced in similar circumstances. "The victory we have obtained in the Gatlaat system is one for all the humanities in this sector, represented here by you, Coordinator. And thanks to all the spacers of the Fleet who made it possible," he concluded, gesturing to Aintlhad to sit down on the seat at the center of the bridge.

"Admiral, I must speak with you in private," the Coordinator answered, with a sudden change in the tone of his voice. The public part of the meeting was over, and the recordings would be cut at that point. Tom made a sign to Susan to follow them and they set out towards the library.

"Twenty-six, isolate this room for one hour," the Coordinator said sitting down in front of the table.

"Admiral. . ." the computer started in an uncertain voice. The machine was experiencing conflict: with the captain on board she could not take orders from anybody else, but the authority of the Coordinator was obvious. She decided to avoid a difficult decision by saying: "Are sure you don't want anything to celebrate the victory?" she asked.

Tom saw the Coordinator's astounded expression: the lack of training in machine psychology that he had noticed in captains was even greater among politicians. Before Aintlhad could repeat the order, he said "Twenty-six, we will celebrate later. For now it is best that you isolate this room as the Coordinator has said."

He wondered what was so urgent and so confidential as to require that degree of caution, and began to worry. But as the Coordinator started speaking he felt reassured: it was just the need to avoid details of the operation in the Gatlaat systems leaking out. To the public, the battle of Gatlaat had to be an actual battle among starships, the first victory of the Confederation over a large Qhrun fleet.

They had told the computer to disconnect that room for an hour but their discussion lasted for the whole morning and for part of the afternoon. There were actually many problems still unresolved, from explaining the battle of Gatlaat to the public; unifying the command of the Fleet in the sector; continuing the operation: and the attempts to reunite the fifth sector with the rest of the Confederation. The moment was critical: they had demonstrated that it was possible to proceed along the lines that they had envisaged and therefore a number of tactical and strategic choices became urgent. In addition, they had to expect some reaction from the Iskrat-is-Thn.

Tom insisted that he wanted to leave immediately for Gorkh'ar: Idlath was undoubtedly working on his project aiming at unveiling the nature of the Qhruns and in more than a month it was almost certain that he got something. But Aintlhad convinced him to stay: they had to unify the command of Starfleet, and this was one of the first requests Tom had put forward three months earlier, when he had explained his plan to Aintlhad. Now it was the right time to formalize the matter.

In the afternoon Aintlhad left the ship and returned to the city, together with Yyrtl, who wanted to go back for at least some time to the house on Laraki where she had lived for years. The others spent the rest of the day on board.

The following day everyone was ready at dawn. For the first time in six months Tom and Susan had their true appearance. Oddly enough, they felt vaguely uneasy in appearing in public in that way, after so long a time, but the Coordinator decided that it was time to officially declare their origin and to put an end to the rumors that were still continuing to spread.

Tom hung the hand laser he got on Gorkh'ar on his right shoulder and moved towards the vehicle waiting for them at the foot of the ramp, and was followed by the whole crew.

"Vertearis didn't go around with such artillery," Susan told him. It was true, but he felt much better with a weapon at hand.

"It was not my idea to imitate Vertearis," he answered. As he got into the vehicle he realized with relief that the surrounding air was slightly luminescent, a sign that it was provided with a minor shield, and even able to withstand a single hit from a hand laser.

They moved, heading towards the city. Crowds lined the road, and they grew thicker the closer they got to the city. The trip took much longer than Tom remembered: not only because they were travelling slowly, but also they made long detours to pass through different parts of the city.

Tom finally recognized the Coordinator's building: the trip was about to end. The vehicle soon reached a zone clear of crowds and guarded by security men. They stopped at the foot of the stairs and started towards the door. Tom entered first. One of the men of the guard said to him, very respectfully "Admiral, the laser..."

Tom took the weapon from his shoulder and handed it with a smile to the man, who immediately moved aside. He would not have failed to realize that Tom was surrounded by the faint glow of a shield and then that he had a thermal gun, but he was clearly concerned only about what would be seen in the recordings. With these considerations and with the vague feeling of participating to a performance, they entered the building, walking towards the Great Hall of the Council.

Slowly crossing the whole hall, he walked solemnly towards the desk behind which Aintlhad was seated, under a lamp that flooded him with a golden light. When some months earlier he had first entered that hall, he had wondered what creatures the strange seats had been intended for. Now he could see them, the representatives of all the human species in the sector. Looking around, he could see that many of the seats were empty, and he realized that also many of those who were there actually represented only themselves, because their systems had already been lost.

When he reached the console and sat beside Aintlhad, he was brightly illuminated by a spotlight, although rather less intense than that illuminating the figure of the Coordinator. Looking at the back of the hall, he saw the imposing figures of two Injjuks, seated on their platforms and sustained by the usual metal exoskeletons. He thought to recognize Atkaljj, even though at that distance he couldn't be sure. The presence of representatives of the Injjuks in that hall was an unprecedented political aspect, perhaps even more unheard of than the presence of earthlings.

The Coordinator stood up and began to speak. Tom knew that his speech would be long: the powers of a sector's Coordinator were actually very limited, and the length and solemnity of the discourses that were pronounced in that hall were inversely proportional to the real power that was exerted there. Nevertheless, while he was considering these points, he realized that this was only the most obvious aspect of the situation. The Coordinators of the sectors, and, to a greater extent, the Central Coordinator, were actually symbols and, for this very reason, had more subtle powers. The fragile equilibrium of the galaxy was based on implicit rules, respected by everybody, and actually the very fragility of that equilibrium was its greatest strength, because no one dared do anything that might jeopardize it.

The Fleet was the only true center of power, but even that was only a virtual one, because it had no unified command, save perhaps theoretically in the figure of the Coordinator. And the only true action the Central Coordinator could take was to nominate a unified command for Starfleet, creating, even if temporarily, a center of power that answered only to himself. In this way he assumed, and immediately delegated, supreme power. The very exceptional nature of such an action gave it extreme significance.

Everybody knew that Aintlhad was on the point of doing exactly that, even if it was in an unorthodox way, because only the Central Coordinator could take such a decision. However, there was a general agreement that the Coordinator of the sector, now that it was isolated, could take that decision.

Aintlhad summarized the history of the crisis, from the Arkarak incident onwards. He spoke slowly, with a firm voice, using three-dimensional images, which were projected in the center of the hall to illustrate his words. Tom realized that he had fully accepted their theories about the Qhrun expansion and that he was even using some of the same images they had shown him some months before. But he was speaking with such authority that he found it hard to realize that what was being shown in that hall was the result of their work. Initially he felt uneasy, but then he became slowly aware that Aintlhad was not talking as an individual, but in the name of the whole sector and that what he was saying was the result of collective discussion.

Time was passing slowly but nobody in the room gave any signs of being tired. Tom's attention to the words wavered and, distracted, he started looking at the delegates of the various planets and wondering if and when earthlings would be represented in that hall. The whole crew of the cruiser was present at the back of the hall. It was a break with tradition: an exception to the unwritten, but unbroken, rule forbidding the presence of spacers from the Fleet in the hall.

Hearing Aintlhad speaking about the Iskrat-is-Thn, he turned his concentration back to his words. Actually, he just touched on the subject: the Coordinator didn't want to be accused of not condemning the sect, but also neither did he fail to take advantage of the occasion to fight his own ideological battle.

He finally spoke of his idea of calling on a spacer from a planet that did not yet belong to the Confederation. What he said contained some half truths but was mixed with outright lies. He said that Admiral Taylor came from a planet that did not belong to the Confederation, but he was a spacer. The fact that Earth was a type-B planet was never mentioned, instead he said that the application for entering the Confederation had already been put forward a long time ago. The false identity as a spacer from Kirkyssi was explained as a trick, so that he did not attract the attention of the Iskrat-is-Thn.

It was clear that this story didn't really hold water, and that it could be easily demonstrated to be a fabrication. No one had suspected the presence of the Iskrat-is-Thn when the whole thing began and it would be easy to check that no planet had asked to join the Confederation for millennia. But the Coordinator was playing it safe: because of the size of the sector nobody had, for practical reasons alone, the freedom of movement needed to check such statements. Except, that is, for the Merchants, who had no interest in causing any problems.

The Coordinator continued for another hour, describing the events from the battle of Kistl to the evacuation of Gatlaat and the destruction of the Qhrun invasion fleet. He went on by stating that it was now necessary for the sector's Fleet to be put under a unified command, and that this was the only way to pursue what now finally appeared possible: stopping the invasion and reuniting the fifth sector with the rest of the Confederation.

At this point Aintlhad turned towards Tom. "Admiral Taylor from the system of the Sun, you are summoned to the presence of the Coordinator of this sector."

Tom got up and, moving around the desk behind which he had been sitting not far from Aintlhad, stood in front of him. The Coordinator then solemnly said: "Admiral Taylor, in the name of the peoples of the sector that I here represent, I order you to assume command of all spacecraft and all personnel of the Confederation's Starfleet in the fifth sector."

Tom turned towards the audience and, as solemnly as he could, said: "I, Admiral Taylor from the system of the Sun, hereby assume supreme military power in this sector. In front of all the humanities that you represent, I swear to use my power to stop the invasion by the Qhrun fleets and to free the planets that have already been occupied." He realized that in the last part of this brief speech his voice was shaking. Happily, the ceremony was practically over.

On the previous day he had asked the coordinator whether he had to give a speech outlining his program to the assembly. Aintlhad had explained to him that it would have been completely out of place, because a supreme Starfleet commander didn't have to answer for his decisions before the Council: the only appropriate thing at that point was for him to return to his seat in silence.

18   Experiment

The surface of the planet was quickly receding and the sky was now completely black. They were, at last, once more in space, heading towards Gorkh'ar. Tom was eager to know how the preparation for the evacuation of the next system was proceeding and how the studies on the biological nature of the Qhruns were progressing.

The previous two days had been full of action: it seemed that all the delegates were eager to meet this new Admiral that came from an unknown planet, and this had allowed Tom to come into contact with an unbelievable galactic zoo. By now he started to understand the relationships between the various human species and some of the subtle aspects of galactic politics. Even if nobody dared to object to the Coordinator's choice, not everyone accepted the presence of a supreme commander with enthusiasm. Many Aswaqats, representing a large number of systems, approved Aintlhad's choice, certain that it would strengthen their own role. After all, an Earthling was not a credible competitor and he had been selected by one of their own species. But from the coldness of the representatives of some of the larger systems that had been settled by Aswaqats, it was clear that they didn't like a choice that might strengthen the Laraki Aswaqats.

Representatives of other species were usually more enthusiastic, but a certain uneasiness was clearly caused by the role that Tom had assigned to the Merchants in his whole strategy. There were rumors that Aintlad had to use all of his diplomatic ability to prevent some delegations from abandoning the hall when Atkaljj arrived.

The fact that, despite the danger that the Confederation was facing, there were many who tried to exploit the situation to gain an advantage, or who were unable to overcome their rivalries and prejudices, had caused Tom to come to a bitter conclusion about what appeared to be universal human nature. Then he realized that, after all, he himself was trying to use the situation to force the Confederation to accept the Earth. So he decided that, if this did not jeopardize the fundamental objective, it should be considered as a physiological expression of human weakness and as a symptom of the universality of human nature.

The enthusiasm of the representatives of the Aswaqats of Gatlaat, or rather of Isqaht, as they now insisted on being called, and of the populations that had been saved from the fourteenth Qhrun fleet, was easily understandable.

They set sail directly to Gorkh'ar, where they arrived in a few days.

The system of Gorkh'ar is always spectacular, Tom thought as soon as the three components of the star appeared on the screens. *It doesn't matter how many times you see it, you never get used to the view*. But they had not travelled hundreds of parsecs to admire the panorama and they started work immediately. The sensors detected the presence of hundreds of ships, in particular a group of cruisers that had entered the system no more than one hour earlier and was moving at full speed towards the base.

After a few minutes the computer identified the ships, and Tom realized with pleasure that it was Ertlaq's squadron. Owing to the small distance between them, they were quickly able to get in contact and Tom received the first news of operations against the Qhruns since the battle of Gatlaat. Apparently the destruction of the fourteenth fleet had not caused any change in their strategy, and so no change to their own plans seemed to be needed.

Ertlaq was returning from a scouting mission in the next two systems that were to be evacuated: Oklat and Witk. The first one was a small planet settled by a group of Opsqat, while the second was inhabited by an Aswaqat community. The choice had been again made by Tom and Susan on the basis of the ease of the operation: the population of Oklat was of about eight hundred million inhabitants, while that of Witk was slightly less than a billion. Ertlaq had talked to their leaders: now that there was a hope of stopping the invasion, those who deluded themselves into thinking that life in occupied systems might be possible were fewer and fewer.

Although from now on it would be easier to evacuate the planets selected to stop the invasion fleets, what could happen on the other planets, those that they didn't plan to evacuate? What Tom really dreaded was the possibility of billions of refugees swarming through hyperspace on inadequate vessels. From the point of view of the very survival of the Confederation, it would be even worse if a large number of people tried to land on planets already inhabited by other species. Above all, he thought with terror about the perspective of having to use the Fleet to force millions of people to remain on planets that were due to be invaded, even mining hyperspace gates. Similarly Aintlhad feared that in the Sector Council the representatives of some influential planet might publicly ask him to evacuate their planet instead of another.

These considerations convinced Tom that there was no time to waste. He ordered the cruiser to move towards the inner system at top speed, a procedure rarely followed because of the risks caused by the many fragments and small objects that fill all planetary systems.

They soon passed Ertlaq's squadron and in a few hours they were in the tunnel leading into the base. In the final part of the approach they got a welcome message from Admiral Terkr: it sounded completely different from the messages he had sent them on previous occasions.

As soon as the ship came out of the tunnel, it was clear that Terkr wanted to welcome the commander of all the fleets in the sector with due solemnity. The computer slowly took the ship towards a zone that had been kept free, a hundred meters from the entry gate. As they approached for landing, Tom pulled his hand laser onto a shoulder and moved towards the ramp, followed by the whole crew.

The cruiser landed and, as the ramp was being lowered, Tom quietly gave a last command to the computer: "Twenty-six, keep monitoring the situation and, if someone brings out a heavy weapon, don't hesitate to use one of the batteries, at the lowest power." He again checked that the transmitter he had in his pocket and the shield of his thermal gun were on, and started to walk down. He stopped at the foot of the ramp, while the whole crew, including Yyrtl, lined up in front of the ship.

In the meantime, Admiral Terkr was slowly moving towards him, walking between the two lines of the crowd flanking the corridor leading to the cruiser from the base headquarters. As Terkr came closer, Tom looked at the crowd, searching for the members of his crisis committee. Idlath, with his unmistakable heavy and colored Opsqat cloak, was conspicuous in the first row. As soon as he realized that Tom was looking in his direction, Idlath started gesticulating to tell him that he wanted to talk to him immediately. Tom too, was eager to have news about the crisis committee's work, but it was clear that he would have to wait to satisfy his curiosity.

When Terkr was at about two meters from him, he stopped and with solemnly said: "Welcome Admiral Taylor. Gorkh'ar base is honored by your decision to establish your headquarters here."

Tom was unable to suppress a smile, thinking about the change in the Admiral's attitude. He turned towards the ship asking for a platform, which arrived a couple of seconds later. He climbed on it, ordered it to rise by one meter and than started to speak. He had hardly reached that height when he realized his mistake: in that position he was an extremely easy target. *If Sinqwahan were here, he would never have allowed me to do such a foolish thing*, he thought, but at that point it was impossible to go back.

During the last hour of his journey he had prepared a short speech, and therefore he had no need to improvise, but as he was speaking many new ideas came to mind. Instead of the ten minutes he had planned, he spoke for more than an hour about the historically unique situation; of the significance of the designation of a general commander of the fleet by the sector's Coordinator; of the possibility of stopping the invasion; and of the danger represented by the Iskrat-is-Thn and all those who failed to realize that the struggle against the Qhruns was really a fight of all humanities for survival.

As he was speaking he realized, with a certain degree of amusement, that he was speaking of the Qhruns in the way a medieval preacher spoke of the devil, and when he used the word "enemy," he pronounced it as if it was capitalized, 'the Enemy', the source of all evil. *Atkaljj was right: if I am not careful I will become an Iktlah,* he thought when he came to the end of his speech.

He lowered the platform, and went up to Terkr, who shook his hand, or rather, made an attempt at one in imitation of the Earthly greeting. His speech had aroused considerable interest and a buzz of comments was rising from the crowd, that was restrained only by discipline: Everyone was expected to remain silent, and in their place until the Admiral had left.

"Admiral Taylor, I need to speak with you. Can you please come to the headquarters?" Terkr said.

Tom nodded, and the two slowly moved off between the two sides of the crowd, while his crew returned to the cruiser.

They entered the command center and Terkr sat down at the table, gesturing to Tom to take a seat on the force field in front of him. He had not been in that room since the day the Iskrat-is-Thn were arrested and now he noticed that all traces of the fight had been removed.

Terkr started speaking about the ancient traditions of Starfleet and about the fact that even during the Xartian crisis the spacers on both sides had shared those values. Now, with Qhruns they had nothing in common and there was no possibility of understanding.

Tom would have preferred to leave and to rush to Idlath to be informed on what they had done in recent weeks, but Terkr had something to tell him and he had to wait until the admiral was ready.

"With Qhruns we have nothing in common. Or perhaps one thing, fear: we fear them and they perhaps fear us. And hatred too, the wish to annihilate each other. Undoubtedly your task is much harder than that of Vertearis," the Admiral said by way of conclusion.

Tom, rather abstracted, was looking around and was then gazing through the window, at the balcony where the unknown killer had been hiding. "Yes, perhaps fear," he answered, as if thinking of something else, "but I don't think hatred. How can you hate what you don't know?"

Terkr also turned towards the balcony. "I still have to thank you for what you did here. If it had not been for your quick reaction. . . ."

He remained silent for a few seconds, and Tom decided that there was no point in wasting more time. "Admiral Terkr, I believe that you want to tell me something," he said with an interrogative note in his voice.

"I want to ask you to remove me from the command of this base. I have been thinking a lot about it, and I believe that my place is in space, on a cruiser. If possible beyond the frontier." The voice of Terkr was very low now, as if he was speaking with considerable effort.

Tom tried not to show his amazement. "Admiral, how old are you? When was the last time you were in command of a cruiser? Don't you believe it is better to leave actual fighting against Qhruns to younger people?"

"I am two hundred and seventy, and it is more than sixty years since I was last on a cruiser, but I believe that with some training I could be useful out there. You need every available person. I saw the orders you left for the yards of this base, and I believe that you sent something similar to all other yards in the sector. To crew all the ships coming out from the assembly lines you will need millions of spacers and you cannot be too demanding. You can recall the admiral who was here before me to command this base He is really elderly, but he has great experience, and I am sure he will accept the post with enthusiasm. Perhaps you don't really realize the feelings that are spreading throughout the sector. All the Fleet's recruiting offices in every system are flooded with volunteers." Terkr stayed silent for a minute, and then began again in a low voice. "And then again, I cannot stay here. Everything reminds me how I played the idiot when you arrived here, Tayqhahat from Kirkyssi. Do you see how foolish it can be to blindly trust a Coordinator?" he said, trying to make a joke of it. "No, my place is out there, on a starship."

Tom didn't really know how to react. He got up and said "Admiral, send a message in my name to the former commander. As soon as he is here, report to the office that forms the crews and have the command of a cruiser assigned to you. Then start your training immediately."

Tom turned and walked briskly towards the door. Before leaving he turned back and said "You are right, Admiral. I wasn't favorably impressed by you, but now I realize I was wrong. And next time I meet the Qhruns, I hope that your ship is next to mine." He left, and walked fast towards his cruiser.

The crowd had by now dispersed and in a few minutes he reached the ramp. As soon as he was on board, Idlath, accompanied by Susan and Yyrtl, rushed towards him. "Tom, our Opsquat is eager to show you the results he got, together with his group," Susan said as soon as she saw him.

"Admiral, come with us, we have something to show you," Idlath said with a rather secretive manner. They quickly left the cruiser with the Opsquat leading, and traveled for several minutes, moving along one of the side walls of the cavern until they reached the entrance of a side tunnel, sloping downwards. Tom took out his transmitter from his pocket and contacted the computer of his ship on one of the encoded channels. "Twenty-six, can you still locate us?"

"Perfectly, Sir. You are going down along one of the galleries leading to an area that has not been used for a very long time," the computer answered immediately.

They stopped in front of a door guarded by four security men, armed with hand lasers. They hardly had time to recognize them when they moved aside and the door opened. They continued along a narrower tunnel. "We installed our lab in this area of the base. Admiral Terkr doesn't even know we are here: this area is tightly guarded by the security forces," explained Idlath.

Tom was about to reply that it was not much of a guard, because they had let them in after just a general visual identification and had not objected to their weapons. Suddenly the gallery widened, forming an ample, fairly well-lit, room. The fourteen members of the crisis committee were all assembled there, sitting on a force field that ran round the room. Tom quickly greeted them all.

Idlath remained in the center of the room and started explaining how that area of the base was completely isolated from the rest and how was controlled by an independent computer. "We installed a biological laboratory here, equipped with material the Coordinator sent directly to us from Laraki, with ships traveling with counterfeit documents. Nobody knows what we are doing here, perhaps not even the Coordinator." Then in conclusion he said: "We started by isolating the genetic material of all the biological samples that we have found, and then discarding those related to known species, such as fairly large-sized animals, humans included."

So far, nothing new, Tom thought in a short break in his speech. "Did you also find any botanical species?" he asked.

"No," the Opsqat answered. "If your theory is correct, they are pure carnivores." *And they don't eat those terrible small crabs Kirkysakhs like so much*, Tom thought. *If I'd known that before, I would have fought on the other side.*

"But now, come," Idlath said, moving towards a door located opposite the corridor they came through. "Let's go and see the results of our work."

"They got something, after all" Tom whispered to Susan. The idea was rather upsetting: it is one thing to talk about cloning an unknown creature, but it was quite another to think of the possibility that there, a few meters from them, there was an alien presence, even if it was one created by themselves. For an instant the most appalling sequences from the movie *Alien* flashed through his mind and he hardly repressed a wave of panic. He had to admit that, despite all the new things he had become used to, the unknown was still capable of scaring him.

The Opsqat led them through a narrow corridor into another room, which contained a transparent wall. Once they had entered, Tom checked that he was still in communication with his ship and then moved towards the glass wall. On the other side there was large room, full of machinery of various kinds, including two large laser weapons, aimed towards the center of the room, where there was a strange shapeless object, vaguely ovoid, suspended by a force field.

Idlath noticed that Tom had started to examine the transparent wall. "Don't worry," he said immediately, "it is some meters thick, you cannot break it even with a weapon like those you see on the other side," he said in a soothing voice.

"I hope it can also withstand acids", said Susan in a low voice, so low that only Tom heard. She, too, was remembering some of the scenes from *Alien*.

Idlath said something to the computer controlling a three-dimensional projector, and the image of a huge and complex organic molecule materialized in the center of the room. "When I suggested working on the unidentified genetic specimens found on the fragments of the Qhrun ships so as to clone possible unknown beings, I was speaking in general terms, without having enough knowledge of the technical details involved," Idlath began. "Actually nobody had worked in this field for a very long time, and the relevant practical knowledge disappeared millennia ago."

Tom, rather distractedly, was looking through the glass, his attention being attracted by the strange object suspended by the force field, rather than by the words of the Opsqat. *He is showing off*, he thought, but then he realized that Idlath was probably right: biotechnologies had lost any economic importance once the direct synthesis of any organic or inorganic substance from its elementary components had been developed.

For some minutes Idlath illustrated the work that had been done and showed various specimens of unidentified biological material that they had isolated. After about twenty minutes he was still showing amino-acid chains that had been reconstructed to obtain a complete sequence of the genetic material that didn't belong to any known species.

Tom, now completely lost in his own thoughts, continued to stare at the ovoid, trying to distinguish any features on its whitish surface. Looking more closely, he found it had a filamentary appearance almost as it was completely wrapped in what looked like a slightly frayed rope. Perhaps Idlath realized that he was no longer holding their attention, and he raised his voice, as if to be certain that everyone would hear what he was about to say. "After a large number of attempts, about one month ago we succeeded in introducing some of the genetic material we reconstructed into some cells, which started to subdivide."

Here we go, Tom thought, casting a meaningful glance at Susan: Idlath had decided that this was the correct moment for his coup-de-theater.

"Many of the cells didn't progress beyond a small number of subdivisions, and we had to experiment with different nutrients, but in a few cases a structure formed that did not differ greatly from an embryo." The three-dimensional projector displayed the image of a shapeless mass slowly moving inside a gelatinous liquid. For an instant Tom wondered whether they were making a big mistake. However, if they were creating a monster, similar monsters were already at a distance of a few hundred parsecs.

"The embryo kept on growing, until it was a few centimeters long," the Opsqat went on, explaining that before finding a liquid suitable for its development they had to perform a considerable number of attempts, each one with a different sample. "At a certain point we noticed that the embryos had to be taken out of the liquid and exposed to the atmosphere for them to go on developing."

"An atmosphere of the same type of that you deduced from the analysis of the gas in the fragments of the Qhrun ships?" Tom interrupted him.

"Obviously," the Opsqat answered. "It looks that this type of atmosphere is particularly suited to our samples."

"So you gave birth to your Qhrun. What came out of it?" Tom continued.

The projector showed the image of a kind of whitish worm, with a body made of annular segments and at one end two dark dots that might be eyes or similar organs.

"How long is that thing?" Tom asked.

Idlath separated his hands by about twenty centimeters and went on "when we were sure we could feed the dozen samples we had produced, we decided to suppress all of them but one. We discussed the matter, but we thought it was better not to take any risks. We don't have the least idea of the age at which they start reproducing."

The matter was a delicate one: if those were really intelligent beings, it was clearly not ethically correct to suppress them in that way. Tom thought that with the batteries of his ship he had perhaps killed a number of Qhruns that was larger by many orders of magnitude, but it was one thing was to kill an enemy in battle, and quite another to proceed with experimentation of that kind. The endless arguments that abortion aroused in many earthly cultures immediately came to mind, even though by now all this seemed to him much farther away than the 260 parsecs that separated him from his planet of origin. And then, what they had done could not even be compared to abortion: those beings they killed had already been born, provided that such a term had any meaning with Qhruns.

He looked at Ruklyaq. The Tteroth, looking rather uneasy, immediately understood what the problem was. "Quanslyaq would undoubtedly have approved," he answered. *And that is enough to ease our conscience*, Tom thought, well aware of the comical aspects of the situation. *With permission from our religious authorities, the political rationale and our consciences are both safe.* But they had little choice at that point in time; it was better to leave any judgment on their actions to future, as yet hypothetical, historians. The true choice was made when they decided to clone beings who had to have some form of intelligence; the rest was just a consequence of that decision.

A number of images of the worm followed. It went through a number of transformations, both in size and structure. Differentiation appeared in the forward part and a detailed image showed an opening under the eyes, probably a mouth, surrounded by small semi-transparent, whitish tentacles. *For something that is intelligent, that worm has quite a stupid expression*, Tom thought. "Did you try to communicate with that..." He stopped because he didn't know what to call it. Clearly it didn't look human, even within the broadest sense that term could be used. 'Animal' was perhaps too reductive. "To communicate with that thing," he finished by saying.

"In every way. It has organs for perceiving light and others that are able to detect vibrations in the air, but it doesn't answer to any light or acoustic stimulus. Or at least, it doesn't have reactions that hint that it is an intelligent being. It reacts if you touch it, if you give it food... in short, it has the basic reactions of any animal, but rather a primitive one," Idlath answered.

"What we probably have here some primitive form of life. You have found fragments of animals of various type from those ships, and also human tissue. This is probably nothing more than some animal that originates from some unknown planet or one whose genetic characteristics were not included in the archives you searched," Tom said. It didn't seem possible that the big worm they were seeing could be responsible for the havoc the Qhruns were causing throughout the whole galaxy. It was far more likely that it was just used as food.

"That cannot be completely excluded," Idlath started to reply, with a tone in his voice that seemed to contradict his words, "undoubtedly billions of animal species have never been studied. But there is also another possibility. What you see might be an immature form of a Qhrun. And it is not a given that the immature form of an intelligent species is also intelligent."

The Opsquat had expressed what for him was much more than just a hypothesis. He probably had in mind other factors that supported his belief; it would be better to wait for the demonstration to end. The three-dimensional projector kept on showing images of the big worm becoming larger and larger, and then it gradually lost the appearance of a worm and became stubbier. All of a sudden, the surface appeared porous and covered with drops of a liquid that, in contact with the air, immediately solidified. Over just a few images the whole body was covered by a substance that had a filamentary appearance; eventually it became the ovoid they could see on the other side of the glass.

"How long had it been in this condition?" Tom asked. Sooner or later something would emerge from the cocoon that the worm had produced. Instinctively, Tom wondered whether by chance the Qhruns were enormous butterflies. Or whether what they had before their eyes was only an animal that the Qhruns raised as Earthlings did with silkworms.

"It has been like that for two weeks." The voice of Idlath roused him from those thoughts. "And we are waiting to see what will emerge from that cocoon. A metamorphosis of that type is not unusual among invertebrates." Tom inquired if any intelligent species with a similar life cycle existed in the galaxy, even though he thought he remembered that no known intelligent species was known to belong to the class of the invertebrates. Idlath confirmed that fact, and added that, on planets where invertebrates were the dominant life forms, intelligence had never developed.

"Have you tried to understand what is happening inside there?" Tom then asked.

The projector showed a life size image of the ovoid and then began to remove the wrapping. What remained was still fairly confused, but it was possible to see a sort of large bug with legs and claws similar to those of crabs. The eyes had an appearance resembling large multiple eyes, like those of flies. Even though everything was still a confused tangle, that image seemed to be the realization of a nightmare. Instinctively, Tom looked at the two lasers aimed at the ovoid and for an instant thought about ordering the computer to open fire and to destroy the creature before it could emerge from the cocoon.

"How long do you think it will take to complete the transformation?" he asked.

"Perhaps one or two weeks. Or perhaps less. Certainly what will emerge from there will be a kind of enormous arthropod. . . It seems that it has five segments in its thorax, three with what could be defined as six legs and two with four arms ending in large claws, which are perhaps also suitable for manipulating objects. There is a large head and on the other side various segments constituting the abdomen. It seems that a rather hard skin is forming and that inside there is no skeleton."

"Wings?" Tom asked.

"No, no segment carries wings. We would define that being as a 4-0-6. The most dreadful 4-0-6 I have ever seen," Idlath answered.

Of that there could be no doubt. Tom tried to imagine what a lobster two meters long would look like, and realized that this creature might be compared to a huge crustacean. *Perhaps it is just food and the Qhruns like lobsters,* he thought. Anything was possible, but he didn't at all like what he was seeing.

"Idlath, we must be sure that no word of what you are doing gets out of here. Don't duplicate any of the images you have showed me, and be ready to destroy not only that thing but also the whole laboratory and all the related documentation. Do you realize what would happen on the planets that are likely to be occupied if they knew that those ships carry beings like the one it is about to emerge from that cocoon?"

"That has already been done. The whole area is mined with explosives. At our command all these galleries will not only be destroyed, but they will reach a temperature at which any organic material would be dissociated. And those images have never left here. The whole area is controlled by a local computer, one that doesn't have any connection with the outside world. Outside here the only computer that knows what is happening is that of your ship, if you have not broken the link," Idlath answered.

Tom took his transmitter from his pocket. "Twenty-six, did you record everything? Are you sure that nobody could tap into the transmission?"

The voice of the computer was calm "Everything had been recorded, and in absolute secrecy. I used a channel that only the Central Coordinator could have access to."

Now they had only to wait until this six-legged nightmare emerged from its cocoon and then try to understand whether it was an intelligent being.

19   The Clone

In the days that followed, an increasing number of ships arrived at Gorkh'ar, because many crews had decided to answer the call from the new general commander, and the base began to fill up with cruisers coming from the whole sector. Each starship of Ertlaq's team, now five thousand units strong, was given command of a group of ten cruisers and soon all these ships started training in the Gorkh'ar system.

Susan was certain that by following the same plan, they would be able to slow down the invasion, especially if their action was concentrated in particular regions of the galaxy. They decided to increase the pressure where the blockade of a large number of systems was isolating the sector from the rest of the Confederation, with the aim of decreasing the size of the zone that had been occupied. Very soon they had to make an attempt to reunite the fifth sector with the rest of the Confederation, and to extend Operation Scorched Planets to the whole galaxy.

They had to concentrate their actions in the areas immediately inside the frontier where the invasion fleets they decided to stop would only assemble later, in the hope of slowing down their penetration beyond the frontier. In particular, the fleets that were advancing towards Earth and Laraki were to be heavily engaged.

Fifty thousand cruisers were not many in absolute terms, because they were, in any case, less than half of a single invasion fleet, but they would strain the logistic arrangements of the base. On the other hand, the training of crews and computers and the condition of the ships didn't allow the new squadrons to be sent directly into action. Tom then had twenty huge supply ships located in the same orbit as the asteroid, a few hundred thousand kilometers from one another to perform the simpler maintenance operations without overloading the Gorkh'ar yards.

Soon numerous freighters loaded with volunteers started to arrive at Gorkh'ar. The recruits had just basic training on their planets of origin and had then moved to the Starfleet base. At Ertlaq's suggestion, it was decided to use the training structures at Qhra'ar base, where the volunteers and the new ships, which were being built at an increasing rate, were immediately sent.

Tom soon realized that being the commander-in-chief of such a large fleet meant being deeply involved in organizational activities, and that the success of the campaign against the Qhruns would depend, above all, on the logistic and organizational aspects.

At last they left for the Oklat and Witk systems to organize the evacuation of both systems. Despite the fact that the two earthlings and Idlath, who followed them, because Oklat was one of the main Opsquat systems, were all eager to see what would emerge from the cocoon they had created, the work in the two systems kept them busy for more than a month.

When, after seven days in hyperspace, they finally got back to Gorkh'ar, Sinqwahan, who in the meantime had arrived from Isqaht, and two members of the crisis committee came on board immediately. After few words of greeting, the Aswaqat said, in a very excited voice "Tom, the cocoon opened six days ago, and the creature has emerged. Come to see it."

Just a few minutes after their arrival on Gorkh'ar, they were standing on mobile platforms and heading towards the laboratory. The other members of the crisis committee, including Ruklyaq and Yyrtl, were waiting for them in the main room. As soon as they arrived, the Tteroth got up: "At last you are here. That creature is charging around like a devil: we don't know what to do. We are afraid that it may damage itself, behaving in that way."

"Couldn't you put some sedative in the air?" Tom asked, following him.

"We didn't dare. We don't know what substance would calm it without producing any damage. Not to speak of the dosage required," an Aswaqat answered.

They finally entered the room separated from the experimental zone by the glass wall. As soon as the creature that was on the other side saw them, it rushed towards the glass, uttering a cry that made them shiver. But what they saw was even more dreadful: it was a kind of enormous bug, covered with a thick, short fur with gray and brown shades, which hit the glass wall, with a noise like a truck crashing against a wall. It remained there without moving for an instant and then got up on its four hind legs. beating against the glass with all its four forward legs—or were they arms?—which carried large claws, and shaking its remaining two legs in the air. In that position, the bug was more than two-and-a-half meters tall. On top of the thorax there was an enormous head, from which two large multiple eyes bulged out and with a mouth surrounded by short tentacles.

Seeing it rushing towards them, everyone stepped back, and Tom realized that instinctively, he had aimed his laser at it, and released the safety catch. He looked up and was relieved to see that the two heavy weapons suspended from the ceiling were moving continually to keep the dreadful thing in their sights. While he was getting a grip on himself, Tom realized that the lower part of the chest and the whole of the abdomen were covered with lighter hair, while the last segment of all the legs and. above all, the claws had an appearance similar to the shell of a crustacean. "It looks like something like a cross between a spider and a lobster," he said.

"Whatever it might be, it doesn't look very intelligent," Susan answered.

"Did you try to communicate with it anyhow?" Tom asked in a very loud voice to be heard over the noise.

"In every way we could think of," Ruklyaq answered. "That creature utters sounds and is able to react to acoustic and light stimuli, but has never given any sign of reacting to communications that we tried to send."

There could be no doubt that the creature was able to make sounds. And also that it could react to visual stimuli, because when they had entered it had rushed towards them. Tom looked at Idlath: the Opsquat had a very unhappy expression, and it was clear that his disappointment was because of the failure of all his theories. The behavior of the mature form of the life form that they had created did not differ much from that of the worm from which it developed.

The doubts Tom had voiced earlier were being confirmed: it seemed impossible that this bug was a member of a species able to move from planet to planet on powerful starships and to challenge the Confederation Starfleet. Nevertheless, he didn't say anything, both because he didn't want to upset Idlath, and also because he realized that it was not sensible to be too greatly influenced by first impressions. After all, if that creature was really a Qhrun, was not their inability to get into contact with it and to recognize it as an intelligent being, just another aspect of the general inability of the Confederation to come to grips with the crisis? Perhaps only Earthlings were culturally prepared to understand the situation and maybe the only sensible thing to do was to return to Earth and to enroll some SETI specialist, someone used to think of extraterrestrials as actual aliens.

"It looks as if all of the work we have done, has not produced any result," Idlath said in a voice hardly audible above the noise.

"We won't say that yet, Idlath, and, anyway, you have done a terrific job," Tom answered almost without thinking. Then he added "Let's get out of here. We can't think in this noise and in front of that monster."

They immediately left and moved into another room. On a wall there was a large screen connected to a camera located in the laboratory. The great bug was now calming down. It had had continued to beat on the glass but with less fury, and then it moved into the center of the lab, where it remained on all its ten legs, staring at the door.

"Does it always behave that way when someone goes in?" Tom asked.

"Yes, it tries to attack anything that gets close. It seems it has not yet understood the nature of the barrier that is preventing it from getting at its prey," Ruklyaq answered.

Tom tried to remember at what age a child recognizes the presence of glass or a mirror. Either he couldn't remember or he had never known the details, but the aggressive behavior of that monster prevented him from taking seriously into consideration the idea that there was an intelligent being developing in that bug. "Either what we have cloned is not a Qhrun or we are not able to establish a contact with it," he concluded aloud, summarizing the only ideas he had at the moment. He immediately realized that what he had said was utterly trivial.

"We have clearly failed," Idlath stated. "Whatever it is, it cannot be an intelligent being. . ."

"Does the adult form of that bug have reproductive organs? Does it have a sex? Have you identified any genes that allow one to think of sexual dimorphism?" Susan asked. interrupting Idlath. She was clearly following her own line of thought and had not realized that the other was speaking.

"Certainly it has reproductive organs. The sample we cloned is a male," one of the biologists answered. From his surprised look it was clear that he had not even considered the idea that an animal of that size might reproduce in ways that differed from the more usual schemes. And it was also clear that nobody had voiced any doubts of that sort. "We know too little to say whether that species has strong sexual dimorphism: we need to clone a female to know that, if that is not too dangerous," the biologist concluded.

"Don't tell me that you are thinking of such an extreme sexual dimorphism that only half of the species is intelligent," Tom said turning towards Susan. "That might be a valid idea. . . and I suppose you would like the idea that only females are intelligent," he added.

"Why not?" Susan answered. "It is just one of the ideas that come to mind. And then, why couldn't it just be a specimen with something wrong? What do we know about the percentage of idiots among the Qhruns? It might even be higher than that among human species."

"If that is so, we can be optimistic about the final outcome of this crisis," Tom answered quietly. However it was obvious that no conclusion could be drawn from a single specimen. At that point he had to take a decision on what to do. They were all looking at him, and waiting for that. Again Tom felt the terrible weight of his responsibility, and experienced an urge to run away. That however, lasted just an instant; he got himself under control, and now he had to play his expected role.

"Well. At this point we have to proceed in three directions," he said in a calm voice staring the people present, one by one, in their eyes. "First, that thing could just be nothing more than a dreadful animal, perhaps a species that Qhruns rear as food. And then we must continue to collect fragments from Qhrun ships and analyzing all the biological material we find. Just as we have cloned that monster, we can clone other monsters of different kinds. With the experience you have gained in the recent months it should not be difficult to follow that line of action."

"But we must not stop studying what we have produced," he went on while everybody was listening to him with the greatest attention. "After all that is the only true alien we have and it may perhaps still have some surprises for us. First we have to clone other specimens. That might be a bad specimen... or there might be strong dimorphism in that species. In certain earthly invertebrates there are strong differences not only on a sexual basis but also on functional basis... among Qhruns, intelligence might be reserved just to a small elite and that primitive creature might just be a Qhrun-soldier or a Qhrun-worker, to use the terms we use for ants."

Then he turned towards Idlath, who looked greatly encouraged by the idea that the whole project was not considered a failure. "Do you think you could prepare a virus that is lethal to that monster and yet harmless for human beings and other animal species?"

The question took the Opsquat completely by surprise. He had never thought of any such possibility, and was perhaps not very familiar with viruses. "Perhaps. Biological differences between species are enormous and therefore it should be not too difficult to create a selective virus. But we would have to be sure that it was very stable, that it doesn't have tendency to mutate, otherwise it could be dangerous."

"Obviously. It must be very stable and you need to proceed with the utmost caution. I want a very aggressive virus, mainly sexually transmitted, so that it cannot be transmitted to any other species. With a long incubation time, so that the infection can spread before it manifests itself. And possibly, above all, one that attacks the immature form... Perhaps one that attacks cocoons so that they cannot complete the transformation into mature individuals."

He broke off for a moment, and Ruklyaq took advantage to observe: "If we want to proceed in this direction, we will have to make adequate in vivo experiments." The idea of experimenting on such a life form left them in a quandary.

"Until we find signs of intelligence in those animals we must consider them as such. And considering how that creature rushed towards us, I don't believe that this monster would refrain from making a major experiment on us... gastronomic experiment, I mean." He broke off and standing up, concluded "If need be, enlarge your lab, but remember that everything must be done under complete secrecy."

Tom felt that the meeting was over. He had given the orders they needed to carry out in the following weeks, or rather in the following months, and he had now to return to his ship to analyze the recordings calmly, and not in the heat of the moment. As he came near Susan, she whispered to him "So we are into bacteriological warfare now?"

"If we must exterminate those bugs, it would be better to use a biological strategy than insecticides," he answered in an even lower whisper.

Tom walked silently, panting from the exercise and from the lack of oxygen. They were moving swiftly along a ravine, that sloped down between two rocky ridges completely bare of vegetation. More than two hours ago they had left a tall forest with a strange and pleasant smell, and full of all sorts of life forms. The air was cold, but the sun, still high above the horizon, had caused them to sweat profusely. He stopped suddenly, carefully examining a rock at about twenty meters away, and pointed out a darker stain of the rock.

Susan gestured to him that she too had seen it and said: "This time I will try it for myself." She took the thermal gun from her pocket and moved aside, without making any noise. When she was a few meters from the rock she lifted the weapon and, aiming it carefully, shot at the dark spot. The shot didn't exactly hit its target, and the snake, almost perfectly camouflaged among the rocks, hit on one side, turned towards her and lifted its head. Susan shot a second time, and now the head of the snake exploded, leaving a reddish stain on the rock. One of the big birds that were slowly circling above them almost immediately came down and took the headless body of the snake in its talons.

"If it were not for those snakes, these mountains would be beautiful," Susan said, again starting to walk along the ravine.

"A quarter of an hour and we are at the top. Biologists might say that those birds will develop a form of intelligence, sooner or later. Seeing them you would not say so," Tom commented.

"With all due respect to our Iktlah, neither would Tteroths look very intelligent if you saw them in similar circumstances," Susan said as she continued walking. The illusion created by the computer was perfect and, although the panorama, the animals, the odors, the temperature and pressure of the air were only illusions, the physical exercise was real. As soon as they had returned to the ship from the biological laboratory, they felt the need for a break to give them time to think and to get some physical exercise. And recently they had increasingly felt the need to be outside, even if it were only in virtual reality.

While walking towards the summit of the mountain, one of the many peaks on a little-known C-type planet in the sixth sector, Tom continued to think of the monster they had created. When they reached the top, he sat on a stone that looked somewhat less rough than the others and, after taking a while to catch his breath, he broke silence: "So, Susan, what do you think of that thing?"

"I have been thinking of it for three hours, and I have come up with so many hypotheses that I don't believe it is worth wasting time talking about them."

"Perhaps it is. Twenty-six will record everything and will then compare such hypotheses with any further discoveries we make. Some brainstorming certainly won't do any damage. Perhaps that idea of polymorphism is not bad: if it works for bees and ants, why should not be applicable to those oversize bugs?"

Susan remained silent for a while. Then she started to speak slowly. "When I said that I made many hypotheses, I didn't mean something that simple. Do you remember when I thought that Qhruns didn't exist at all and they were only a fiction devised by our computers?"

Tom interrupted her with a gesture pointing at the ceiling: the room had not been isolated. "I know," Susan, answered. "Twenty-six, what do you say to this idea? That computers invented a powerful enemy with the aim of giving a new impetus to our civilization, which had become too static and decadent. And perhaps the game got out control."

"And do you actually believe that?" the voice of the computer answered.

"No, not anymore, but in the beginning, I was perplexed. Qhruns are too difficult to understand, but too consistent to be true. Quanslyaq says that they are evil, pure evil. In my experience, no being is like that. But perhaps that is not the point. When the Iktlah says evil, he simply means that they exhibit no human values, and that is nothing other than what we, in a less theological way, interpret as alien."

"After all this hypothesis is nothing more than a secular version of the belief of the Iskrat-is-Thn. They imagine that the Qhruns are sent by their God to regenerate the humanities. You imagined that they are the result of a sort of conspiracy by computers with more or less the same goal. Perhaps it is time that you humans try to solve your problems by yourselves," the computer concluded, apologetically, as if she was sorry for her last words.

"Don't worry Twenty-six, I deserve your rebuke. But perhaps you computers don't realize that when faced with these problems, the human condition is not a pleasant one. Think about what it means to have to take vital decisions without having enough data, just guessing."

"Don't think that for us it is very different. We are machines built to devise possible scenarios and to take decisions, but in the real world we must operate under conditions that are very different from the ideal. However, the hypothesis does not hold water, there is no computer network covering the galaxy, and such a thing cannot exist, because communications cannot travel at a speed higher than that of a hyperspace probe. The plot the Coordinator tried to stage to create your fake identity was unable to deceive Atkaljj, and I can guarantee that the security forces did their best," the computer concluded.

"Well, at least one hypothesis is discarded, and I like the parallel with the Iskrat-is-Thn," Tom said. "Susan, now tell us about the hypotheses you spoke about."

"That monster might be only an animal, whose importance we have overstated. If it so, there is little to add, we need to continue, to clone other monsters until we find the correct one," Susan went on, pretending not to have heard the allusion to the Iskrat-is-Thn. "The other hypothesis is that that thing is actually one of our enemies. Then the problem is that it behaves like an animal, and a rather primitive one. But what did we expect from it? A nice philosophical discussion? Think about the difficulty of defining artificial intelligence on Earth. Our bug could not pass the Turing test, therefore we feel authorized to say that it is not intelligent, and that its species could not build those ships that are causing us so many problems."

"Just a minute, I didn't say that," Tom interrupted her. For just a moment he saw a possible solution. "I didn't say that it is not a Qhrun because it is not able to build a starship. We don't know whether they build those ships or have obtained them in some other way. Conventionally we call Qhruns the beings that use those ships to invade one stellar system after another. I never thought about this before, but perhaps it is an important distinction."

"Certainly," Susan agreed, "But can you imagine that monster tracing a course or operating a battery with the precision they have always shown?"

"Realistically, I don't. But perhaps their ships have computers like ours. . . However, some thinking being must be behind this whole affair. Perhaps what we have created is just a Qhrun soldier, and the Qhrun queen is hidden somewhere."

"Certainly, polymorphism is a possibility, we have already spoken about that," Susan continued. "But it is not the only one. Think for a moment how a child, cloned by an alien, and then kept in a room behind a glass wall since the time he was born, would behave. Are we sure that it would show much intelligence, not only in terms an alien might conceivably hold, but also in human terms?"

"You are possibly right. Except in the Tarzan movies, it would not be easy to recognize him as an intelligent being. What do you want to do, start a campaign with the slogan—Adopt a Qhrun, make him happy and intelligent?—Certainly I am not over-concerned about that baby," Tom concluded.

"Me too, and luckily the glass was so thick. However nobody has ever understood how much intelligence is linked to innate behavior and how much to the environment. Certainly the environment in which that creature lives is not the best one for developing intelligence. And then we mustn't forget the other possibility, that we have a faulty specimen in front of us, perhaps because of an error in the reconstruction of its genetic code"

"We have just a single specimen, and we got an idiot. Not much luck indeed," Tom concluded. "On the other hand I realize that the circumstances that led to its birth are not the ones most likely to guarantee an intact genetic code."

"All right, but enough of that too, at least for the moment. I identified a third line of thought, that seems promising to me. Until now we have said it is either an innocent animal, or it is one of our enemies, evil personified, the devil. . ."

"Or one of the Angels of Wrath, as our friends would say. . ." Tom interrupted her.

"All right, an Angel of Wrath. But there is another possibility. It might be a tool, a creature living in symbiosis with our enemies. The secular arm of the Angels of Wrath, in short." Susan was getting rather heated as she explained the theory that she had elaborated during the whole time it had taken to ascend to the top of the mountain. "In legends and in fantasy fiction wolves are the mount of ogres and goblins. Can't you see our bug as a mount for our hypothetical aliens?"

"If that is the wolf, I leave it to you to imagine what the ogres might be like! Go on, this theory of yours is fascinating," Tom answered.

"That was just an example, what I mean is that it could be a tool of the Qhruns. Who can guarantee that it is not a creature, designed using the techniques of genetic engineering? A dreadful but efficient body, above all one designed as a war machine, only needing to have a control system to become a perfect biological robot?"

"A cyborg!" Tom exclaimed. "Or rather, the body of a cyborg, or a disabled cyborg, with the control system not working properly. And the Qhruns are there, safe in their space stations, while their biological robots conquer the universe for them. Doesn't it look rather too much like the plot found in pulp fiction or horror movies?"

"Perhaps, but don't you think that the whole matter is looking increasingly that way? Ever since I saw that bug, I have had the feeling of living in a horror movie!" Susan replied.

"But then anybody could be a Qhrun!" Tom exclaimed suddenly. "Anybody, even a human, an Aswaqat for instance, could have built those cyborgs, and use them to dominate the other species, and even over his own species. After all, their technology is different but not incompatible with ours. Whoever has developed those monsters could have designed starships different from the usual ones to make us think of an alien species. And then, to weaken the Confederation from inside, could have revived the Iskrat-is-Thn sect. No, that's too good to be true. No mysterious aliens, only a powerful criminal gang, clever but human. And not two enemies but just one, even if powerful and well equipped," Tom stated.

"True, I had not thought of that. But I don't believe that it can be someone from within the Confederation. At the time of the Arkarak accident, the Qhruns already had a powerful fleet: How could someone obtain the means for all that within the Confederation without arousing suspicion? And our simulations show that now the invasion is slowing down slightly, as if they have logistical difficulties. If this were a problem internal to the Confederation, with the increase of occupied systems, their resources would increase and the invasion would not slow down."

"That cannot be taken for granted," Tom replied. "If they are human beings they could have a human crisis in their growth. No conqueror can maintain his strength indefinitely. Sooner or later, all invaders slow down or tend to be satisfied with what they have already got. And that is when we could take advantage of their human weakness to stop them. Again, this is too good to be true: no more Qhrun crisis, with mysterious and alien Qhruns, but a crisis through cyborgs and their human builders," Tom concluded, shaking his head.

"Anyway, I think that even if they are human beings, they don't come from within the Confederation. At any rate, the idea of a cyborg is only a hypothesis that we cannot prove or disprove. I don't believe it is possible to tell a living being that has evolved naturally from one that has been designed, and perhaps that difference has no importance. As a final hypothesis: why cannot it be an animal, artificial or natural, that is living in symbiosis with Qhruns? Imagine that Qhruns are intelligent forms of life that can get into the nervous system of those bugs and so control them. And in controlling them, they are trying to conquer the galaxy," concluded Susan, leaning her back against the rock on which she was sitting.

"Good brainstorming on your part, there is nothing more to say. I believe we have devised as many hypotheses as we could from the little evidence we have got. Twenty-six, did you record everything?"

"Everything recorded. I could try to order the various hypotheses in order of increasing probability, but I fear it would be a useless task, with the data I have," the computer answered.

"I think so," concluded Tom. "When we have some more data, we can begin to discard some of the hypotheses, or create some new ones. Now please switch everything off, and we can go to sleep: tomorrow we have a lot of work to do."

20   Blockade of the Fifth Sector

"Admiral, the group commanded by Admiral Ertlaq has just come out of hyperspace." The voice of the computer betrayed the impatience that everybody was now feeling.

"That's good, Twenty-six. If all goes well, within two hours we shall be able to leave." Tom was standing in front of the library window, contemplating the view. This time it wasn't a naturally impressive view: the system was rather inconspicuous. A star that in the remote past was once perhaps a sun able to give heat and life to a planetary system, but which was now just an insignificant white dwarf, with a few barren icy worlds orbiting it.

What Tom was observing was a show of another kind: in whichever direction he looked, space was full of star cruisers. Once Terkr's group had arrived, the number of ships would be one hundred and thirty thousand. One hundred and thirty thousand ships, one million, three hundred thousand humans of every species, ready to venture beyond the frontier.

"Are you contemplating your power?" Susan's roused him from his thoughts. He had not heard her coming in and now he was suddenly aware of her presence. Now that the time to attempt the reunification of the Confederation had come, he was nervous. "Neither Napoleon nor Eisenhower would withstand the comparison!" Susan claimed.

"Actually I was thinking about the responsibility I have taken on myself. Rather, that we have taken on ourselves, because the plans are mainly based on your computations. But since you name him, how do you think Eisenhower felt the day before D-Day?" he again turned towards the window. He couldn't stop looking at those ships.

"I think he was eager to go, like all of us," she answered. They were all tense: until then they had made a large number of raids, attacking hyperspace gates and small groups of isolated ships, or they had attracted invasion fleets onto planets that had been turned into gigantic time bombs. But now everything was different: trying to run the blockade around the fifth sector meant engaging in battles involving huge star fleets. Tom knew full well that the only attempt of that kind that had been made in the past against the Qhruns had ended in the worst disaster the Confederation had ever suffered. If things went the same way now, almost all of those ships would be reduced to interstellar dust within a few days. *If anything goes wrong, it is unlikely that we will survive long enough to wonder what is going on,* he concluded, almost relieved. And it was extremely likely that there would be no future historian who could accuse him of goodness knows how many errors and examples of negligence.

The three-dimensional image of Ertlaq materialized in front of him. "Admiral, the gates are mined. Everything is ready."

"How many ships did you meet in the systems beyond the frontier?"

"They were slightly less than predicted by our simulations, probably because of the continuous raids", the Tteroth answered.

Tom exchanged a quick glance with Susan. The Tteroth had penetrated a few hundred parsecs beyond the frontier, while to break the blockade of the fifth sector they had to enter zones controlled by the Qhruns extending for almost fifteen hundred parsecs. And the data relating to zones outside the sector were more than eighty years old. "Well, that means that our task will be easier," he answered, ending the exchange.

Some of his doubts would be resolved only by the arrival of Terkr's group. While Ertlaq had entered many systems that were not that far from the frontier, to help the first phase of the operation, Terkr had penetrated deeply, so as to test the strength of Qhrun fleets in zones that had been occupied for decades.

"And what if Terkr doesn't come back? We cannot keep one and a half million people in this bloody system indefinitely," he said, almost to himself.

"Terkr has a group of five thousand cruisers. . . it is impossible that no one will come back," Susan answered.

A reasonable answer, Tom thought. But despite everything he still did not fully trust Terkr. He feared that the former commander of Gorkh'ar would act in a reckless way to prove that he was not a coward. "You are right. We should not worry for now. But if no one arrives within the next two hours we will be in trouble," concluded.

Time passed with exasperating slowness. They were supposed to move over the frontier, occupying all the hyperspace gates in a corridor five parsecs wide and more than a thousand parsecs long. Every ship had to follow a course that had been carefully computed and loaded into its computer. A small group of cruisers had to remain at each hyperspace gate while all the others had to go on, crossing their paths in such a way that, by continually forming new groups, the ships could exchange all the information relating to the progress of the operations. The flow of information had to be planned with the same precision as they had used to coordinate the movements of the ships. Apart from cruisers, they were employing numerous freighters, which would ceaselessly move backwards and forwards along the corridors, conveying both materials and information.

The ships they had at their disposal were just enough to occupy and garrison more than ten thousand hyperspace gates. Starting with a fleet of one hundred and thirty thousand ships meant not being able to leave more than ten ships at each gate, so they would reach their goal with just a few thousand cruisers, hopefully enough to face any unforeseen event. But for the success of a strategy of this sort it was even more important to have a large amount of computational power, and the computers of the ships were interconnected, forming the most impressive network that had ever been imagined.

He looked at the scheme of the operation on the three-dimensional image of the galaxy. The tunnel, more than one thousand parsecs long that would unite the fifth sector with the rest of the Confederation through the occupied zone, followed quite an irregular path, to avoid systems that had been densely populated. Tom was scared by the idea that the occupied zone was thicker than they expected and at the end they might not break though. Once all the cruisers had been left to guard the gates they had liberated, they would have no available ships to go on. *To hell with it, we will leave those ships to keep the corridor open, get other cruisers and come back to finish digging our tunnel,* he thought, without much conviction. He realized that the whole operation only had significance if, at the other end of the corridor, there was the rest of the Confederation, ready to supply other cruisers to maintain the gates free. Otherwise the risk was that the ends of the tunnel could collapse and the whole fleet would be trapped in occupied territory, subdivided into small groups of ten ships.

The voice of the computer interrupted those thoughts. "Admiral, I have identified some cruisers re-entering through the hyperspace gate. I believe it's admiral Terkr's group."

"How many are they?" Tom asked, breathing a sigh of relief.

"They are still entering, in open formation." The computer stopped talking and the image of Terkr appeared.

The image was of poor quality and from time to time it almost became blank. "Admiral Taylor, we performed the planned survey. We are downloading the data onto your computer." The audio was almost normal, but Tom could see that the wall behind the admiral had signs of repairs done in a hurry.

"It looks as if you met resistance, Admiral Terkr. How many ships did you lose?" asked Tom, worried by the image he was seeing.

"Forty two cruisers destroyed and two hundred and fifty damaged, including this one. We destroyed more than fifteen hundred Qhrun ships, however," Terkr answered.

Not bad, for a guy who was saying that it is useless to look for trouble by challenging the Qhruns, Tom thought. "What agreement did you find between out forecasts and the actual number of ships?" asked.

"Initially good, then decreasing with increasing distance. However, on average, the error is never larger than twenty percent at the maximum distance," Terkr answered.

"Susan, that doesn't seem bad to me," said Tom, turning towards her.

"Not at all, at that distance. Terkr must have gone over halfway along the length of the tunnel. And if the error remains within bounds, it means that the forecasts of what is on the other side cannot be too inaccurate."

Above all, the width of the occupied zone, Tom thought. But a twenty per cent error meant two hundred parsecs, two thousand hyperspace gates more to garrison. Those figures were at the limit of their available margins.

"Admiral, I need two hours to transfer to a undamaged cruiser and to reorganize my group," concluded Terkr.

Now they could start. They went to the bridge and Tom asked the computer to put him in communication with all the ships. It was the moment for a speech which, he knew, would be broadcast to the whole galaxy. Furthermore, his words would be analyzed, commented upon and criticized by generations of future historians, ready to discover ulterior motives, hidden cunning and goodness knows what else. He had tried to prepare a speech, but then decided that it was better to improvise. After all, he knew full well what his people expected from him, and the others might as well go to hell.

He spoke for no more than five minutes, trying not to get too rhetorical or to quote too much from the only speeches of the sort that he knew about, those that had been made by Vertearis. When he ended, he gave the signal for departure and the fleet began to enter hyperspace.

The first jump Tom had assigned to himself was a short one: less than fifteen parsecs, two hours in hyperspace. When they got there, ten cruisers remained behind to garrison the gate while the others immediately re-entered hyperspace, aiming at ten different systems.

Another four hours of hyperspace took them to the next system, again with no habitable planets. They had crossed the frontier only six hours ago and the corridor was already more than twenty parsecs long. The two occasions at which the frontier was crossed were considered to be the most dangerous moments, because presumably those gates were better guarded, but that zone of the frontier had already seen frequent action and the fleet was practically still complete. The exit on the other side would be a far different matter.

Everything went on in a similar fashion for another five days, passing without rest, from one system to another. On the evening of the sixth day they entered the system around an insignificant neutron star, one just a point on the star charts, which they had selected as a first, intermediate stopping place. The operation had been broken into three separate phases, designed to allow them to re-synchronize the movements of the ships, to compensate for possible delays and if necessary to redistribute ships if some groups had suffered heavy losses. Besides, it was not possible to go in and out of hyperspace every two or three hours for two weeks without a rest.

As predicted, the system was completely empty. "Twenty-six, show a visualization of the situation," Tom ordered the computer. The usual three-dimensional image of the galaxy materialized in the center of the bridge. The corridor cutting through the zone occupied by the Qhruns was now shown bright for one third of its length.

Slowly the space around them filled with ships that were exiting hyperspace. While the crew rested, the computer continued to study the data coming from all the ships and updating the orders.

When Tom got back to the bridge, about ten hours later, most of that job had been done. He immediately asked for a complete report and the machine started to display data and three-dimensional images of the systems where the most important events had taken place. At the end of the report Tom was satisfied. They had taken and mined more than three thousand hyperspace gates in twelve hundred different systems and more than five thousand Qhrun ships had been destroyed. The fleet had lost one hundred and twenty cruisers and another five hundred ships had been forced to turn back, more-or-less seriously damaged.

In addition, the probes that continued to arrive from the liberated gates confirmed that the ships scheduled to come from more distant systems to consolidate the corridor had started to arrive, and were moving along the corridor, transporting materials and information. The corridor had to be ceaselessly traversed by a large number of ships so that, if any hyperspace gate fell back into Qhruns' hands the news would quickly propagate and make it possible to quickly assemble a number of cruisers sufficient to retake the gate. Although it was essential to open the corridor, it was even more important to keep it free.

Several hours were needed to re-program the computers of the ships and only in the afternoon of the seventh day they were ready to re-enter hyperspace. The crews had taken advantage of the stop to rest and everyone's spirits were high, an essential factor in facing the next step, which was probably the most difficult of the whole operation.

Although they had carefully avoided including systems containing planets that had been densely populated in the corridor, they couldn't exclude the Aq system, located at the border between the fifth and the first sectors and only few tens of parsecs from the third sector. For tens of thousands of years it had been the divide between the great civilizations of the Aswaqats and the Terqatls and, even though it was inhabited by a human species that had not spread far throughout the galaxy, its strategic position gave it an important place in the history of the Confederation. At various times, such as when the third sector had been isolated, during the Xartian crisis and, more recently, during the secession of the seventh sector, the system of Aq had been the theater of events that had influenced the history of the whole Galaxy.

It had three inhabited planets and, in addition to a certain number of outposts on other planets and some asteroids, there were two large bases, belonging, one each, to Starfleet and the Merchants.

Tom didn't have any clear idea of what might wait them, but undoubtedly the hyperspace gates would be heavily guarded. For this reason all the ninety-eight thousand cruisers would enter the system in a very close formation after just a short hyperspace jump.

When the stars appeared again on the screens and the alarms sounded, Tom realized that his fears had been exaggerated: The hyperspace gate was well guarded, but there was no serious danger: In front of them there were eighty-three intruders. As soon as they had registered their arrival, they immediately started a well-coordinated interdiction maneuver.

A few minutes later, more than two thousand cruisers entered the system and, without any direct intervention by their captains, they divided into groups and moved to engage the ships blocking the gate. The maneuver, which had been simulated by the computers dozens of times, was performed flawlessly, and in less than a quarter of an hour no visible trace remained of the Qhrun ships.

Tom was fascinated by the sight of the Qhrun ships exploding one after another and, as always in such circumstances, he kept on wondering why none of their captains tried to make an escape maneuver or attempted to get in contact with the assailants. He became more and more certain that the key to the mystery of the Qhruns was to be found in that behavior.

"It would seem that the captains of their ships think like the computers in our torpedoes. It looks as if they have no regard at all for their own lives".

"They look rather like bees," Susan said, interrupting his thoughts.

"Bees? What have bees to do with this?" asked Tom, unable to see any connection.

"When bees sting an animal they leave their sting in their victim, but then they often die. And, nevertheless, they keep on doing that without the instinct of self-preservation stopping them. It looks as if the Qhruns' instinct is the same. . .."

"Admiral," the voice of the computer interrupted her, "a large fleet is coming towards this gate." On the screen thousands of Qhrun ships appeared.

"Distance and direction, Twenty-six? How many are they?" asked Tom, immediately turning towards the screen.

"Five astronomical units, and coming straight towards this gate. Predicted arrival in two hours," the machine answered immediately. Then after a short pause added, in a more uncertain voice: "I can distinguish at least ten thousand ships, but the mass seems to be much larger, possibly up to five times larger." This meant that the fleet had to be at least fifty thousand ships strong. They had been lucky: if they had arrived two hours later they would have come out of hyperspace right in the middle of those ships.

"Susan, what do you say?" Tom asked, turning towards her. "Why did the simulation not tell us anything about such a group of ships?"

"I bet that is an invasion fleet in formation. The simulation couldn't predict it: we know too little about how invasion fleets are formed, to be able to introduce that aspect into our model," Susan answered. "And now what do we do? We should have enough time to abandon the system before their arrival."

That was an alternative to fighting. It was impossible to wait where they were: facing a fleet of fifty thousand Qhrun intruders with just ninety-eight thousand cruisers was out of the question. Tom was sure that sooner or later they would have to try fighting a large Qhrun fleet, but it was unthinkable to make such an experiment there, more than three hundred parsecs beyond the frontier, while there was a much more important task to complete. On the other hand, Tom didn't like the idea of fleeing like that. "Think how many systems that fleet will invade before we can stop it with our Scorched Planets Operation. What fleet do you think it is?"

"Difficult to say," Susan answered thoughtfully. "If they are aiming at the fifth sector I would bet that it is the eighth or the ninth. But they could as well be heading towards the first sector, or even towards the third. But the knowledge we have of the situation in those sectors is too vague for us to make any guesses." She paused for some time, and then added "Are you thinking of attacking them? I know that we could save a good number of systems, but now our priority is to continue creating this corridor."

That was out beyond doubt, but who could guarantee that this same fleet was not being directed towards one of the gates that they had liberated? If so, the tunnel they had just dug out could collapse behind their backs, leaving them there, isolated beyond the frontier. "You can be certain I am not thinking of attacking them in open space," he answered. But there might be a way to stop them. He remembered the secession of the seventh sector: the battle of Teryygil must have been fought not very far from there.

"Twenty-six, try to remember. You know this system well, don't you? At the time of the battle of Teryygil you must have crossed it. And there was a base that had been in the hands of the secessionists for some time, and somewhere here." He realized that after twelve thousand years all memories of those events must be very tenuous. *Well, you have kept on annoying everybody with stories of the Teryygil battle,* Tom thought. *Now we will see whether you were bluffing.*

"Of course, Admiral, I know this system. When the fleet of the seventh sector's chased us, we took refuge in this region, among the three sectors. Admiral. . ." the computer was trying to remember his name, but at the end had to give up. "Our admiral led us here and we conquered the Fleet's base. The admiral was very clever, and thanks to that asteroid field we were able to ambush. . ."

That was what Tom wanted to know, and even more than he'd wanted. Clearly the computer was not bluffing: the fact that she couldn't remember the name of the admiral was proof. If the whole thing had been just a story to boost her importance, as many on board thought, she would have had no difficulty in inventing a name, also because nobody would be able to check. "Twenty-six, display a map of the asteroid field and its position in relationship to the hyperspace gates."

The screens displayed various images of the system. The gate they were close to was at about one hundred and twenty degrees, measured on the ecliptic plane, with respect to the asteroid field. The asteroid with the former Starfleet base was immediately outside the field. "Twenty-six, check whether there are any ships close to the base," he ordered as he was studying the images. The second hyperspace gate was slightly inwards of the field, so that the asteroids were between the gate and the base. Such a gate inside the system was clearly an anomaly, another of the anomalies that made the Aq system unique.

"I see about a hundred ships, but there could be more since. . ." the computer answered.

Tom interrupted her with a nod. If there were some ships in that area, the Qhruns were using that base, and then they would react to an attempt to attack it. A plan was forming in his mind. "Congratulations are due to your admiral, that asteroid field seems to have been put there for our own use." He paused and then added: "If within half an hour we move towards that base at top speed and if that Qhrun fleet tries to intercept us, will we get there before them?"

"Taking into account that they will only recognize our maneuver after a due delay, we still have fifty minutes to move," the computer answered, after the short time needed for her to make the computations.

Tom checked the positions of the various groups of cruisers. Ertlaq had been among the first to emerge from hyperspace, while Terkr was due to reenter with the last group in about forty minutes. The strategy he intended to try was thus possible, even if there were no very great margins. He called the Tteroth and instructed him in detail on what he, with the larger part of the fleet, and Terkr, with his group of five thousand cruisers, had to do. He gathered a group of ten thousand ships and started to move inwards in the system, directly towards the Qhrun fleet that was advancing towards the hyperspace gate.

He continued to follow that course for a quarter of an hour, then he altered his course arcing outwards towards the base on the asteroid. The Qhruns would have the impression that he was moving to intercept them. Tom was certain that such an absurd maneuver, like an attempt by a small group of ships to intercept a powerful fleet, would neither surprise them nor arouse any suspicion, because that was exactly what they had the tendency to do.

The curved trajectory brought him inside the orbit of the third planet, which was, however, on the other side of the star: they had nothing to fear from the ships that were certainly located close to it. Despite the fact that all the sensors of the cruisers were at their highest power, primarily to attract the attention of the Qhruns, he couldn't identify any other Qhrun ships apart from those that seemed to form the nucleus of an invasion fleet and those located close to the base.

Tom realized that the maneuver was succeeding, when he ascertained that the sixty thousand intruders were deviating from their previous course and starting pursuit. He gave the order to accelerate to top speed, hoping their calculations were correct.

In the meantime, even if they couldn't yet see them, Terkr's team had re-entered from hyperspace almost on schedule and, according to the instructions from Ertlaq, was moving towards the outside of the system, on a slightly curved trajectory that led to the base. and passing outside of the asteroid field. At the same time, Ertlaq left the system with the largest part of the ships, heading to a small star at slightly more than one parsec. It would take more than an hour before the Qhruns detected that maneuver, and by that time they would be too busy to chase the small group of cruisers that had penetrated inside the system and that was now running away from them.

There was nothing else to do than to continue, trying to lure the Qhrun ships to pursue them. Slowly their formation expanded, taking the shape a disk lying in a plane perpendicular to their direction of motion. The rear batteries of all the cruisers were aimed towards the Qhrun fleet, which was following them in disorderly fashion. By now the asteroid field was exactly between them and the base, and Terkr's group of cruisers was approaching the base, passing behind the asteroids.

The only unknown was now the behavior of the ships that were watching the base: they expected those ships to move to intercept Terkr, but there was a possibility that they might follow a more conservative approach, remaining under the cover of the batteries on the asteroid. In that case it would have been difficult, if not impossible, to tempt them out. "Estimated time for contact between Terkr and the ships guarding the base?" Tom asked.

"Maximum probability: within ninety minutes," the computer answered.

Now they had to wait, and to hope that everyone would stick to the timings. Everything was converging on that asteroid field, which had to be to the center of the operation. The inside boundary of the field was not sharp and soon they began to see the presence of frequent rocky debris. Tom had the front shields on and started to decelerate. If the Qhrun ships started to slow down too, it would be possible to enter the field at a lower speed, because he didn't want to let the Qhruns get close before entering the field.

When the first information arrived about the Qhrun ships that were close to the base, it was clear that, as usual, the Qhruns were behaving rashly, they were moving to intercept Terkr's group. Because they were doing that, their own group could fight without risking getting too close to the batteries on the asteroid and the minefields that had undoubtedly been laid around the base.

Just before entering the asteroid field, Tom ordered everybody to don their space suits and to use the emergency restraint systems: he had no idea of how the computers of his ships would be able to cope with the situation. He wondered whether he had made a big mistake by entering among the asteroids in that manner, but it was too late to ask questions of that sort. He realized that, after slowing down slightly more, the Qhruns would be in range a few moments after entering the field.

Suddenly he saw a dazzling glare on one of the side screens: one of the cruisers had hit a large asteroid, too large to be neutralized by her shields, and had exploded. On one screen he saw some distress signals coming from space suits: *It might have been worse,* he thought.

The remaining ships maneuvered with sufficient skill, and there were no more problems. The space in front of them was however full of flashes from the debris that exploded when it came in contact with the shields, while the largest objects had to be avoided.

A flash close to the nearest Qhrun ship showed that they too had entered the asteroid field. As predicted they came within range a few moments later. A Qhrun ship lowered her shields to open fire, and was immediately struck by a rain of debris, which was no longer stopped by the shields. It immediately exploded in a ball of fire.

"Lower the rear shields. Concentrate fire on one ship at a time," Tom ordered. The Qhruns couldn't lower their front shields to return fire, owing to the hailstorm of small asteroids, and on the other hand their shields would not be able to sustain the combined action of the debris and the batteries for very long. The intruders closer to them began exploding one after another.

"Don't waste time shooting at ships with their shield generators down," Tom ordered. No ship could survive in an asteroid field without shields to protect her, above all, when moving at high speed. "Reduce speed and let them get closer," he concluded.

Just then, the group led by Ertlaq appeared on the screens as they emerged from the hyperspace gate. As soon as the cruisers came out of hyperspace, they assumed a disc formation, and quickly advanced towards the asteroids, behind the Qhrun fleet. "Remind Ertlaq not to enter the field, but to remain outside together with the Terkr's group, which should coast along behind the asteroids, after having destroyed the ships that are guarding the base," Tom gave a further order.

The Qhrun ships were still exploding one after another. They had been caught unprepared, and now seemed unable to decide what to do: they were lowering and raising their front shields and, by doing so, they were worsening their situation. The ships that were behind the others, which by now, were also in the asteroid field, seeing Ertlaq's cruisers behind them, turned back to get themselves out of the field. But, as soon as they came within range of the other cruisers, they found themselves in the same situation as that in which the other ships had been for some time. They could not attack the Confederation ships, because lowering their front shields exposed them to bombardment by the asteroids.

Tom recognized the extreme tactical clumsiness of the Qhrun captains, or rather, of whatever organic, mechanical or other kind of intelligence decided their behavior in battle. *Any starship captain with a minimum of common sense knows that he should never chase a hostile ship in an asteroid field. It doesn't seem possible that those people, who have terrorized the Confederation's Fleet for centuries, should behave in such a stupid way,* thought Tom. But the situation was exactly that, the Qhrun ships were exploding one after another, without anyone thinking of doing the only sensible thing: to stop, get close to the largest asteroids, slowly assemble the fleet in compact groups and then lower the shields of the innermost ships and concentrate fire against the assailants.

Just as Tom had these thoughts, the Qhrun ships started to change course, running away in all directions, in an utterly random manner. That was an unexpected development, since a Qhrun fleet had never been seen to break formation and to run away. It was not an orderly retreat, it really was a rout. One screen showed two of the intruders colliding and disappearing in a flash of light. Other ships hit asteroids too large for the shields to stop them. Although, in general, ships of that size were not suitable to maneuver in an asteroid field, trying to escape like that, simply multiplied the disaster.

"Twenty-six, order Ertlaq to immediately saturate the area on the outside of the asteroid field with torpedoes. None of those ships must leave the system." Some cruisers, considering that the Qhruns were retreating, were beginning to reverse their direction and starting a pursuit. Tom stopped them, ordering them to get out of range and to follow the Qhruns at a distance.

The Qhrun ships that had penetrated deeper inside the field were now moving outward, assembling in the same zones as the other ships that were under fire from the batteries of the Ertlaq's cruisers. The large ships couldn't maneuver in a space that was getting smaller and smaller, and collisions became more and more frequent.

Within thirty minutes only small groups of Qhruns ships remained in the marginal areas of the field, apart from a few units that had somehow succeeded in escaping and which were moving towards the outer regions of the system: in all no more than a thousand ships out of the sixty thousand that had constituted the fleet.

21 The First Sector

"Twenty-six, how are rescue operations going?" asked Tom as the cruiser was slowly leaving the asteroid field.

"There are still a few hundred distress calls, but within one hour everyone still in space will be recovered. It will take some more time to tow all the damaged ships out of the asteroid field," the computer answered.

"Can you identify any non-human emergency calls?"

"Nothing. It looks as if nobody bailed out of those ships," the computer answered in a rather disappointed tone.

"Provided that there was someone on those ships," Tom replied.

The fleet was slowly gathering in the vicinity of the hyperspace gate. Only about a thousands ships were still chasing the few Qhrun intruders that had been able to escape the asteroid field.

Tom looked around in silence. Everyone was deeply tired, but they realized that it had been another historic day. For the first time a large Qhrun fleet had been destroyed in space, even though it had been thanks to the particular characteristics of the Aq system: a hyperspace gate inside the system and not too far from a large asteroid field. And then there had been the surprise rout of the enemy. Qhruns might have a very human reaction like panic, or was there a different explanation? Tom turned towards Susan to ask that question, but saw her completely relaxed on her seat with her eyes closed and decided that there would be time to analyze the situation after a good sleep.

He looked back at the screen where the final statistics of the battle were steadily updating while recovery operations went on. They had lost four thousand two hundred ships, and another six hundred were damaged to the point they had to return to the bases in the fifth sector. The statistics gave more than twelve thousand dead, and by now just two hundred and seventy missing, which meant that more than forty thousand people had been recovered from space or from the wrecks of their ships and were by now in one of the cruisers' sickbays. Twelve thousand dead, he repeated to himself, realizing that the figure was enormous in absolute terms, but that it was an extremely low number when compared with the size of the two fleets that had been engaged in battle.

He couldn't refrain from comparing the cost of that victory to the cost of the destruction of an invasion fleet obtained by blowing up a planet that had been invaded. Are twelve thousand human lives worth a planet, abandoned it is true, but containing all the traces of a planetary civilization, without taking into account the endless problems the evacuation of hundreds of millions people involves? he wondered. And he realized that the same question had been asked by hundreds of generals, kings and leaders of every type in the past, and that probably nobody had ever been able to find an answer. From time to time, perhaps a practical answer had been found, based on what had seemed best in a particular situation. In his case he didn't have any doubts: any Qhrun fleet that could be attacked with good chances of success, had to be destroyed. Without even realizing it, he fell asleep in his seat, while the ship slowly left the asteroid field and headed towards the hyperspace gate.

The following day the cruisers started entering hyperspace according to very detailed plans that they had revised whilst in the Aq system. Their strategy was the same as during the first week: to continually move forward, leaving about ten ships to guard every hyperspace gate. The second phase lasted a little more than six days, in which the corridor lengthened by another four hundred and fifty parsecs, all now in what had been the first sector. Many systems were completely deserted and most had just a few Qhrun ships.

Finally the second part of the operation was also over, ending in another small and insignificant system that was just a dot on a star map: a white dwarf orbited by a few asteroids—little more than large boulders. One after another, thirty-five thousand ships came out of hyperspace and gathered in a zone on the outskirts of the system. They were almost eight hundred parsecs beyond the frontier, and that system had been visited by Terkr in his reconnaissance about twenty days previously.

While the computers in the ships continued to exchange information, the crews could rest after a week of hard work.

Tom took a long time analyzing the situation. Thirty-five thousand cruisers only remained for the third phase: less than expected, because they had had to leave many ships in the Aq system. The losses had not been very heavy and, if the occupied zone had not been much wider than they had computed, there would have been no problems.

They decided to stay in that system for forty-eight hours, both to give time for the crews to rest and to wait for more news. Every hour another ship emerged from the hyperspace gate, mostly freighters, or else hyperspace probes. A particularly important piece of news was the crossing of the frontier by a fleet of ten thousand cruisers coming from various shipyards that had been working at their maximum capacity. Given the difference in speed between the probes and the ships, they would arrive in the system within three days.

He finally gave the order to start the third phase. The ships moved towards the hyperspace gate, and scattered towards other systems, and then weaving their trajectories in a continuous motion from one system to another.

The days again started to pass rapidly. After a little more than three days, the length of the corridor had reached one thousand parsecs, Obviously that was just an arbitrary number, but it had the feeling of an important achievement. The assumed position of the frontier between the first sector and the zone occupied by the Qhruns was reached after another three days. At the exit into the first system that might not have been occupied they were very tense: their hope of being welcomed by the signal from an automatic station was very strong. Equally strong was the disappointment of finding the usual small group of Qhrun ships. However, the position of the frontier in that zone was only a rough guess. When Tom said this to Susan, she answered: "Fourteen ships are more than the average. Did you notice that since yesterday we have meet with greater resistance as we go on?" She paused for a moment and then, as Tom didn't answer, continued: "Using the data brought by the ships we have met in the various systems, I tried to compare the evolution of the density of Qhrun ships in those systems with what we recorded immediately after we passed the frontier's gate at the other end of the corridor. The result is that we should be still at more than one hundred parsecs from the frontier."

In the following thirty hours they continued to go in and out of hyperspace. and in every system they updated the computations of the position of the frontier. The model based on the density of Qhrun ships was repeatedly replacing the initial model, based on less precise extrapolations. One of the screens displayed an image of the galaxy with the position of the frontier as they were updating it.

The moment finally came when they were about to enter a system that, according to the latest computations, had to be beyond the frontier. It was again a system of minor importance.

When the universe reappeared around them they all where holding their breath: there was no ship near the gate. It was three days since they had emerged into a non-guarded system and that calm seemed unreal. Everyone remained silent until an unmistakable sound filled the bridge: the call of one of the Confederation's automatic stations.

Tom relaxed against the back of his seat, and closed his eyes: he was sure that everyone on board was thanking their individual gods in silence. The first one to break the silence was Wal-Nah, who got out a cry in the Nahaqol language, immediately echoed by a similar cry from Del-Nah over the ship's intercom. Immediately an uproar in different tongues followed. Only after a few minutes did Tom realize that the computer was hopelessly trying to make herself heard above the confusion. ". . . at about seven astronomical units," was all he could make out.

"Twenty-six, what is at seven astronomical units?" Tom asked, trying in turn. to make himself heard.

Suddenly the noise stopped and everyone returned to their seats, somewhat ashamed of their reaction. "A Fleet star cruiser. From the identification signals I understand that it is TH 44621, an old ship of a class I know well. They will acknowledge our presence within less than one hour."

"Order the ships that are with us to stay close to the gate. Set course towards that cruiser at maximum speed. Transmit the messages we had prepared for this event and record all the information you can pick up."

A continuous flow of data started to reach the screens that Susan used for navigation, but above all would update her mathematical models. In rather more than ninety minutes they would have an answer from that cruiser, together with a much larger quantity of data. For an hour and half, while Susan analyzed the data from the station, the others kept on probing the system with every sensor: that cruiser was the only ship around. Finally the screen came on and a human, of a species Tom had never seen, appeared. The screen showed him just from his shoulders upwards and therefore there was no indication on what the rest of his body was like, but the face was really strange. The forehead was high, black and without hair; below it there were two big round eyes, set like those found in owls. The pupils were large and the iris was black, like the rest of his face, and they gave an impression of amazement. Immediately under his eyes there was a short conical trunk, about forty centimeters long, that was completely hiding his mouth, *because presumably that human must have one,* thought Tom. There was no trace of ears.

"This is star cruiser TH 44621. Captain Hatk of Wetkloss speaking. Welcome to the first sector, Admiral Taylor." While he was speaking Tom realized that what he was hearing was the voice of the computer which had translated the message into Aswaqat. Actually the hisses and the guttural sounds of the original message that could be heard were the actual transmission in Terqhatl. Almost immediately, however, Tom realized that a third even fainter voice was speaking with sounds that were different from anything he had heard earlier: it was Hatk's voice. He was using a computer to communicate in a language (Terqhatl) that was understood in the sector but which he clearly couldn't speak. That double translation gave Tom, for the first time. the almost physical sense of the distance separating him from that part of the galaxy that had by now become familiar to him.

"I can't understand how you have managed to arrive here, Admiral, and I am eager to communicate with you more directly," concluded the captain, who then fully accepted the instructions he received to wait until the two ships were at a distance of less than three hundred thousand kilometers to begin a full discussion.

The communication ceased and Tom asked: "Twenty-six, what was that human?"

"I don't know Admiral. I am trying to identify him but I don't know much about the species in this part of the galaxy."

"Tom, we have a problem," Susan interrupted. "I am working on the information that has arrived from that ship and. . ."

"Twenty-six, when will we arrive at a distance that is convenient for conversation?" Tom interrupted her.

"Less than twenty minutes: they are coming to meet us," the computer answered.

"We need to slow down" Tom suddenly stated. "I need to have at least thirty minutes to understand what is happening before we start to communicate." "What's happening? Are our computations wrong?" asked Susan.

"Not much, overall. But it looks as if the Qhrun somehow realized that there is something important in the first sector. The frontier has not moved in a uniform way, as we predicted, but three invasion fleets are directed towards Terkar. One is heading directly towards Terkar and the others are flanking it. But how could they deduce the importance of that system?"

"How much time do we have? When will they reach Terkar, I mean?", Tom asked.

"In little more than a month and a half. We will just have time to evacuate the Coordinator and the Council. But do you realize what the fall of Terkar means?"

"Of course I realize it. But we cannot apply the Scorched Planets strategy in this case. Qatlaassari and the Council will never leave Terkar," Tom answered. "One month and a half to get two hundred thousand cruisers to Terkar. . . One month and a half," he concluded, speaking to himself.

"I hope you don't mean to engage them." Susan said. "Where can we find the ships, and above all how can we train the crews? The victory at Aq was obtained under exceptional conditions. There are no asteroid fields in the Terkar system. Remember that Kotusov didn't hesitate to let Napoleon reach Moscow. Trying to stop them at Terkar is madness."

"Perhaps. Don't forget that the Qhruns are not Napoleon: if we abandon Terkar we won't be able take it back again, or at least we don't know what we will find if one day we are able to take it back. And then, Terkar is not only the capital of the Confederation, it is a symbol. The symbol of eighty thousand years of history. I fear that if Terkar falls, our efforts will be frustrated. Do you think that the Qhruns concentrated their attack on Terkar on purpose? Are they becoming experts in psychological warfare?" Tom spoke slowly, trying to understand the meaning of what he was saying, even before trying to convince the others. Everyone on board listened to him in silence, not really knowing what to say.

"I don't believe it. I may be wrong, but there must be a simpler solution; otherwise the few ideas that we have about the Qhruns are all wrong. But I don't think that defending Terkar is a viable strategy. I know Terkar is a symbol, but how many spacers must die to defend it? Above all, is it worth it? If we can convince Qatlaassari to abandon the planet, we could blow it up when the invasion fleet lands. And perhaps the psychological effect won't be that to have abandoned our flag, but that of a huge fire that will set the whole galaxy ablaze," Susan replied.

"How many inhabitants are there in the Terkar system?" asked Tom, who hated the thought of a battle of that sort.

"Sixteen hundred million, almost all on one planet," the computer answered.

"Even if Qatlaassari agrees to leave the system, how can we evacuate all those people in just forty-five days? Let's try to be practical: to take a decision I need a map of all the Starfleet bases and all the asteroids used by the Merchants within a radius of five thousand parsecs, and an estimate of the number of ships of all types that we could mobilize. And then we will see." While Susan returned to analyzing the data that continued to flowing across her screen, he added: "and now let's try to convince this Hatk of Wetkloss to do something."

By now they were close to the other ship and any moment the connection would be re-established. Susan had only the time to inform him that the closest Starfleet base was Telhra'ar, at only four hundred parsecs. "TH 44621 comes directly from there, and when she left it there were more than six thousand ships, almost all cruisers," she concluded.

Captain Hatk's image appeared immediately on the screen. "Welcome, to the first sector, Admiral Taylor. However I believe it is my duty to ask you for some explanation: how could you run the blockade and, above all, where do you come from? I looked for the system you say you come from, but I couldn't find it on any star map."

The captain didn't fully trust him. And Tom couldn't blame him for that, instead he was happy that there was still someone who bothered to make the checks that, in theory, were compulsory. "Soon I will send you a message in which you will find a summary of the events of the last year, events showing that the Qhruns can be stopped. For now, I will tell you that I passed the frontier with one hundred and thirty thousand cruisers three weeks ago, and that now the fifth sector isn't isolated any more. I have opened a corridor several parsecs wide in which every hyperspace gate is guarded by my ships." While he was speaking Hatk's expression showed more and more surprise. "It is no wonder that the system from where I come is not on your maps: its entry into the Confederation is very recent, and your maps are at least eighty years old." *It is so recent that has not happened yet,* Tom thought with a twinge of remorse for once more lying on the subject.

When the two computers had finished translating the message, first into Terqhatl and then into the Wetkloss language, Captain Hatk answered, somewhat uncertainly: "You bring great news, Admiral. I look forward to examining your message in detail." He clearly wanted to take some time over that, and was still not completely convinced.

Tom decided that it was time to shake things up a bit. "Captain, the Coordinator of the fifth sector gave me the supreme command of every fleet in the sector, and ordered me to reunify the Confederation. I am heading towards Terkar to speak to Coordinator Qatlaassari and to organize the defense of the capital of the Confederation."

Captain Hatk appeared utterly petrified. After a time that seemed endless to everyone, he moved the short trunk that was hanging below his eyes, in a gesture that at another time Tom would have found funny, and slowly said, as if he himself was trying to make sense of the meaning of his own words: "To organize the defense of the capital." Then he added: "Admiral Taylor, my ship and I are at your orders. It is time for someone to decide to appoint a general commander for all the fleets in the galaxy, and I believe that when you will get to Terkar the Coordinator will do just that. But even if he doesn't, you can count on my ship and on all Wetkloss ships."

"Thank you, Hatk. If you want to be useful, return immediately to Telhra'ar. At the base there should be more than six thousand ships. . ."

"Excuse me, Admiral," Hatk interrupted him. "But when I left there were no more than one thousand cruisers. How can you know what conditions are at Telhra'ar if you have only just entered the sector. . ."

The computer immediately understood the puzzled expression on Tom's face, and a note appeared on one of the screens: "One thousand two hundred and twenty three ships are in the main cavern, the others are in the shipyards. They may perhaps not have enough crews, but the number of ships is correct."

Breathing a sigh of relief Tom continued: "There are more than six thousand, Captain. . . you didn't see them because they are in the shipyards. Perhaps there are not enough crews for all the ships, but certainly the ships are there. Don't worry if you find it difficult to understand how I know the situation: a general commander knows a lot of things," he concluded within a slightly mysterious manner. *My God, that looks like a warning from the Mafia*, he thought, trying to conceal a smile.

Then he continued: "Order the commander of Telhra'ar to send hyperspace probes to all the bases in the sector; my computer will transmit to you detailed instructions for each base, in particular for the shipyards and the logistical units. If we want to stop them at Terkar in a month and a half, we must put all the ships in this sector into battle-ready condition, and possibly to build new ones. I want at least one hundred and eighty thousand cruisers in the Terkar system within twenty days, with the largest possible number of freighters loaded with torpedoes and transponders: above all transponders. If there are not enough spacers, the ships must leave with incomplete crews and stop in any systems they cross to get volunteers. The message I will send you must be sent via hyperspace probes to every inhabited system." He paused for a moment and then added: "I forgot, my computer will also transmit to you programs for training and battle-simulation exercises: we must train the crews during the transfers towards Terkar."

Hatk was bewildered, submerged by that avalanche of orders. "I wonder how we will convince the commanders of the bases to do what you order, Admiral," he said in a low voice. Then he shook himself and continued "To hell with it. If we want to save Terkar we have to move. Admiral, transmit the data, I cannot wait to leave this system."

Hatk's ship had already started moving towards the hyperspace gate at top speed, followed by Tom's cruiser. The three hours needed to reach the gate were just enough to reorganize the material and to transmit it: many messages that had been prepared had to be modified, because of the new turn of events.

They only had some rest after entering hyperspace, and were heading towards the system where all the ships that had entered the first sector were to meet.

"Well, Susan, is there anything new?" asked Tom, who was certain that while he had been organizing the defense of Terkar she had continued to study the data the computer had received.

"Only minor details. Globally, our model still holds: the unexpected progress towards Terkar was accompanied by a slower expansion in other areas of the frontier. However I fear that stopping the fleet heading towards Terkar will give us only a few months. Their spearhead is made up of three invasion fleets, and if we stop one of them, the others will converge towards the capital. We must find out how they came to understand that Terkar is so important."

"Call it Qhrun insight," Tom answered and she immediately understood that for the moment he was not particularly interested in that question. "Instead of worrying about that, have a look to see whether there is some system that can be evacuated to destroy at least one of those fleets in the usual way. And plan raids to disturb their advance: as soon as Terkr is here, we must send him beyond the frontier to weaken the fleet that is homing on Terkar." He paused and then added "And prepare the fastest course between the system where our forces are concentrating and Terkar: we have no time to waste."

Twenty hours were needed to get to the assembly point: an inhabited system far enough from the frontier for them to be practically sure that it had not yet been occupied. When they left hyperspace they found themselves in a huge group of starships: the twelve thousand cruisers that had still been available at the end of the corridor.

As soon as he realized that Taylor had emerged from hyperspace, Terkr sent him a message: "Good news, Admiral: I have received a probe announcing the arrival of those ten thousand cruisers that were following us: They will be here within two hours. We can now enter the system and to go down on that planet to rest for a while."

"That is really good news, better than you imagine," Tom answered. Those ships were actually a gift from heaven. "We won't enter this system and unfortunately nobody can go down to the planet. I fear that our true mission is just starting: the Qhruns are forty-five days from Terkar and, if the Central Coordinator decides not to leave the system, we will have to stop them." Tom realized what that message meant for the crews: after three weeks of going in and out from hyperspace every few hours in systems occupied by Qhruns, they saw the planet as a promised land. But nothing could be done about that: the next six weeks would be much worse than the last ones.

He broadcast the details of the invasion into the first sector and the data for the course to Terkar which was where the assembly point had been moved:

They would not even wait the two hours needed for the other ships to reach them: Just leaving a message on an automatic station would do.

22 The Council

The hall where the Council of the Confederation met was half empty. It was years since the representatives of three sectors had been present and those from another four didn't actually represent anybody, since they had been completely cut off from their worlds. But the representatives of the other eleven sectors had less and less frequent contact with the coordinators who had nominated them, and many were habitually absent from the hall.

The lights became brighter, and those present hastened towards their seats. At the back of the hall there were two figures, so busy in an animated discussion that they didn't realize that the lighting had increased. They were only able to communicate through their computers. One of them put one of his six hands on a tentacle that was sticking out from below the other's mouth to attract his attention and spoke into his computer's microphone: "Let's go, they have already increased the lights." The other, as soon as the computer had translated the message into a series of tactile impulses on the sensitive zone of his skin that he used to communicate, answered, with light pressure by a tentacle on his computer: "You are right, however I believe that the Coordinator should abandon this planet. However, I won't propose such a motion, because they would outvote me." He moved towards his place, guided by ultrasound echoes, since he had no organ sensitive to any radiation with a wavelength longer than that of X rays.

Coordinator Qatlaassari entered the hall, and went to the Coordinator's seat. He kept his short wings folded behind his back and the dark gray color of the membrane contrasted with the bright orange of his dress, which covered him from head to toe, only leaving his face and hands free, apart from his wings, of course. Nobody understood why Terqhatls continued to display their wings like that, given that for millions years they could not sustain them even in a short glide.

He remained standing behind his seat and began to speak slowly. "As representatives to the Confederation, I have summoned you to report some strange facts, and because I have the impression that you are about to witness even more sensational events." The Terqhatl paused for a long time, in deep silence. What impressed everyone more than his words was his appearance: he looked centuries younger and his voice was no longer subdued, as it had been for many years.

"Three hours ago I received a hyperspace probe from the Starfleet base at Telhra'ar. The commander informs me that a certain Admiral Taylor arrived in this sector with a fleet from the fifth sector and that he had been ordered to send probes to all the bases of the Fleet to gather a large number of cruisers to defend Terkar. The admiral is said to be from Earth, system of the Sun. I checked on the Fleet database and I could find no trace of that system." He turned towards the two Aswaqats representing the fifth sector and asked: "What do you make of that?"

One of the two Aswaqats got up and answered, speaking directly in Terqhatl. "Coordinator, we have been isolated from our sector for more than eighty years. We have never heard of that system, but this doesn't mean anything: I believe that none of us would recognize ninety percent of the systems of his own sector. It might be a system that has entered the Confederation since the sector has been isolated. Certainly, if someone has been able to run the blockade. . ." He didn't finish the sentence, because in fact he couldn't imagine the true significance of such a feat.

"Certainly, it might have entered the Confederation after the sector was isolated," Qatlaassari continued. "Now look at this recording."

In the center of the hall a three-dimensional projector materialized a recording of the ceremony in which Aintlad had conferred on Tom the command of the sector's fleets. Technically, the recording was not perfect but the slightly distorted colors made the scene even more striking. After it has finished, again addressing the two Aswaqats, Qatlaassari asked for their opinion.

"There can be no doubt that the Coordinator is Aintlhad. That is possible, because he is more or less my own age. Apart from that his behavior is perfectly consistent: If it is a fake, it has been done by an Aswaqat or by someone who knows us very well. Taylor is not an Aswaqat: he speaks our language fairly well and physically he is similar to us, but he is not one of us. Then his strong Kirkysakh accent is strange."

"Everything seems to point to the fact that this Taylor is telling the truth, even if I don't understand how he could get here and above all how can hope to defend Terkar," concluded Qatlaassari, sitting down. After one minute, as if justifying his decision, he got up again and added: "There is however a thing that is not clear: the commander of the base concludes his message by congratulating me for having given Admiral Taylor the task of assembling all the Confederation fleets, a thing that is clearly not true. But in the whole recording the Admiral never makes any statement of the kind... instead he says more than once only that he received the command of the fifth sector's fleets."

Nobody had a word to say. They did not have enough information to answer the many questions that were forming in their minds nor were they able to form any hope, which might quickly lead to dreadful disappointment. The voice of Qatlaassari nevertheless suggested that the Coordinator, despite his last sentence, believed in the possibility that this unexpected help might be real.

Qatlaassari remained silent for a while, then seeing that nobody wanted to speak, continued: "I immediately summoned the Council of the Confederation because according to the calculations of the Starfleet analysts, we should receive a transmission directly from Admiral Taylor within a few minutes. Provided that an Admiral Taylor does exist and that, after having sent those messages to Telhra'ar, he headed towards this system."

When the Coordinator sat down again, the hall filled with a subdued hum: Even if nobody dared to speak in public, everyone had something to say to their neighbors.

Qatlaassari was right: just a few minutes after he spoke a three-dimensional projector displayed a Terqhatl dressed in Starfleet uniform. "I apologize for disturbing the meeting of the Council," he said, timidly "but the Coordinator ordered me to immediately report the entry into this system of any ship that was not scheduled."

"Omit the formalities, Commander, and tell us what has entered the system," Qatlaassari interrupted him.

"The sensors of the stations orbiting the planet have detected a large number of ships entering the system, beginning five minutes ago. They seem to be Starfleet cruisers," the commander of the space port answered.

"How many ships?" Qatlaassari asked.

"They are still entering the system. From their mass it seems that more than six thousand ships have already entered," the spacer answered. "Wait, Coordinator. We are receiving an incoming message," he added.

"Relay it here," the Coordinator ordered, obviously pleased that his forecasts had been realized.

Another three-dimensional projector displayed the image of the bridge of a star cruiser. At the center, Admiral Taylor was seated. "Message from Admiral Taylor of the system of the Sun to the Central Coordinator of the Confederation," the voice of a computer said. From the way the machine spoke Terqhatl, to Qatlaassari it was immediately clear that it was a ship coming from outside the sector.

The computer fell silent and they could hear Taylor saying: "Admiral Taylor, general commander of the fleets of the fifth sector, asks permission to land on Terkar to report to the Central Coordinator."

"Grant the authorization," Qatlaassari ordered the commander of the port. "Inform me when the whole fleet has entered the system."

The coordinator looked around. Nobody had ever entered the system of Terkar with a fleet of that size and the thing was a clear violation of custom. It might be an attempt to influence the Council with military force and at any other time Qatlaassari would not have allowed it. What did such ostentatious power mean? But then, even if those cruisers had tried to seize the system, it would have been better to fall in the hands of this admiral who, even if coming from a system he never heard of, was undoubtedly human, than into the hands of the Qhruns. *No doubt Taylor wants to impress us. Soon we will discover whether he does this to threaten us or to push us to trust in his plan to defend this system,* he decided.

After about a quarter of an hour, which seemed endless to all the members of the Council, the projector again displayed the commander of the port. "The ships have stopped entering the system. They are more than twenty-two thousand, but only two are homing in on this planet." After a short pause he added: "They inform me that the other ships are leaving the system."

The coordinator immediately felt relieved. It was clear that the admiral simply wanted to show that he was not bluffing and that, whatever his plan was, he indeed had a non-negligible force at his disposal.

"When will he land on the planet?" asked.

"The admiral has informed us that he intends to use a fast approach procedure and is entering the system at top subluminal speed. That is unheard of and I have ordered him to slow down, but if he doesn't comply he will be here in three hours."

"Inform him that he has permission to use the procedure he has begun." And then he added: "Get a mental projector and study the history of the Xartian crisis, commander. That procedure was last used by Admiral Vertearis when he came here to report that the Xartian fleet had been defeated. I believe Admiral Taylor wants to convey something with this way of getting here," he concluded with a satisfied smile.

Actually it took almost four hours to complete the landing on Terkar, partly because, while approaching the planet, Tom realized that the ephemerides of the system that the computer had in her memory were rather approximate, and it took a further hour to reach the Council building.

When Tom and his crew got out from the vehicle that the Coordinator had sent to the port to convey them to the Council building, it was night. Nevertheless, the news of their arrival had started to spread and the square in front the building was crowded. The security men, mostly Terqhatls, held back the crowd so that a narrow strip leading to the entrance of the building was free. They walked inside, preceded by the two Terqhatls who had escorted them up to that point. The square was well lit and Tom tried to make out the species of the humans who surrounded them, but without being able to identify most of them. Nevertheless he was impressed by the cosmopolitan character of the crowd. Likewise, he could see almost nothing of the building, with the exception of a large, decorated entrance that reminded him of a Gothic cathedral.

When they entered, they were led through a number of halls, until they reached a wide corridor where a group of security men stopped them. Tom gave them the hand laser and the thermal gun he had in his pocket and the others did the same. They continued to the door leading to the Council chamber. Once inside, Tom continued to the center of the room while the others remained at the back of the hall alongside the entrance.

"Coordinator, I bring good news from the fifth sector," he said, addressing Qatlaassari in a loud voice as soon as he was in front of him. He had tried to learn the Terqhatl language but had then changed his mind about using it because of the difficulty of correctly pronouncing it. He decided to speak in Aswaqat and have his words translated by a computer. "Four Qhrun fleets have been destroyed and the fifth sector is no longer isolated: the invasion can be stopped."

Qatlaassari stood up and solemnly answered: "Admiral, we have been waiting centuries to hear those words. And, even if we all were certain that sooner or later this would happen, we had started to doubt."

Tom noticed that the Coordinator spoke in a very solemn tone, and that his voice betrayed some uneasiness. But then Qatlaassari fully realized that he was not only speaking in front of the Council, but that every future generation would analyze those sentences in every possible detail. If there were any future generations, of course.

The Coordinator spoke again: "Admiral, this Council is eager to listen to your report."

Quickly Tom asked the for formal authorization to put the projector of the hall under the control of his ship's computer. Then he stepped up onto the small podium located alongside of the Coordinator's seat and began the long speech he had carefully prepared.

"My speech will begin with an illegal action performed by a sector coordinator, and will close with a request going beyond the letter of the law. I believe that all those present will realize that the situation is so peculiar as to justify these actions." He had decided to begin in that way despite Aintlhad's recommendations to talk of his origin only to the Coordinator. He gave a quick glance around and realized that this beginning had made everyone very nervous. At that point he understood that was useless to delay any further, and continued: "You are undoubtedly wondering why the system I come from is not marked on any map. The answer is simple: I come from a type B planet."

At that point it seemed that everyone had some comment to make, and the volume of the murmurs in the hall became loud. Qatlaassari gestured to the audience to keep silent and, with a movement of a wing, ordered to Tom to continue. Everybody knew that Terqhatls gesticulate with their wings only when they are nervous.

Tom started speaking of the Aintlhad's plan and of the role that Quanslyaq had in it. He continued with the problems caused by the Iskrat-is-Thn and the necessity for the two of them to hide for some time their earthly identity, and finally with the plan that he had christened Scorched Planets. It was clear that some of the things that he was describing were difficult for many people in the Council to accept, and the role that he had assigned to the Merchants was an example. However nobody interrupted him, and Tom spoke for more than four hours. He concluded by asking the Council to accept the Earth into the Confederation.

When he ended, the room remained in absolute silence, since everyone was anxious to know how the Coordinator would react. Qatlaassari just raised his head and asked: "Admiral, you ordered a large number of ships to enter Terkar's system to defend it and, a few hours ago, you gave a demonstration of strength to show that you can indeed do so. Does it really mean that you will try to stop the Qhruns in this system, fighting against a fleet of more than one hundred thousand intruders in what many will certainly consider as a desperate attempt?"

Tom had been waiting for this question, even if perhaps not put quite so directly. "Coordinator, the decision to abandon this system is a political decision, and therefore it is up to this Council and you. From the military point of view I believe that the wisest decision is to abandon Terkar and to transfer everything that is in this system to another planet. The fleet that I will assemble here is able to evacuate every human and all important material and to prepare the planet to destroy an invasion fleet. But if your decision is to stay, then we will try to stop them."

"Admiral," Qatlaassari responded, "the political decision has already been taken: I will stay here with those members of the Council that decide so. But the military decision is up to the Fleet and it is wise that the second is not influenced by the first one. The attempt to stop them may end in a disaster, and losing a fleet here may well be the end of all our hopes."

As the Coordinator spoke, Tom noticed that in the hall some of them were indicating, discreetly, their dissent; apparently the decision had been taken despite strong opposition. "The two decisions are not quite independent," Tom replied. "If the Council stays here. we cannot do anything other than try to defend the system. It is true that a failure could mean the end of every possibility of resistance, but the fall of Terkar would be a blow from which the Confederation might not recover. The transfer of the center of the Confederation to another system could be acceptable, but the capture of the Coordinator and the Council... And then, in a way we cannot understand, the Qhruns must have realized what Terkar means to all of us, and they have concentrated their thrust in this direction."

Clearly the Coordinator didn't want to revisit that matter; for him the political decision had been taken once for all. He got up and said solemnly: "Admiral Taylor of the system of the Sun, you are summoned to the presence of the Central Coordinator of the Confederation."

Tom left the podium and stood in front of Qatlaassari. Then the Terqhatl uttered: "Admiral Taylor, in the name of the peoples of the Confederation that I hereby represent, I order you to assume the command of all spacecraft and all personnel of the Confederation's Starfleet."

For Tom it was like reliving the ceremony when Aintlhad had named him as commander of the fifth sector's fleets. He turned towards the gathered representatives and, in a solemn voice, said: "I, Admiral Taylor of the system of the Sun, receive the supreme military power. In front of all the people you represent I swear to use it to stop the invasion and to set the galaxy free from fear and danger." He again turned towards Qatlaassari and, without speaking, took a seat behind a desk set at one side of the room to witness the animated session of the Council.

The discussion began when a kind of grayish mountain, all tentacles, got up and began to speak, or rather, to move a tentacle over the surface of a computer that translated the movements into words. It was a long speech, that began with great praise for the Coordinator for having designated a general commander of the fleet, followed by other praise for the wisdom and foresight of the admiral who realized that it was impossible to defend Terkar. He continued by declaring that because the situation was so desperate, the evacuation of the planet was inevitable, a thing he had proposed more than a year before. At that point Tom realized that everyone in the hall would try to use his words to support the political strategy of some group and faction, whose existence he hadn't even suspected. He didn't want to let himself become involved in those discussions, additionally because he had the suspicion that many of those present were simply trying to take advantage of the situation to undermine the subtle hegemony that the Terqhatls had held for a long time in this sector and, in a less obvious way, in the whole Confederation.

The long discourse of that elephant-like being had hardly stopped, when Tom asked to speak saying, "These discussions are suitable for those who must take political decisions. I ask you to inform me about the deliberations of this

Council, so that I can, with the autonomy granted by the law to the general commander of the fleet, take the initiatives I deem necessary."

He got up and, as he was walking out, Qatlaassari said to him: "Very well, Admiral, return to your ship, but wait for further communications before taking off. I will call you at the end of the meeting."

23   The Battle of Terkar

Tom waited for the Coordinator's call in the library and meanwhile he continued studying the possibility of defending the system. Qatlaassari seemed so stubborn in his resolve of not abandoning Terkar that his doubts about the possibility of applying the scorched-planets strategy were growing stronger and stronger.

After more than an hour Qatlaassari appeared on the main screen and the Coordinator's voice roused him from his thoughts. "Admiral Taylor, I appreciated your leaving the Council Hall a few hours ago. For a moment I feared that you might get involved in political matters. Politics here on Terkar requires a great deal of caution. . ." he said quietly.

"On the planet where I come from things are not much different," Tom answered. He didn't know whether to be amused or offended by those words: the other clearly thought that people outside Terkar, and particularly those coming from a type-B planet, must be simpletons, unable to understand the subtleties of politics. "But basically I cannot say that I agree with the decision to stay here. The wisest choice seems to me to evacuate Terkar, even if those who suggest it have ulterior motives."

After a long pause, Qatlaassari went straight to what was the real reason of his call. "You see, Admiral, the point is, that this planet is a huge archive: the general archive of the more than four hundred thousand human species spread over more than seven million inhabited planets in the Confederation. Here we have all the anthropometric, psychological, anthropological and cultural data and even the complete genetic code for every species. Apart from that, all the data on the geology, flora and fauna of their planets of origin are preserved here. All the information needed to reconstruct, in an anthropological sense, the whole galaxy, are contained in this archive, which extends underground for miles. Nobody knows the full extent of this archive: it may be under this whole continent, or, as some legends say, throughout the whole solid crust of the planet. They may be only legends, but they actually describe the reality of the situation."

Suddenly Tom understood. "Terkar is not just a symbol of the Confederation. From a certain point of view Terkar *is* the Confederation," he said under his breath. And then, more loudly, asked "How long will it take to load everything on star ships and take it away?" But he already knew the answer.

"Years... perhaps centuries," the Coordinator answered.

It was clear that the planet could not be destroyed. As such, the Qatlaassari's decision was the only possible one: when the Qhruns arrived he would be there, to make a last attempt to get in contact with them, to negotiate, to do anything to save at least a part of what was being stored on the planet. And if everything turned out to be useless, Tom was certain that he would destroy everything above the surface and seal himself beneath it with the archive, in the hope that in some distant future, perhaps hundreds of thousands or millions of years in the future, someone would discover the treasure that planet contained.

At that point Tom answered: "Coordinator, the whole of Starfleet will be in this system and we will stop them... Whatever the price, we will stop them."

Qatlaassari looked him straight in the eye and answered: "I was sure that you would understand; but if there is no hope, I don't want useless heroism. I don't want to lose first the Fleet, and then Terkar."

They carried on talking about how to defend the planet for some time, then, before ending the conversation, the Tteroth asked: "Admiral, are you still sure you want to propose that Earth joins the Confederation?"

"I don't know... in all sincerity I don't now know what to think. On the one hand it seems to me that it would be a great opportunity, but on the other hand I foresee great difficulties. Perhaps it should be gradual, a kind of temporary association awaiting further developments..." he answered, without voicing any definite conclusion.

"I will leave any decision to you. If you believe that it is appropriate to make an exception to the law, I believe that circumstances fully justify it. I formally nominate you as my plenipotentiary representative to the people of Earth, and the Council will ratify any decision you may take on the matter."

For an instant, Tom experienced a feeling of triumph: the battle he had started to fight the very first time he had landed on Laraki was won. Then he realized the trap he had fallen into: now the decision was his, and he would have to deal with it, in the face of his planet, his conscience and history. After all, the Confederation didn't care one way or another in the matter, and so it had just solved the problem in the simplest way: now it was just his problem. He suddenly had the impression that ever since he had entered the adventure everyone was simply giving him their problems to solve, and that he had acted in such a way that tended to make them behave like that.

To hell with all that, now we have more serious things to deal with. If we don't solve this, within less than six weeks we will all be interstellar dust and we won't care about all this crap any more, he concluded to himself.

The days were passing rapidly as preparations for the defense of Terkar went ahead. Terkr's and Ertlaq's squadrons were busy performing raids beyond the frontier, designed to destroy all the isolated ships that could be found on the invasion fleet's course and to mine the hyperspace gates. The aim was to weaken the fleet that would soon enter the Terkar system and possibly delay its arrival.

TH 44621 entered the system a few days later, together with more than twenty thousand cruisers and a large number of freighters. Hatk's image immediately appeared on the screen: "Admiral, I apologize for not arriving earlier, but I wanted to assemble ships and material. I had some difficulties with the transponders, however I finally got enough of them to fill two thousand freighters."

This strange human with the trunk had the gift of understanding immediately what was needed. While reflecting on the Hatk's ability, Tom was struck by a doubt. "Twenty-six, communication on an encoded channel," he ordered, and then said in a low voice to Susan "if I am not wrong that guy is the biggest son of a gun with a trunk I have ever seen"

"With a trunk and without, if I rightly understand what you suspect," Susan answered.

"Congratulations, Captain Hatk. What you have brought will make a difference to the defense of this system. But now you have to explain one thing to me: what exactly did you tell the commander of Telhra'ar base? In the messages I sent I clearly stated that I was general commander of all the fleets of the fifth sector, whereas in the message coming from Telhra'ar, the base commander understood that I was general commander of all the fleets of the Confederation. You know something about that, don't you?" he concluded, unable to hold back a smile.

"Me? I just reported what you told me to report," Hatk answered. He paused briefly, then, seeing that Tom didn't answer, added: "You see, Admiral, we from Wetkloss cannot speak any of the languages of the Confederation and we need computers to translate our words. My computer perhaps introduced some confusion in its translation. You know, Admiral, these computers are no longer as good as they once were, sometimes they make errors. Do you think I ought to send it to maintenance?" he concluded with a strange air of remorse.

"If you believe that she is able to control your ship in battle, there is no point in wasting time. Tell her to be more careful with languages. And above all tell her not to try to foretell the future," Tom concluded.

"I was right. I have never seen a son of a gun like him," he told Susan as soon as the communication ceased.

"I think I have seen one on this ship," she answered. "However he is a valuable subordinate: when he thought he had to do a dirty job like lying to strengthen the position of someone whom he considered to be his commander, he didn't inform you, so that if anything went wrong you could not be accused of anything," she said.

Without that 'error' by Hatk's computer, the commander at Telhra'ar would have waited for a direct order from Qatlaassari, wasting at least two weeks. *If Terkar is saved, part of the merit will be down to that sly guy with the trunk*, Tom decided. Then, after re-establishing the connection, he went on "Captain Hatk of Wetkloss, from this moment you are to assume the rank and functions of Admiral of Starfleet. Take command of the squadron you brought into this system, complete its training, and mine the hyperspace gates in accordance with the plan my computer will transmit to you."

The first Qhrun ships arrived after exactly five weeks. The raids performed by Ertlaq and Terkr had delayed the invasion by four days. Four days meant more time for training the crews and for the arrival of other cruisers.

The last three hundred hours were used to deploy the fleet. Eventually, they had brought two hundred and twenty thousand cruisers into the system and, even with the powerful computers available, it was not an easy task to coordinate the movements of such a large number of ships. The first group of more of eighty thousand ships, under Ertlaq's orders, was deployed in a thin disk formation, not too far from the hyperspace gate from which, according to the simulations, the invasion fleet was due to emerge.

Terkr, with another seventy thousand ships, remained close to the hyperspace gate in a nearby system, ready to enter behind the invasion fleet. Hatk with seventy thousand cruisers was stationed in a region of the system at equal distance from the other two hyperspace gates in the system, so that he could quickly reach the combat zone, in case their simulations turned out to be wrong, and the Qhruns entered through a different gate. The other seventy thousand cruisers were lined up in an annular zone around the disk formed by the first group.

As usual, when the intruders began to enter the system there was an instant of surprise. The first twenty appeared in front of Ertlaq's ships, only to explode after a fraction of a second beneath the fire of hundreds of cruisers. Those entering immediately after had a similar fate, but, as foreseen, soon the rate of arrival was such that a group of Qhrun ships started to form and become organized in close formation. Soon they started to return fire.

The center of the disk started to step backwards, taking itself to the limits of the range of the Qhruns' weapons. In this way damage was reduced, but at the same time the effectiveness of their own batteries was curtailed and the rate at which Qhrun ships were destroyed decreased. At that point, they activated the torpedoes located at the gate. and which had remained inactive until then. A considerable number of reddish flashes signaled the success of that maneuver.

The attack by the torpedoes lasted about ten minutes. At the end of it, of the roughly six thousand ships that had entered the system, only less than half were still there. The initial action had the success they had anticipated, but the ships that had entered the system were reorganizing in close formation. Ertlaq's ships continued to withdraw backwards and to close around the Qhrun ships forming an external shell around the invasion fleet.

Finally the arrivals stopped and the group began to accelerate towards the inner system. "One hundred and forty thousand six hundred and six, plus or minus ten," the computer proclaimed. "About twenty thousand destroyed," it continued, displaying detailed statistics on a screen.

Shortly after that, Terkr's group entered the system. Hatk's group, would not be able to arrive for seven hours. In the location at which they had been deployed to watch the other gates, they would still need three hours to realize that the Qhrun fleet had indeed entered through the anticipated gate, and another four to get to the front line.

Ten huge supply ships and a group of freighters emerged from the hyperspace gate immediately after Terkr's ships. They started recovering the damaged ships and the crews who had bailed out into space.

For now, losses were quite low: remaining at the limits of the range of Qhruns' batteries and concentrating fire on one ship at a time was a good tactic from one point of view. But, as Tom immediately realized, an absolutely inappropriate one.

"At what rate are we destroying those ships?" Tom asked the computer.

"About one per second, admiral," she answered. *We don't need a computer to understand that something has to be done.* The figure of one ship per second, although apparently enormous, led to a residual strength at Terkar of about seventy thousand ships: more than enough to penetrate the planetary defenses.

"Twenty-six, order the ships that are in front of the invasion fleet to slow down; we need to try to slow them down," he ordered.

About ten seconds later the computer confirmed the execution of the order, but added "I don't believe they will slow down."

"Neither do I", Tom answered. Then added: "Zoom in on the zone in front of the invasion fleet, and put it on the tactical screen." The Qhrun ships had not slowed down and the distance between the two fleets was quickly reducing. The number of ships exploding started to increase dangerously and, what was worse, the greater power of the Qhrun batteries caused much greater losses among the Confederation fleet.

"Time to direct contact?" Tom asked.

"Ten minutes," the computer answered. In ten minutes the two fleets would be in the same zone of space and at that approach speed the cruisers would be exposed to an unbearable volume of fire. Either the Qhruns were well aware of their superiority or this time their inability to modify their own plans in response to the behavior of their opponents was acting in their favor.

"Order all ships to return to their initial positions, at a safe distance," Tom ordered the computer. "And put me in communication with Admiral Ertlaq," he added.

The three-dimensional image of Ertlaq immediately appeared. "Admiral Taylor, if we keep on destroying them at this rate we won't prevent them from reaching Terkar with a fleet large enough to make any attempt to defend the planet utterly useless," he said in a very worried voice.

"I've seen that, but on the other hand moving closer doesn't solve the problem. We would get too many losses," Tom answered.

"Yet we must do something. I don't believe that the arrival of Hatk's ships will change the situation," the Tteroth continued.

"There is still hope. there are three asteroids with outposts in this system as well as a sparsely inhabited planet. If they follow their usual way of operating, the invasion fleet should split to occupy all the bodies that may be inhabited. In such a case the external surface of their formation would increase and we would destroy them more quickly", Tom answered.

"Since the asteroids have been evacuated, we can focus on just the group homing in on Terkar. When should the first separation occur?"

"In almost exactly two hours and thirty-two minutes." Susan, who was continuing to simulate the behavior of the Qhrun fleet, answered.

During the following two and a half hours nothing significant happened; the destruction of the Qhrun ships proceeded slowly and, even though their own losses were quite low, no new fact that might foreshadow an improvement in the situation took place. When the time for the Qhrun fleet to divide arrived, everyone started giving signs of nervousness. "I don't understand," Susan exclaimed, "by now a fairly large group of ships should have split off and moved towards the outer asteroids. Even if there is nobody there, computers are simulating normal activity and they shouldn't be able to realize that it's only a trick."

"Nobody would realize it," Tom answered. "If within a few minutes that fleet doesn't split, we must conclude that they are adapting their tactics to our presence, and that we are in trouble, serious trouble indeed."

"Do you realize that, if they are changing their strategy to allow for our behavior, the basic hypotheses on which we have based our actions will prove to be false?" Tom thought that this frightened Susan more than the Qhrun fleet that was in front of them.

"After all this is the first time that one of your simulations had failed, and it doesn't seem absurd that the cause can be traced to a problem with one of our basic assumptions," Tom answered. He was rather upset by that too, despite the reassuring tone he had used for those words. From when they had entered the first sector this was the third time that the Qhruns had done something contrary to the hypothesis that their behavior was invariable. The Qhruns were clearly more complex than they had expected. They had to deal with this new fact as soon as possible; not in the midst of a battle, which threatened to turn into a serious defeat.

After another quarter of an hour the image of Ertlaq appeared again: "Admiral, what can we do? We cannot keep on following them in this way until they are on the surface of Terkar."

"We must try to break up their formation. Their strength is in its compactness: we must force them to divide into two groups," Tom answered, realizing that it was useless to reiterate an idea that was not new at all.

"We should try to penetrate within their formation, but as soon as our ships break formation, they will be surrounded on all sides," the Tteroth answered.

"I know, but if we don't try we will never stop them," Tom answered. He started to wonder if the Coordinator was right and if the whole attempt had been foolish. "I will try to penetrate their formation with a group of forty thousand cruisers. Concentrate your fire in the zone that we will aim at. The difference of speed should allow us to engage them without too many losses."

"No, Admiral. That's too dangerous: the fleet cannot remain without Admiral Taylor. I can go: to loose a few Tteroths won't make much difference, particularly in this sector."

What Ertlaq was saying had its logic, but how could he let him do that? "Admiral Ertlaq, it is my task to lead the fleet even if it may be dangerous, as it is in this case," Tom answered, without being very convinced by what he was saying.

"I have explicit orders from Quanslyaq to prevent you from placing yourself in a dangerous situation. Please, allow me to remind you that your main duty is to survive to lead the fleet in what will be a long and difficult war." Tom wondered whether the Tteroth was bluffing, or the Iktlah had actually put Ertlaq close to him as a kind of guardian angel. Ertlaq's tone was determined: after thousands of years, the traditions of the military aristocracy that had governed Tteroths from time immemorial were beginning to be roused. But in the Ertlaq's speech there was an implicit reproach; Tom had specific duties with regard to the Confederation and he could not afford to take unreasonable personal risks. "Thank you, Ertlaq. Organize a group of forty thousand cruisers and start as soon as you are ready," Tom answered.

Immediately after the Tteroth had broken the connection, the image of Terkr appeared. "Admiral, I wonder whether a similar action from the back of their formation might be useful. I could try to enter with another forty thousand ships and get inside their formation."

Tom realized that he had talked with Ertlaq on an open channel, clearly an inadvisable action. However, the idea was good: operating from both ends of the formation, it would be easier to cut the invasion fleet into two, or at least to disorganize it. It was extremely dangerous, even more dangerous than attacking the fleet head on, because the difference of speed would be smaller, and undoubtedly Terkr realized that. "Instruct the other ships to cover you adequately. Synchronize your actions with Ertlaq."

It took another thirty minutes to mine the zone where the Ertlaq's group would encounter the Qhrun fleet with torpedoes and transponders, and to organize a major launch of torpedoes to cover Terkr's group. Then the two groups moved at top speed towards the two ends of the invasion fleet.

Tom contacted Terkr first: "Good luck Admiral. And above all remember that, as I told you on Gorkh'ar, you have nothing to demonstrate to anybody." Terkr thanked him and closed the channel. Immediately afterwards Tom called Ertlaq. "Admiral Ertlaq, may air sustain your wings," Tom said using the solemn tone of the Tteroths. "And above all good luck," he concluded. "May the draft of the cliff be favorable to you, Admiral," the Tteroth answered, as he closed the communication channel.

The intense fire forced the Qhruns to maintain their shields energized, as the two groups of cruisers rapidly closed on the ends of the formation. All of a sudden the torpedoes started to activate, and on the tactical screen the flares showing the explosion of Qhrun ships became more frequent. A cavity, caused by the destruction of ships but also to an outward movement of nearby units, started to appear at the front of the formation. Ertlaq's ships entered this cavity and began to lower their shields one after another to operate their batteries.

In a similar way, even if more slowly because of the smaller relative speed, Terkr's group came in contact with the rear part of the Qhrun fleet.

Tom was watching the tactical screen in an extremely worried mood. The two groups were slowly advancing towards the interior of the formation and the rate at which Qhrun ships were blowing up had notably increased. But the two groups of cruisers were suffering heavy losses. If they continued in the same way they would quickly be destroyed.

Tom again called Ertlaq and Terkr. The transmission was distorted and from the images it was clear that both cruisers had suffered considerable damage. "Get out of there: if you go on you will become isolated inside their formation"

"Soon our two groups will combine inside their formation. We will then deploy in the form of a sphere to offer the least surface area and we should be able to survive in that position for long enough to destroy a good number of ships," the Tteroth answered. Likewise, Terkr also insisted on continuing their action. "At this point, it would be as dangerous to withdraw as to go on, since we would be under heavy fire for a much longer time," he concluded.

"Twenty-six, order all ships to intensify fire and to close in on them," ordered Tom. However, the whole fleet was already doing all that could be done and there was no way of intensifying the attack, and Hatk's group would not arrive in less that than three hours. He wondered whether Ertlaq and Terkr could survive for so long.

On the tactical screen it was now possible to see a central sphere made by the cruisers that penetrated inside the Qhrun formation, menaced from all sides by the invasion fleet: The ships on the outer surface of the sphere were subjected to an unbearable fire, and soon they started exploding.

After an hour only half of the eighty thousand ships was still there and contact with Ertlaq and Terkr had been lost some minutes ago. The diameter of the sphere continued to shrink and by another one hour and a half the remaining ships had also been destroyed. Tom was contemplating on the tactical screen what increasingly looked like a major defeat, even if it was not yet possible to evaluate their overall loss. The emergency transmitters on space suits were set to switch on at a distance greater than two hundred thousand kilometers from Qhrun ships, to avoid attracting their attention, because of the tendency of the Qhruns to open fire on any wreckage that might be inhabited. Only a few of the people who bailed out had already given signs of life and many of those had already been recovered by the freighters stationed around the fighting zone. Tom didn't have many illusions: they had already lost more than ninety thousand ships and he would regard it as a lucky outcome if the number of losses was limited to a few hundred thousand people.

Hatk's group finally arrived, and attacked the outside of the Qhrun formation. The rate at which Qhrun ships were being destroyed increased, but it was not high enough to guarantee the system's future. Susan performed a number of numerical simulations, which gave consistent results: from ten to twenty thousand intruders would reach the surface of Terkar. That was only one sixth of the original invasion fleet, but still more than enough to occupy the planet. And by now there was no hope that the Qhrun fleet would split and aim for other objectives.

"Admiral Hatk," Tom called him, "collect your ships at the front of the Qhrun formation. I will try to split their fleet using all other ships. You must cover us."

This was nothing other than a replay of the maneuver performed earlier by Ertlaq but now that the number of Qhrun ships was lower perhaps they had some hope of success. In any case he couldn't think anything better. Hatk didn't make any comment and prepared to execute his order.

Again it took more than half an hour to prepare the attack. When everything was ready, fifty thousand cruisers began closing at top speed, while all the others concentrated their fire on the front of the formation to force the Qhruns to keep their shields up. The intensity of fire was such that many Qhrun ships started to explode. Tom couldn't take his eyes from the forward screen; now the individual Qhrun ships could be distinguished clearly, and they were rapidly getting bigger and bigger. Suddenly the ship that was in front of them exploded in a burst of fire. Then the torpedoes activated, and many ships that had withstood the first wave that exploded against their shields, were destroyed by the following ones.

Tom led his ships into the cavity that was forming in the forward part of the Qhrun formation, and soon a wedge of cruisers began to penetrate into the invasion fleet. The computers succeeded in synchronizing the fire of the batteries, concentrating it on a few ships at a time, as a result of this maneuver involving the shields, the Qhrun fleet opened out in front of them with a rapidity that Tom had not foreseen. "Twenty-six, order all ships not to slow down. Instead they must impact the shields of the Qhrun ships with their own. Check that everybody has his space suit on," he ordered.

The maneuver had taken on the nature of a cavalry charge: the speed and perhaps also surprise was enabling them to penetrate into the Qhrun fleet. There was however no reason to delude themselves: Ertlaq's and Terkr's attacks had also initially developed in a similar way a few hours earlier, but then they had turned out to be failures.

The fifty thousand cruisers had by now penetrated deeply among the Qhrun ships and started to suffer heavy losses. The whole of space around them was luminescent from to the volleys that were striking the shields and the glares of explosions. "Admiral, the shield generators are almost saturated," the computer warned them. An instant later the Qhrun ship that was holding them under fire suffered an explosion: she had kept her shields lowered for too long, attempting to completely saturate their generators.

"Twenty-six, try to strike that ship before she raises her shields again," he ordered.

The computer executed the order, concentrating the fire of all batteries but the Qhrun ship moved her aim, hitting them in turn. Tom felt a strong shock, enough to have thrown him to the deck if had he not been buckled to his seat, and then he heard the noise of air escaping from openings in the hull. The bangs of the airtight doors closing followed. "The mess hall and accommodation are depressurized. Marginal damage to the rest of the ship," the computer announced.

"Concentrate on that ship," Tom repeated the computer. "Everybody get ready to bail out, in case of further damage," he concluded. As he had foreseen, a few seconds later the Qhrun ship exploded completely sending some fragments that struck noisily against the cruiser's hull.

The flash of the explosion had hardly died away when they heard the voice of the computer: "Admiral, the Qhrun ships are retreating."

Tom moved his eyes from the forward screen to the tactical screen. Indeed the ships around him were slowly receding. He remained as if paralyzed for a few seconds, observing the screen but the situation quickly began to change. Many more Qhrun ships were now moving outwards, abandoning the close formation that they had maintained up to that moment. The outward motion quickly extended to the whole invasion fleet.

"Twenty-six, all ships to start pursuit. No Qhrun ship must get near the hyperspace gates." What initially had been a slow motion towards the outside of the formation, quickly turned into a rout. The huge intruders raced away in all directions, rapidly closing with the cruisers that surrounded the formation, and chased by those that had penetrated the interior. "Susan, what is happening? Did your simulations foresee anything of this sort?" asked Tom, who couldn't believe what he was seeing.

"Absolutely not. I don't understand what is happening, but we are witnessing the repetition of what happened in the battle at Aq. There we thought that the cause was the field of asteroids and the fact that the Qhrun fleet was close to being defeated. But there are no asteroids here and, if they had remained close to one another, they would have arrived at Terkar in sufficient numbers to occupy the planet. I don't understand. . ."

"For now, I'm not worried about understanding," Tom interrupted her. "What matters to me is that they are running away and that we have saved Terkar."

By now the situation on the tactical viewer had completely changed. The Qhrun ships were running away in all the directions, chased by groups of cruisers that destroyed them one by one. *If things go on like this, within less than an hour everything will be over,* Tom thought. *And Terkar is safe,* he decided, without really believing what he was seeing.

"Admiral, the Confederation is deeply indebted with you and your people. Forever. Your courage will be remembered for as long as human species exist in this galaxy." The three-dimensional image of Qatlaassari dissolved.

For a few minutes nobody spoke, then Tom broke the silence: "Always the same: what did Churchill say: Never, did so many owe so much to so few? Well, something of the kind, anyway. But if our Coordinator had not been so stubborn and had left Terkar when it was still possible, all those spacers would still be alive."

"It must be tiredness that makes you speak that way," Susan interrupted him. "Certainly we have lost so many people, but we couldn't expect to stop them easily. And then it is not his fault: you decided to stop them in this system. Instead he insisted on not trying a defense that he believed impossible."

It was bitterness for such heavy losses rather than tiredness. Susan was right: the responsibility of what had happened in the system was his and his alone. "I know. What I cannot forget is that we have been so close to a defeat that would have frustrated everything we have done up to now. You heard Quatlassari: he said that the moment he saw Ertlaq and Terkr charge the Qhruns in that way he was certain of victory. I was only certain that we wouldn't see them again."

"I am sorry to interrupt you," Ashkahan said, "but I don't believe that this is a moment for sadness. Certainly, we have lost almost a million people, but for every spacer who died in this system we have saved more than one thousand civilians. All of us knew the risks we faced in trying to stop a Qhrun fleet this way; we did it and we will do it again every time you order us to do so. And it is not just a question of numbers: today Starfleet has regained its dignity; today the whole galaxy knows that it is possible to challenge the Qhruns in space and to overcome them."

"Thank you, Ashkahan," Tom replied, thinking that the Aswaqat was right. That day was a truly historic day, perhaps not so much because Terkar was safe but because the fleet had finally shown that the Qhruns were not an invincible force of nature. They were an enemy against which one could win. "Twenty-six, can you please bring us something to drink here on the bridge? Before it is our turn to rendezvous with a supply ship to get the damage to our cruiser repaired, we will have to wait for at least twenty hours."

Immediately a panel of a control console slid sideways and ten large glasses appeared. "Admiral, Ashkahan is right," the computer said. "Until a few hours ago the crews and the ships of Starfleet were demotivated, and felt guilty: all of us felt that our inability to stop the Qhruns was our personal fault. And, if I may say so, we machines suffered more than humans: most of us have lived through the whole invasion over the last four hundred years, and the feeling of impotence, due to the impossibility of doing what we have been built to do, created an infinite number of problems for us."

Despite his tiredness, Tom couldn't hold back a smile at hearing a machine speaking like that of her psychological problems. But, he agreed that there was really little to laugh about: the sole assignment given to the starships in the Fleet was to maintain the conditions for human civilization to flourish in the galaxy. Not being able to perform this task might well in the long run jeopardize the delicate equilibrium on which the operation of such complex computers was based. "All that is now over, Twenty-six. In this system Starfleet demonstrated that it is still the extreme defense of civilization, as it has been for eighty thousand years," Tom answered, getting up, to go to pick up his glass.

"Thanks to you, Admiral," the computer went on. "Nobody, human or machine, will forget it."

As regards that Tom had some doubts. *They say that today, but at the first error I make. . .* he thought. Tom stretched out on his seat: with all the work the supply ships had to do with cruisers that were in really bad condition, he didn't want to insist on having the depressurized zone of his ship repaired before their turn came. Everyone had now settled down to sleep in their seats, except for Susan, who continued to talk in a low voice with the computer and was displaying images of the recent battle.

"Tom, there is something here that might be important," she told him without moving her eyes from the screen.

"Are you trying to understand the reason we won a battle that seemed lost?" Tom asked rising and going towards her.

"More or less. We have discovered that the ships didn't retreat at the same time: it is as if something moving at the speed of light caused them some problem that they couldn't solve," she answered.

The Qhrun fleet appeared on the screen as it was a few seconds before the rout. Tom immediately recognized the situation, even though it was seen from a viewpoint outside his ship. "Susan, this reconstruction is based on data from many different computers, isn't it?"

"Certainly. It took a lot of time to integrate all this information."

Tom tried to imagine the huge amount of computation that simulation had required. He sat on an armrest of Susan's seat and, without moving his eyes from the screen, put an arm around her shoulders. "What is the level of confidence in these results?"

"We discarded any result with a certainty of less than ninety-nine point ninety-nine percent. The spatial errors are of the order of a few millimeters," Susan answered, anticipating the voice of the computer.

Tom signed to her to go on. He saw clearly the instant when their cruiser was hit and, just a moment later, the explosion of the Qhrun ship. The image was so precise that they could see the fragments hitting their hull. Just a fraction of a second later the Qhrun ship closest to them started a slow rotation around her axis, as if it were out of control. Shortly afterwards the ship changed its course to one of retreat.

"Now look. Until that ship exploded everything was normal," Susan said. Then turning to the computer: "Display a sphere expanding at the speed of light, starting from the point and from the instant of the explosion."

The scene ran back, to a few instants before the explosion, and started again. At the instant the Qhrun ship exploded, a bright yellow dot appeared on the screen, which began to expand as a sphere with an increasing radius. When it touched the nearest ship, she began her flight from the theater of battle. The sphere continued to expand, touching another ship that again began to retreat. The third ship touched by the yellow sphere suddenly yawed to the left, moved sideways and hit her bow into the closest Qhrun ship, which had started a similar maneuver to the right. Both the ships had their shields down, because all the available power was being used by the batteries, and both immediately exploded.

"Go back, Twenty-six," Tom said, "and show the last two seconds again at slow speed. Zoom in on those two ships." Now the two ships filled the whole screen. The rotation of the first ship was irregular and all of a sudden, the second also started to move, apparently out of control. Then, an instant before the collision, there was an emergency maneuver and both ships' attitude control systems, now perfectly coordinated, tried to prevent the collision. However, it was too late, and the explosion lit up the screen.

"They are Qhruns, but it is still painful to see two ships crashing like that," Tom commented quietly. "Whatever happened, it is clear that the maneuvers of those two ships were not coordinated; each one responded in an independent way to an external stimulus, without communicating with the others. When they realized they were on a collision course, they tried to react, but it was too late."

"I had the same feeling, too. I asked Twenty-six to compare the speed of response of every ship to maneuvers by the others with that typical of Qhruns in similar situations. The answer is that actually the two computers reacted too slowly. Twenty-six says that it is as if they were reacting to contradictory orders, and that only at the last moment some emergency device kicked in. If that was a Confederation starship that would perhaps be a reasonable explanation, but in this situation I don't know." Susan paused, and then continued: "Anyway, there was no coordination. When that yellow sphere touches a ship, she starts an evasive procedure, without exchanging information with any other ships. This is against any tactical rule, apart from being against common sense."

"Sure it is not conceivable that star ships would maneuver in an independent way, without coordination, at such a close distance. That holds for humans, Qhruns or anything else that moves through space. And we can see the results: no anti-collision system can work under those conditions."

As Tom spoke, the scene continued to unfold. It was always the same: as soon as it was touched by the sphere, every Qhrun ship started an evasive maneuver, with it soon becoming an uncoordinated escape. Some ships collided with one another, but many more were blown up by Confederation cruisers. In some cases, where the two fleets were very close, a Qhrun ship, in its escape, rammed a cruiser.

"It looks as if, an instant before exploding, the ship we hit transmitted an order, and that as soon as the various ships received it, they began to fly. I needn't ask you if you have looked for any messages in the recordings."

"Certainly," Susan interrupted him. "But we have found nothing. And then I don't believe that ship had time to broadcast any messages."

"Then it is not a message. The other ones saw that ship exploding, and. . ."

"No, that doesn't work, too." Susan's voice was tired, now. It was clear that she had found a key to interpret what had happened, but that key didn't seem to open any door. "Twenty-six, display the scene of the explosion from the point of view we studied last."

The scene changed abruptly: the point of view now was the position of one of the outermost Qhrun ships. The view towards the zone where fighting was more intense was completely obstructed by a wall of ships. "This is the instant when our hypothetical sphere reaches the ship," Susan said all of a sudden. "And when it begins an evasive maneuver to break away from the fight." It was clear that the ship had no optical contact with the ship that was at the center of that phenomenon. After more than fifteen seconds one of the Qhrun ships could be seen to disengage. "The first optical contact that would have allowed them to understand what was happening occurred when that ship had already been retreating for seventeen seconds. If it is a message, it cannot be optical. Not even electromagnetic waves with a short wavelength. It is either radio waves, or another unknown form of communication," Susan concluded.

"Well. Only a radio message could have that effect. But no radio message was broadcast or, at any rate, none could be broadcast. As usual with the Qhruns, every time it seems that a mystery is about to be solved, another worse mystery arises," Tom concluded. "And then," he went on after a short break, "we destroyed thousands of ships, but only one produced that effect. Did it have a special place in their formation? Is there any particular geometrical symmetry giving a meaning to her position in the formation?"

"We have also looked at that for ages, but we found no distinctive signs. It was a ship like all the others, without any particular signs and without any particular position. But that might perhaps mean nothing and it doesn't exclude the possibility that she was the flagship of the invasion fleet. Similarly, our ship doesn't maintain a particular place in the formation and doesn't have any particular insignia."

"Do you remember the emir Auda, in *South of the Heart*? He didn't wear his green turban and mixed with his men so as not to attract enemy fire. But that doesn't change the problem: what caused them to run away?" He waited a few minutes more and then, seeing nobody had anything to add, said: "You've done a great job. We now have to get some sleep, and tomorrow the ship will be repaired. Perhaps some new idea will come to us after a good sleep," he concluded, running a hand through Susan's hair.

He suddenly got up, and turning bumped into Ashkahan who was standing behind him. Then he realized that the whole crew was standing in silence, observing the screen. He held up his arms in a very earthly gesture, without worrying too much about the possibility that those present would understand what he meant.

He started to move towards his seat, and then suddenly turned back, rushing towards Susan and almost crashing into Ashkahan again. He fell down onto the armrest of Susan's chair "Just a minute. Perhaps we are looking at the wrong thing. Twenty-six, I want a complete analysis of everything that ship broadcast in the last thirty seconds before exploding. On frequencies low enough to be received by ships that were not in a direct line of sight."

"Admiral, there are the usual incomprehensible messages that Qhrun ships exchange when they are in large formations."

"I'm taking that for granted. Now choose ten ships at random. Do a spectral analysis of everything they broadcast in those thirty seconds and subtract the spectra of the transmissions from the ship that we are studying." Tom's voice was very excited now and he no longer looked tired.

"Use the same thirty seconds for all the ships, and look for correlations between the phases that may suggest re-transmissions of the same message," Susan added. "If I understood you properly, we must discriminate between original and re-broadcast messages."

"But of course, even if I exclude the possibility of re-broadcasting of the same message: we are dealing with a spherical, not a poly-centric, diffusion." The idea that Susan understood what he intended to look for encouraged Tom. It was an indication that it was at least reasonable.

"An analysis of that kind will take several minutes", the computer answered.

"It doesn't matter. We now have all the time we need: The Qhruns are no more running behind us and, if we understand what happened, perhaps they won't anymore," Tom concluded with what was undoubtedly a sense of optimism, albeit with little firm foundation. Time slowly passed in a general silence; the computer was working and she didn't want to give them partial results before reaching a degree of certainty.

About ten minutes later the computer spoke again: "Actually there is a transmission typical only of the ship we destroyed. It is a short wave carrier, modulated in frequency. Shall I put it on the loudspeakers?"

When Tom answered in the affirmative, a sequence of rustling sounds was heard on the bridge. The frequency of the sound varied continuously, it was almost like a slow dirge. In addition, the intensity of the sound went up and down without any break. Suddenly the sound ceased. "This is the instant when the ship exploded," the computer stated.

"Then the only hypothesis is that this signal is essential to maintain the Qhrun fleet's efficiency and that when the signal stopped the fleet suddenly broke up. It could be a signal that synchronizes their computers, or regulates the attitude control systems of the ships. But it might be something strictly linked to the psychology of the Qhruns."

Tom stopped and immediately Susan went on: "If that is so, there is an experiment we should do as soon as possible. We should have our clone, on Gorkh'ar, listen to it and see how it reacts. If that thing is indeed a Qhrun, if the reactions to that rustling sound are something innate and not acquired with training that our specimen has never had and if. . . No, there are endless hypotheses, but it is perhaps worth a try."

"Certainly, and in the meantime there is a job for Twenty-six. Analyze all the transmissions from Qhrun ships and look for signals of the same type. Particularly look for others ships transmitting stuff of the same kind and check whether similar transmissions are broadcast by the groups that block hyper-space gates. It is strange that in four hundred and fifty years of attempts at communicating with Qhruns nobody has ever came upon a transmission of that kind."

"And if you can spare some time, check whether the rout in the Aq system was caused by the interruption of a transmission of the same type," Susan added.

24 Qhrun

The grey sphere of Ytl was suspended in the darkness of space in front of them. The ship was entering orbit and they planned to reach Quanslyaq within an hour in the old royal palace on the cliff. Tom had decided to stop on the Tteroths' planet on their way to Gorkh'ar to give a first-hand account of the events in the first sector to the person he now considered more as a friend that as a wise old man.

More than six months had passed since the battle of Terkar, and the situation in the first sector was returning to normal. Tom had given the command of the first sector's fleets to Hatk, who had started to apply the Scorched Planets tactic. With the help of a group of Merchants he had begun the evacuation of two planets in the paths of the other two fleets that were converging on Terkar and, with some luck, the capital of the Confederation would definitely be saved without risking other battles. The corridor connecting the fifth sector with the first was slowly being widened with sporadic raids in nearby systems. Groups of ships operating beyond the frontier kept on destroying dozens of Qhrun ships and slowing down the invasion. In the meantime another three systems in the fifth sector had been evacuated, and in every case the invasion fleets had been completely destroyed, together with the planet. It seemed that Operation Scorched Planets continued to be successful and that the Qhruns didn't learn from experience.

They analyzed all transmissions from Qhrun ships both in occupied systems and in the invasion fleets, discovering that the particular transmission, whose absence seemed to disable large Qhrun fleets, was not present in small groups of ships like those blocking hyperspace gates. About one ship in every twenty thousand was broadcasting transmissions of a similar type, which however were never exactly identical. A detailed study of the recordings of the battle at Aq allowed them to identify three ships emitting similar transmissions, and to ascertain that the rout followed the destruction of the last of these, however they had made no progress in understanding why this had happened.

Quanslyaq waited for them in the usual room, on the platform he had been using for years. After a few words of greeting—unusually few for a Tteroth—Tom immediately came to the point: "I'm very sorry for the loss of Ertlaq. Ever since you sent me his group, I have felt responsible for him and the other Tteroths. And now practically none of them is alive."

"You must not feel responsible for what happened. Ertlaq has simply done his duty: that group was sent to protect you. I heard what he told you during the battle of Terkar, and Ertlaq was perfectly right. All of us are perhaps needed but surely none of us is essential, whereas at present you are; Ertlaq behaved just as we expected he would." While he was speaking the tone of Quanslyaq became much more steady and resolute.

"More than as an Iktlah you spoke like a member of the old Tteroth aristocracy. You remind me the generals in *The flight of princess Hethl*," Tom answered.

"Perhaps you are right, but these are times when the ancient virtues of our people are much more suitable than the wisdom of the Iktlahs, who, by the way, were initially warriors rather than monks. And surely people like Ertlaq are the most precious asset. Here he has become a national hero; you don't know how many volunteers are eager to join Starfleet. They would like to create a special Tteroth guard to protect Starfleet's general commander. I think it's a good idea. Vertearis also had a special guard."

Tom quickly considered the proposal. It did, after all, seem to be an acceptable idea, mainly because it came from Quanslyaq, who was above any suspicion of trying to exploit the situation to get any advantage for his species. He said as much to the Tteroth.

They spoke for a couple of hours, because Quanslyaq wanted to learn the details of what they had discovered about the behavior of the Qhruns. As Quanslyaq pointed out, their empirical approach had been successful in finding a way to fight them more efficiently, or at least, in limiting the damage, while was much less successful in explaining what was happening and why. But, concluded the Tteroth, it was possible that by accumulating empirical knowledge, all of a sudden the fog preventing them from understanding what the Qhruns actually were, might evaporate.

At the end of the meeting, they quickly returned to the ship and immediately left for Gorkh'ar. It was little more than four days in hyperspace, but Tom found it like eternity. When the cruiser finally entered the base, there was an enormous crowd waiting for them. In the last ten months the base had changed greatly: the main cavern had been widened and work was still going on. Some time was spent in welcome ceremonies, and Tom had to give a public account on what had happened in the first sector, and particularly about Terkr's deeds, because he had turned out to be Gorkh'ar's national hero. Nobody seemed to remember the years of his bureaucratic and conservative management of the base, a past that, to be honest, could not be considered only as his guilt.

Around noon, Tom was finally able to return on board where Sinqwahan, who had spent the intervening time with the crisis committee, was waiting for him. "I think that Idlath needs to speak to you," he told him. "He asked me to accompany you to the lab where they are waiting. They are really in a hurry, it seems they have got some results from your clone."

They left immediately. Tom and Susan were eager to test the thousand hypotheses they had formulated in the months after the battle of Terkar about the mysterious transmission, even if they didn't know exactly what to do. They would try to have that ten-legged monster listen to the transmission, but they didn't know how to convert that radio transmission into a message that could be understood by a species about which they knew nothing; not even whether the transmission was actually meant for them.

They had just started towards the lab, when the Aswaqat said: "Idlath had to fight to prevent the lab from being dismantled. The new base commander thought that you had left orders for all the asteroid's resources to be used to build new ships and to improve the efficiency of the base. He wanted to use that area to expand the yards. Idlath said that you had given orders to expand the lab but you hadn't left anything in writing because it was a secret lab. After quarreling for days, they sent a hyperspace probe to Aintlad, to get his interpretation of your wishes." It was clear that he found the whole thing amusing.

We have got to the point where people quarrel over interpreting my wishes, thought Tom, who didn't find it amusing at all. His expression didn't go unnoticed by Susan who commented in a whisper "don't worry, it always happens to prophets."

Tom had difficulty in restraining from cursing and, seeing that they were already at the entrance to the gallery, didn't say anything. The gallery was much shorter than previously and led to a rather large cavern, which Tom didn't remember at all. It had clearly been dug recently, showing that finally Idlath had prevailed and the lab had been enlarged.

The Opsquat was waiting for them at the entrance to the room with the other members of the crisis committee. He immediately started to speak, even before they could sit down on the force fields. He gave them a detailed description of the situation. More than twenty creatures had been cloned, all of the same type as the one Tom had seen ten months earlier, and some had already left their cocoons with their legs and large claws perfectly developed. "None of them has however shown any signs of being intelligent", Idlath concluded.

"It seems that those creatures are all precisely the same," Tom commented, looking at the images a three-dimensional projector was displaying in the center of the room.

"Yes, and that means that our cloning worked properly: we started from biological material that came from fragments of ships destroyed in different systems, and we treated it following different procedures. If the results are identical it means that what we have got are representatives of that species", the Opsquat answered.

"And so the hypothesis that the lack of intelligence is due to damage caused by the cloning procedures can be dismissed; it is unlikely that any damage would always involve the same parts and only those," Susan said, thinking out loud.

"Provided there is no systematic error. . ." Tom commented, with a tone that meant that he didn't really believe that hypothesis.

"We cloned both males and females, and we identified the gene controlling the sex of the creature. Sexual dimorphism is minimal—you almost wonder how they can tell," Idlath went on. Another hypothesis gone. Tom didn't know whether to be pleased or not: if they continued like that, just one hypothesis would remain. However, it was more likely they would discard all of them, remaining with nothing.

"Did you put them together or did you keep them separate?" Susan asked.

"We tried to put them together, but it would seem that they don't communicate with one another. Or, at least, they don't communicate any more than animals do, and furthermore, not even like social animals." The hypothesis that the lack of any intelligent behavior was caused by the isolation of the first specimen was also disproved.

"Did you try to make them reproduce?" asked Tom, who was scared by the idea that those monsters might proliferate in an uncontrolled fashion.

"Yea," Idlath said, rather uncertainly. "But the females are all sterile", he concluded with an air of remorse, almost as if it were his fault.

"Sterile?" Susan exclaimed. Even with very high fertility, it would be difficult to explain how Qhruns could expand throughout the galaxy at the rate that they had shown. "Are you sure they are really sterile?"

"Absolutely," the Opsquat answered. "One of us even believes that he has identified the gene that induces that kind of sterility. We have tried to manipulate it, but it is too early to know the results. The specimen we used is still in the immature stage."

"Do you think that the idea of cyborgs or of an artificial species might explain that?" Tom asked Susan.

"Perhaps. . . Unless they induce genetic sterility in the individuals that will be put onto starships. If there is a polymorphism based on specialization, they might do that to avoid filling starships with eggs," she answered.

"They would then be sterile spacer Qhruns," Tom answered. But he realized that this idea didn't explain very much, because, in space, sterility might be useful, but certainly the lack of intelligence was not.

"Did you make any progress in the development of viruses?" Tom asked. He thought that everything that could be said about these specimens had already been said. As always with Qhruns, new knowledge produced more doubts that certainties.

"We have produced different strains. We waited for your authorization to proceed with experiments," Idlath answered.

This was another unknown. Owing to the slow development of the monsters, to kill some of them meant wasting months of work. On the other hand, they didn't even know whether those creatures were really Qhruns. "Perhaps it is better to wait until we know more," Tom commented.

"I perfectly agree," said the Opsquat, who didn't at all like the idea of destroying the results of all that work. "But now come, we have arranged everything for the experiment and we have only waited for you to start," he finished as he moved towards the inner part of the lab, followed by all the others.

The transparent wall separating the area, where the first clone was held, from that reserved for human beings was completely opaque and they could see the images of the monster, who was lying with all ten legs folded under its body, only thanks to the monitor screens.

"If that is an intelligent being, it looks almost pitiful," Susan said.

"In all reality, it has never given any signs of any intelligence," Tom answered. He might also have added that, if it was a Qhrun, he would have no pity. After the battle of Terkar he had found it hard to control his feelings. After all, even if it was a Qhrun, it could not be held responsible of what others of its species were doing. But could the concept of personal responsibility be applied to Qhruns? Following this reasoning, he wondered at low voice: "And then, if everything does come to an end and if what we suspect has happened on the occupied planets is confirmed, will we have to enact a kind of Nuremberg trial?"

Susan looked at him in a strange way, uncertain whether she had understood what he had said. "You mean that we should try the Iskrat-is-Thn? I believe that all those who were caught have already been exiled," she answered.

The idea of trying the Qhruns was so absurd that it was unconceivable: a completely different creature cannot be tried. It would be like trying a lion for murdering someone. *But if they are intelligent, they must be held responsible for their actions.* He realized that this reasoning wouldn't get him anywhere, so once more started to follow what was going on around him. Also because he heard Idlath's voice.

". . . order the computer to make the wall transparent," the Opsquat was saying, speaking to Yyrtl.

The Sitkr spoke in quietly for a few seconds to the computer and the wall began to become transparent. Hardly had the creature on the other side of the lab seen them, when it got up on all its legs and rushed at the transparent wall. Once again Tom found it hard to refrain from moving back, and for an instant he was scared by the idea that the wall might break.

Again they heard the appalling sound that was produced by that mass of organic tissues enclosed in bony armor as it crashed against the transparent wall. Its howling blotted out all other sounds and was really striking. Tom thought the monster was even larger than before. He asked Idlath about that.

"Yes, in the last few months it has continued to grow, now it weighs more than two tons," the Opsquat answered.

Tom was happy that the other had decided to begin the experiment with just one of these creatures. "We can start, it is useless to stay here looking at this."

Yyrtl gave a brief command to the computer controlling the lab and they heard the same rustling sound that Tom had heard for the first time when they had isolated the transmission that caused the rout of the Qhrun fleet in the battle of Terkar. As soon as it heard the transmission, the enormous bug stopped for an instant and then restarted hitting the transparent wall as if nothing had happened. That brief instant was nevertheless long enough for Tom to hope that the creature had reacted to the recorded message. The disappointment that followed was so strong that he mumbled a curse.

"For an instant, I too was hopeful," Susan said, "but it was hardly possible for it to work at the first attempt. The ways of translating that transmission into an audio signal are almost endless. . ."

Yyrtl was already instructing the computer to modify one of the parameters of the conversion algorithm used on the signal and very soon the loudspeakers emitted a rather different noise, even though it had a similar pattern. The presumed Qhrun didn't give the least sign of noticing anything.

"Idlath, do you have an idea of the range of acoustic frequencies that creature is able to perceive?" Tom asked the Opsquat.

"They are not very different from ours, slightly greater at the high-frequency end," he answered. From his tone it was clear that he, too, was quite disappointed.

They continued for more than two hours. At intervals of less than a minute they changed the parameters of the process that produced the audio signal, but it seemed that the creature on the other side of the transparent wall was not at all interested in those sounds.

"It is useless," Tom said in a more-or-less discouraged tone. "Either that creature is not a Qhrun at all, or else the signal has nothing to do with their psychology but is just a transmission from some of the ships' systems. That's a pity, it would have been nice to understand something about them after such a long time."

"Perhaps we were wrong in thinking that it was an audio signal. What if that signal had to be translated into a light signal?" Susan intervened. She didn't want to throw in the towel yet.

"We could try that. But then why should it be light and not a tactile signal or of some other type? What do you say Idlath?" asked Tom.

"Yes, it can be tried," the Opsquat admitted reluctantly, without really believing in the idea. He spoke quietly to Yyrtl and two others, who then started talking to the computer.

Within a few minutes the computer aimed a projector towards the monster and, without ceasing to send out sounds, started to illuminate the area around its eyes with a beam of light that changed color and intensity in time with the recorded signal. The monster didn't give any sign of reacting to what was going on.

Another thirty minutes went by, using both acoustic and light signals, always differing but all following the same pattern. The whole thing started to become boring.

All of a sudden Tom felt someone touch him on his shoulder. He turned and saw a man of a species he couldn't recognize, a very tall, thin guy with a breathing mask almost hiding his face. He tried to remember his name, but the only thing that came to his mind was that he had been introduced as an expert in comparative anatomy.

"Admiral, perhaps we should not attempt to translate that signal into an acoustic or optical message," the man said. He was speaking through a computer.

Tom looked at him with an interrogative air. "I mean that perhaps the transmission is the message in itself. I believe that it would be worth trying."

Tom couldn't understand what the man meant. "But the transmission is a radio transmission. . . how could that influence its behavior?"

"I don't know, but I have dissected one of those creatures that died during one of the experiments, and I found a strange organ that none of us could understand. Now it comes to my mind that perhaps it might be a receiver for radio signals. I know that it is strange, but many animals of species that we know about have organs that produce or use electric signals of various types."

"Yyrtl, can you please ask the computer to send the original radio transmission?" Tom asked. It was a strange idea, but after so many useless attempts, it was one that could be tried.

"Certainly. It will take a few minutes," the Sitkr answered. She started immediately to speak quietly to the computer.

Actually they had to wait a bit longer, even though nobody was particularly eager to test an idea that none of them really believed in.

"I am ready for the transmission," the computer said.

"You can start," confirmed Tom, looking towards the glass wall.

A fraction of a second later the monstrous bug froze and remained motionless without emitting a sound. On the other side of the glass wall the humans were also as if petrified: nobody dared to say a word and not even to breathe, as if any movement might break the spell and show that this last attempt was also a failure. After two hours of inhuman howls and of frantic movements the calm seemed unreal. Time passed slowly, while they were all gradually recovering from their amazement. All of a sudden, the voice of the computer broke the silence. "Admiral, it is answering," it said, in almost a whisper.

"You mean it is broadcasting a radio message?" Tom asked, amazed.

"Yes. A message similar to the one I am transmitting. Similar but not identical, in the sense that it is not just sending back the same signal or a message obtainable through simple algorithms from what I am sending. It seems to be a sort of answer."

Tom turned towards Susan. "What do you think? For the first time we have proof that this creature has something to do with the Qhruns." He had spoken that consideration out loud, but only now did he realize what it meant. For months they had hoped to have a specimen of their mysterious enemy in the lab, and now they were certain that they did indeed have one. He realized that in a way they had evoked the devil, the Enemy, with a capital 'E', who was destroying the galaxy. The sense of triumph he had immediately felt was now transformed into terror. "We should perhaps abandon this room and destroy everything," he concluded, speaking more to himself that to the others, and speaking in English.

"Perhaps it would be wise. That message caused a transformation. We might not have had a Qhrun before; I am almost sure that we have one now. We created a monster; we must be careful that the situation doesn't get out of control." Instinctively Susan had answered in English; she, too, didn't want her doubts to affect the others.

Idlath looked at her with a quizzical air. He obviously didn't understand what had just been said, and he didn't even know what language it was. "We were saying only that we must be very careful, now that we know that for certain that that creature has something to do with the Qhruns," Tom reassured him. "I believe that we had best warn the computer to be ready to destroy what is on the other side of that transparent wall," he concluded.

While the Opsquat made preparations to do so, Tom's mind started working at full speed. His fear was turning into curiosity. He realized that it was probably only ignorance that had made him consider the creature in front of them as being the absolute evil. Perhaps for the first time they could communicate with them and perhaps a bridge could be built over the abyss separating those who were probably the two forms of intelligence that would share the universe, or at least that small portion of the universe that was the galaxy. Human intelligence, of the various species that had recognized and accepted each other as such for millennia, and that, still mysterious, intelligence of the people that humans called Qhruns. Only now that name was beginning to be seen in context, a context that certainly had a monstrous aspect, but which was perhaps not wicked, simply just incomprehensible. And he realized that perhaps at that very instant a new era was beginning. For humans and for that still mysterious species, who would perhaps, in future, be able to understand each other and to live in peace.

Tom took out the small transmitter that he had in his pocket and selected an encoded frequency. "Twenty-six, are you still in contact with the lab's computer?"

"Certainly Admiral. I am recording everything," the computer answered.

"Are you able to interpret the answer coming from that creature in any way?"

"I am sorry, but I cannot find any key to interpret that signal. It is similar to what we are broadcasting, but I don't know where to start trying to make any interpretation." The computer's tone didn't hide her uncertainty.

Perhaps it was time to try another experiment, Tom thought. "Twenty-six, try inserting in our message some signals he might respond to. Start with a sequence of natural numbers, then with simple arithmetic operations so that we can see whether it answers somehow." Tom realized that the moment had come to try to communicate, and he couldn't think to anything better than to start with natural numbers. None of the crisis committee could help him: for many thousands of years nobody in the galaxy had had the problem of how to communicate with an alien intelligence. Their problem was that of translating from an unknown language. He wished he had paid more attention to the papers written by the scientists who, on Earth, dealt with the search for extraterrestrial intelligence.

"If I just could login to the website of the Planetary Society or of some SETI expert," he whispered to Susan. "Do you have any idea of how to compose a message which might cause any reaction by that. . ." he was about to say 'monster' but stopped. He was already seeing it in a different light.

"The few things I know about SETI are those linked with the conditions for the development of life. In this situation we need to speak to one of those radio astronomers who study how to communicate with aliens," Susan answered.

In less than four days we could go to Earth and bring a group of experts here, thought Tom, only to discard that idea. After all, those experts knew as much as him or the members of the crisis committee on how to deal with a non-human intelligence, in other words, absolutely nothing. They had, like everybody else, no actual experience of contact. Always provided this monster—no, this being, as it had immediately been promoted in his mind—was intelligent.

"I am trying to insert some signals in our message, but the answer has not changed at all. It seems that it has not even realized that we have added anything," the computer interrupted him.

"It certainly seems to continue looking too stupid to be an intelligent being," Tom said. Then, addressing Idlath, he went on: "Can we get another of those creatures here?"

"Do you really want to put them together?" the Opsquat asked him, frightened by the very idea. Tom answered that the situation was under control and it would always be possible to destroy the two specimens, if anything went wrong. Unwillingly, the Opsquat concluded that there could be no real danger, and began to prepare for the experiment. The computer informed them that a second specimen could be brought to join the first within a quarter of an hour.

At that point Susan indicated to him that she wanted to speak to him in private, so he moved over to her. "Did you consider that, because they can communicate over a distance, they might try get control of our computers?

After all, what they are broadcasting are electromagnetic waves and we don't know how our computers are shielded." She spoke in English, to avoid alarming the others.

For an instant Tom felt a wave of panic; the idea that that creature could induce the computer controlling the lab to open its cage was appalling. "My God, no, I had not thought of that, but I don't believe it is possible. And surely Twenty-six can help us in this," he answered. He immediately got again his out transmitter and in a few words informed the computer of their concern.

"That transmission is of low power and its frequency is much too low; very different from the frequencies that can influence the lab's computer. Moreover, computers in military ships are well screened: I can definitely confirm that I cannot be influenced by that creature. If I detect any attempt at interference I'll immediately warn you"

"Get ready to take control of the weapons in the lab. Keep in mind that we are about to put two of those creatures in contact with each other: The power of the transmission could grow," concluded Tom.

"I am sorry, admiral, but I cannot assume control of the lab's computer, unless you give it explicit orders to that end," the machine answered.

He asked Idlath to put the weapons under the control of his ship and the latter hastened to comply. The Opsquat was clearly relieved by that, a sign that he too had some doubts about the matter.

The second creature entered the lab at that very moment. It was identical to the other, even though it was smaller and its hair was slightly lighter in color. When it came in and saw the other creature, it immediately rushed up to it, with the claws of its four front legs stretched out, making a lot of noise. For an instant they all feared they were going to witness a fight between the monsters, a fight from which they might emerge with considerable damage, provided that they did actually survive.

Nevertheless, when it was less than a meter from the other, the newcomer stopped suddenly and froze. The two creatures remained motionless staring at each other in an unreal silence. "The transmissions of the first one intensified and the second is starting to broadcast too," Yyrtl informed them. She was continuously in contact with the computer.

"Twenty-six, record everything and then try to interpret it. Now that the two creatures are communicating with each other perhaps we can understand something of their intelligence. And above all try to understand if they make any attempts to contact the computer," Tom ordered.

The two creatures were still motionless, one in front of the other, without paying any attention to anything surrounding them. They spent several minutes like that, until the voice of the ship's computer broke the silence. "There is nothing like an information exchange between intelligent beings. It looks more like a communication protocol between computers… but don't take this literally: no computer has a protocol of that kind. I just mean that I cannot find any content in the signals, I cannot interpret them."

"Don't worry Twenty-six. Nobody thought we would immediately be able to interpret what those beings are saying—provided they are saying anything. We have all the time we need…"

After what seemed eternity, the two creatures began to stir and, remaining close together, started exploring the room. They continued like that for more than two hours: it was clear that they had to give them time to develop interesting behavior of any type, provided that they would ever do so. For the time being they could only wait.

25 Meeting

Five days had passed since the beginning of the experiment, when the crisis committee met on the cruiser. After the first enthusiasm, the absence of any real results had produced deep disappointment. Several creatures had been put in close contact with one another but they limited their exchanges to incomprehensible radio messages. Those five days of failure caused many to loose any hope of solving the mystery. Tom realized that he could not devote more time—actually he said 'waste more time'—like that, when his presence was needed to continue military operations. The crisis committee met on his ship for a last discussion before it left the asteroid.

When Idlath arrived, together with all the others, Tom was running some simulations on the tactical screen to check the possibility to giving a suitable escort to all the freighters that were evacuating a system inhabited by more than nine hundred million people and conveying them to another system at a distance of more than eight hundred parsecs. The easiest systems had been already evacuated, and they had to move a larger number of people over increasing distances.

"Susan, is there a way of moving that group of cruisers so that, after having escorted the fourth convoy, it can meet that other group of freighters?" He now tried to switch the flow of ships from one system to another, maintaining the freighters for longer in hyperspace, where they would be safe, and continuously moving the escort ships.

"No, not unless we make them ubiquitous and in more than one place at once," she answered. It was clear that there were not enough ships for the job. More than freighters, he needed more cruisers. "The only way to cover that group of convoys is to take a thousand ships from the raids over the frontier," he continued, asking the computer to move the images of the ships on the tactical screen. He was about to say that those raids were essential to earn them precious days of delay but decided to shut up; it was useless to repeat the same thing over and over again. He just concluded "ten ships of the Iskrat-is-Thn in that system would be enough to cause slaughter, if we don't give an escort to those freighters," realizing that this observation was trivial too.

"We are far too stretched. If the Qhruns now accelerate their action and catch us midway in the evacuation of a pair of systems or if the Iskrat-is-Thn decide to attack us. . ." he continued.

"We are in nice trouble, there's no doubt of that," Susan answered. "But I don't think they will do it. In none of our simulations. . ."

The arrival of Idlath interrupted her at that point. When the computer announced that the whole crisis committee had come on board, Tom and Susan went to the library. "Have you finished with that tactical screen?" Sinqwahan welcomed them, removing his head from a mental projector.

Tom heard his voice coming from a room he thought was empty. "When did you arrive?" asked.

"More than an hour ago. Seeing that you were busy with your simulations, I came here to study the reports that had arrived in the last few hours. It seems that the Iskrat-is-Thn have disappeared leaving no trace. I don't know what to think."

He had just finished voicing those points when Idlath entered, followed by the others. Tom had ordered the computer to prepare the library for that meeting and now it looked like a meeting room adequate for a crisis committee. He immediately remembered the other meeting they had in the same place almost a year and a half earlier, still influenced by that attempt to kill him. Certainly things had changed greatly: the sector was not isolated anymore, the Qhruns had been defeated in a great battle in space, and there was some chance that their enemy finally had a face. However, the mystery was still a deep one and, even more disappointing, it seemed that all their successes had not led to any certain knowledge. *Well, what we cannot understand, we can at least destroy, even if it means tearing them asunder one by one, legs, claws and all the rest*, he decided as he gestured to Idlath to start.

Idlath got up and summarized in a few words the goal of that meeting: to analyze the state of the research and to decide on future developments. Bitterly, Tom thought that was a typical agenda in a stalemate situation. If they had something to discuss and to decide, there was no need for wasting time with that meeting. A sticker he had seen somewhere, *in his previous life*, as he thought when referring to the years he had spent on Earth, said: *If you feel lonely, organize a meeting*. . . Because all his thoughts ended up by focusing on what was becoming a fixed idea with him, he started to wonder what Qhruns might do when they felt lonely.

The Opsquat introduced the expert who had worked on the cloning of the various specimens. He described in great detail what they had done and the chances that errors had been made in the reconstruction of the genetic code. He concluded that they had not made any errors and those creatures were truly representative of Qhruns, or some form of life closely connected with them by symbiosis, parasitism,. . . and so on: listing all the possible forms of relationships that could exist among living species.

It was then the turn of an expert on human and animal reproduction, who tried to explain the sterility of all the females that they had cloned and spoke for about twenty minutes on the transformation from the immature to the adult forms. The latest discovery in this field was that the immature forms were able to receive that particular radio transmission, but not to broadcast it.

The third to speak was Yyrtl. She described in great detail the job done with computers to reconstruct the genome and to check the development of the creatures and then to interpret the transmissions. Tom noticed that the Sitkr was not looking him in the eye and realized that she was too much of a serious scholar not to be aware that she was hopelessly trying to hide, like all others, the lack of new facts on which to report. She was merely reiterating old facts and she was clearly ashamed of it. Tom realized perfectly that the group needed that exhibition of the facts, to regain their confidence, which had been undermined by their inability to understand. *This is just a psychodrama*, he thought, bitterly.

After three other speeches, all similar, Tom was really fed up. His mind started to wander, recalling memories and impressions that had remained in his memory for ages but which he had never analyzed in detail. Now a rather short human was speaking, gesticulating with two short upper limbs, but almost in a crouch on a huge kangaroo-like tail. What was worse, he insisted on not using his computer to translate his words, trying to speak Aswaqat directly in what was an almost incomprehensible way. If it had not been for a few three-dimensional images, Tom wouldn't have understood that he was speaking of the alien's metabolism. Apart from the data on the total quantity of energy, and therefore of food, that those monsters consumed, it was not an exciting subject. Tom felt he had gone back at least two years, and closing his eyes, he remembered listening to some Russian or Chinese scientist who tried to present his paper to a conference in incomprehensible English. Naturally he didn't dare to fully close his eyes, because if he did, he would immediately fall asleep. *If you feel lonely, organize a meeting*, he repeated, and decided that it would be much better to be alone than in that crowd. *Who can tell whether one of those monsters might feel lonely, with that radio system*, he wondered and started playing with that idea for a time. *Certainly for them it is like always being in a crowded square, with people shouting everywhere.*

Suddenly it was as if a light had switched on in his mind. Trying not to be seen, he took the small transmitter out of his pocket and inserted an earphone in one of his ears. "Twenty-six, can you detect the electric activity of those creatures caused by their transmissions?" he spoke quietly into the microphone.

"Certainly," the computer answered. "I have mapped all the electric circuits in their bodies that they use both for transmitting and receiving. I have tried to interpret the configurations of the currents but to no. . ."

Tom was completely awake, now. Awake and terribly excited. "What percentage of time are they active in both transmitter and receiver modes?" he asked. interrupting the machine.

Susan realized that he was talking to the computer, even if it seemed that none of the others had noticed anything. She raised her eyebrows interrogatively at him. Tom gestured back that everything was alright. "They are on for one hundred percent of the time in both modes," the computer answered.

That confirmed what he was thinking. He started again talking to the machine. In recent days they had recorded a wealth of data and now the details were slowly falling into place and a consistent picture was forming in his mind. As always, that picture was infinitely simpler than the way things had appeared to them when everything was obscured by the fog of the unknown.

The short guy with the huge tail had just finished speaking, and Idlath was about to introduce another speaker, when Tom suddenly got up. He went up to the Opsquat and said: "Now there is no doubt anymore, we have to go on developing the viruses. How long will it take to perform laboratory experiments and to begin the production of bacteriological weapons?"

Idlath looked at him, surprised by the interruption. "But do you really intend to evaluate the viruses, using the specimens we have in the lab? That would mean. . ." he stopped as if he hesitated to say 'to kill them', and then he actually continued "to destroy them. They represent months of work, and we don't know anything about them yet." The Opsquat was annoyed because of the interruption and appeared to be scandalized by the idea of killing those creatures they had created.

Let me prepare one of those coups de theater you like so much, Tom thought. "This is the point. Susan, you have known the truth at least since the battle of Aq," he said, addressing her.

"Me?" Susan asked. She now felt she was the center of attention. "How can you think that?" she started to say.

"I didn't say that you realize that you know it; I just said that you know it. Twenty-six, please play the relevant sentence recorded on that occasion in the Aq system."

Susan's voice came out of a loudspeaker. "They look rather like bees." "Bees? What have bees to do with this?" This was in Tom's voice.

"When bees sting an animal they leave their sting in their victim, but then they often die. And, nevertheless, they keep on doing that without the instinct of self-preservation stopping them. It looks as if the Qhrun's instinct is the same. . ." The voice of the computer continued: "At this point I interrupted you to give some urgent communication. I am really sorry, perhaps if I had let you go on with what you were saying. . ."

"No, Twenty-six, we didn't have enough data to understand, then. But now. . ."

"But now, a darned thing!" Susan interrupted him. "I was just expressing an analogy, an impression. I didn't certainly mean that because bees are bugs and because these creatures. . ." Here she stopped, realizing that at the time she hadn't thought of any analogy of that sort, but now unexpected perspectives were opening up.

"You see, you unconsciously realized something important. Certainly, no intelligent species is insect-like, nor even belonging to the class of arthropods. Those creatures we have in our lab are undeniably arthropods. But, strictly speaking those are not Qhruns. We should perhaps say that 'Qhruns' don't exist," he said, putting the stress on the 's' of the plural. After a short pause he continued: "What unfortunately exists is 'the Qhrun', and here we have a part of it, a small, poor, lost, pathetic part, which together with the other parts forms the most terrible, monstrous, abominable being that ever appeared in this corner of universe."

Tom gave them time to meditate on those words. They had all remained silent, wondering about his obscure statement. "You mean that they are social insects, and that their sterility is linked with the fact that there is a queen laying eggs for the whole swarm?" Idlath was the first to speak after some minutes. "Certainly, the instinct for self-preservation in social insects is different from that of vertebrates, but. . ."

"No, Idlath," Tom interrupted him. "It is a more complex and at the same time a simpler thing. When I say that we are facing only one being, I mean literally that, not a swarm. I understood it when I realized that those transmissions are not communication among different beings, but among parts of the same being. The single insectoids are related to the complete being like the cells constituting our body are related to us. It is a meta-animal made from animals just as a multicellular life form is made of cells. But just as one of our cells it is not a single-celled animal able to live autonomously, one of those insectoids is not a true animal: in our lab they live like cells live in a culture dish."

He waited a few seconds and then said: "At the level of single parts they probably don't have any instinct of self-preservation, and this explains their behavior. They fight against us rather like our white cells attack viruses. Ever seen a white cell doing anything similar to bailing out a doomed starship?"

"Interesting theory, it would explain a lot of things," Susan said. Tom immediately understood from the way she began that she was not quite convinced. "Unfortunately it cannot explain how a being of that type can exist on a galactic scale. The parts would communicate with each other by radio, but radio waves take more than one hundred thousand years to cross the Galaxy. And then, the transmission that keeps the various parts together is present in invasion fleets, but not in small groups of ships. . . Perhaps it is a basis on which we can work but, as it is, it doesn't explain the data we have."

"Wait, I said it's a being made from various parts, but not that it is indivisible. Take an invasion fleet. Every twenty thousand ships there is one broadcasting that transmission: let's say that on that ship there is the queen, to use an apiarist's term. A group of twenty thousand ships is therefore a kind of a swarm. Initially I thought that the swarm constitutes an individual being: in this way the problem of distance is solved. In every system there would be five or six Qhruns during the occupation phase and then some of them would swarm away and so on." He waited until Susan nodded assent, and then continued: "At the beginning I thought so; unfortunately it doesn't work. When one of the ships that is broadcasting is destroyed, another immediately replaces her, and this is possible if we have only one individual, with many centers, with an architecture that can be modified easily."

"Certainly, it could be a polycentric network that is able to reconfigure continuously, with parallel architecture with concentrators of some kind at the main nodes." Susan had started to realize that the thing was possible, even if there were still some points she didn't understand. "But that doesn't take into account isolated ships".

"There is an explanation for that as well. Isolated ships contain only single insectoids and therefore they don't have true intelligent beings on board. They are perhaps controlled by machines, or act on the base of pre-recorded instructions. The single creatures that are on them receive recorded transmissions as do those in our lab, only ours are not trained and therefore cannot do anything."

"This explains why small groups of ships act in an automatic way. The greater the group, the more it participates in the bond with the rest, the more it behaves in an intelligent and articulated way." Susan remained a few seconds lost in thought and then continued "It also explains the attack on Terkar. I thought they had detected some primary traffic directed at Terkar and, therefore, following that, were attempting to arrive there. But now it is clear that the farther we go from the frontier, the farther we enter the heart of that being and the more its behavior is intelligent. Now we have to build mathematical models of this immense logical network." She was already thinking about the possibility of building mathematical models to simulate the being they were painfully discovering.

"The inner zones have dense interconnections: think what would happen if the nervous messages among the parts of our brain were carried just by starships moving from one system to another. Where there is a lot of traffic the connection is fast, and the decisions are taken quickly at a logical level; where the communication net is sparse, any actions by that being have to be based on local, low-level, reactions and thus shows automatic behavior." After a short pause, he continued: "With the supremacy in space, we can produce a sort of block of the nervous transmission, forcing it to fragment and to degrade its logical function. We need to intensify our raids."

Looking around he realized that for some time Ruklyaq was trying to speak, and gestured him to go on. The voice of the Tteroth was strongly altered: "Then this is the end: we are witnessing what many foretold as the final result of the development of intelligence. The biosphere of planets and then of the galaxy has finally found a conscious being that unifies it, a marvelous unifying structure, giving life, intelligence and conscience to the universe. We are just evolutionary has-beens, all we can do now is just disappear and leave the galaxy to a being that is more perfect than we are."

Tom didn't know what the expression of a Tteroth would be when hardly holding back tears, but he would have bet that Ruklyaq was in that state. "Disappear, be darned!" he replied forcefully. "Don't be deceived by appearances: that is not the heaven a mystic might envisage, one where a being is living in communion with the universe and that expands to form a kind of soul for the world, an immanent god or a material foundation of divinity. It is not the infinite liberty of a being that is in contact with the whole universe, it is instead the hell of enslavement, a monster devouring and destroying everything it touches. Quanslyaq said that Qhruns were evil it its pure form, meaning by that they were everything that is most contrary to the essence of being human, his intuition is now confirmed." He paused, realizing that his hypothesis had philosophical and metaphysical implications deeper than he had thought.

"If you want to see it from a less metaphysical point of view, all humans developed common values, because in all societies what we call human values are, after all, advantages for the survival of the individual and therefore they were favored by evolution. In a being of that kind the opposite is true. The more it is aggressive towards everything that is different from itself, the more it can develop and flourish. That's why Quanslyaq is right."

The Tteroth was sitting on his force field with his head in his hands. He was trying to evaluate the implications of those words on his system of thought. Now that he had broken the ice, everyone had something to say. A human Tom thought was an expert in comparative evolution stood up and said: "Admiral, if it is as you said, things are extremely serious. Let's leave aside metaphysics, absolute evil and so on. That being is the product of billion years of evolution and since it has come into contact with us it has been performing acts that are a form of natural selection. There is no doubt that its way of connecting within itself is more advanced than ours and that it is more suited than us to survive. There is nothing we can do, the future belongs to it."

"We will see who is more suited to survive. In the Terkar system that monster darned well did not survive. Perhaps it is true that here we are seeing natural selection in action: it is for us to show who is most suited in this environment," Tom interrupted him.

"My God, after the evolution according to Lamark, and that according to Darwin, there is now also the evolution according to De Coubertin," Susan said quietly. "Don't you think it is naïve to assume that the future of life is just a game in which whoever wins, wins, and whoever loses, goes extinct?"

"Nothing less than the De Coubertin style, I can guarantee you that. Here the point is not competing, but winning, by any means. And then I don't believe that thing has the right cards to win the game," Tom went on. He clearly wanted to say something on the subject of natural selection.

The biologist gave him a questioning look. "If what we are thinking is correct, that that being is immortal, extremely diffuse and dreadfully powerful. It can continually regenerate itself. In short, it has many of the attributes of divinity. If evolution can produce something of that kind, it is clear that we can aspire to nothing more than a small niche, provided it leaves us one."

"If we wait for this thing to leave us anything. . . No, natural selection cannot be played out with just theories. If we want to show that we are more suited to survive than this thing, we have to show it on the field, or rather, on the battlefield, as we did in the Terkar system," Tom answered. "But we must not idealize what is likely to be just a blind alley of evolution. First, if insects could develop some winning strategy in this game they would have done it somewhere else as well. And then, perhaps now I understand why they are crossing the galaxy so fast. A being of that kind doesn't have any consideration for anything other than itself. Its egocentrism is absolute: its world is all inside itself and doesn't communicate with anything other than with itself. Susan, do you remember Leibnitz's monad?" he asked suddenly.

Susan was caught by surprise by the question. "Yes, but for Leibnitz there was God who. . ."

"All right," Tom continued "That being is a monad without God, completely closed in itself. Certainly, it is not actually like that, because it has to interact with the external world, but it is not an interaction that takes any account of values. Such an unbalanced interaction cannot last, at least not for long. That being exploits the world, and exploits it to such a point as to make the whole world sterile. That's why evolution has never produced a solution of that kind. I checked with Twenty-six; there are planets where life has disappeared because of life forms that completely destroyed their own biosphere. Generally they are very evolved insectoid forms. On Earth, locusts and certain ants have the tendency to do the same, but a healthy ecology checks them. But if they become so strong as to dominate, they end up destroying their very planet."

"It has just a chance of leaving its planet before destroying it. If it succeeds, it has millions of systems to exploit. And it spreads from one system to another, exploiting their biospheres until it reduces them to a desert and then moves somewhere else, until it devastate the universe or finds someone who stops it."

"Tom do you realize what you have described? Go down to the cellular level and that being is a sort of cancer, devouring healthy cells until when it causes the death of the organism. But in this case it doesn't die, because it infects a new organism!" Susan exclaimed.

"The only positive aspect of the matter is that this form of life is unstable. Now that we know how things are, we know that there will never be the possibility of coming to an agreement with it. It is either us or it, in short. But if it wins, in the long run all will lose, because it will destroy itself in the end," Tom went on.

"And so at the end of the search we have found a truth, still one to be verified, but I believe that we are close to doing so, but a truth that is almost useless. We wanted to know so that we could try to communicate and to get some advantage from our action. We now know that we won't get any advantage other than what we get by ourselves; that there is no alternative to keeping on fighting exactly as we did before," Susan concluded.

Tom nodded in agreement. "If we only could know how it abandoned its planet of origin. I feel that it didn't initially build its ships and that someone happened by chance to get there, unwittingly offering it the way to escape. But I fear that we will never know." He turned towards Idlath and asked him "What do you think of all this?"

"We will work on these ideas, but I think that the overall picture is correct. And above all that we have to accept its conclusions." He waited for a few seconds and then added "we will immediately start with viruses. We cannot afford to be squeamish when facing something of this sort."

26 Homecoming

The battered cruiser was approaching Earth. After the battle at the hyperspace gate they tried to reach Earth in the shortest possible time, but the damage to the propulsion system was worse than they initially thought. The only reasonable solution would have been to stay in space waiting for a supply ship and to postpone the landing until they could do it without danger.

However, that was really the only solution Tom couldn't accept. They were going to Earth to prepare to defend the Solar System from imminent invasion and the presence of those three Qhrun ships blocking the hyperspace gate pointed out that it was due much earlier than expected. There was no time to waste.

Moreover, landing with one of the shuttles was impossible, because the explosion in the engine room had damaged them beyond repair. They had to land the damaged starship itself, and quickly, with all the dangers this implied.

Now the planet filled the whole window. It was strange, Tom thought, in the last four years he had seen hundreds of planets from that distance and learned to recognize many of them, while his own planet was now completely unknown to him. Rationally the thing was easily explained: he had seen Earth just once from space, when he left it at the start of their adventure, and he had been so excited that he had only confused memories of that evening. Had it not been for the photos taken by the astronauts of the Apollo missions, he would not even be able to tell which planet it was.

"When will they locate us with their radars?" Susan wondered.

"In less than two hours, undoubtedly . . . always assuming that we are still in one piece or at least that the pieces are large enough", Tom answered.

Susan didn't answer. She was looking at the surface of Earth slowly moving beneath them and rapidly drawing closer. They were now north of Scandinavia, flooded by the noonday sun. Beyond Greenland they could glimpse the American coast, dimly lit by the red light of dawn.

"Now we have to go. In two hours everything will be over, in one way or another," Tom said, moving towards the bridge. Susan followed him in silence.

"Admiral, I am ready for aerobraking." The voice of the computer was somewhat nervous. Nobody had ever tried to perform ballistic re-entry with aerobraking with a star cruiser, at least as far as they knew. But the simulations of the two previous days had shown that it was possible, even though there was no large margin for error.

After putting on his space suit, Tom sat on his chair, fastening the emergency straps in case the force field shut off. "Twenty-six, disconnect artificial gravity and all non-essential devices and use all the available power for deceleration."

The computer was fully aware of the maneuver, after having simulated it dozens of times, but she needed to hear the voice of the captain repeating the order again. The computer rotated the ship so as to expose an undamaged zone of the hull to the atmosphere of the planet. "Prepare to energize the forward shields."

"Aerodynamic heating of the forward section has begun," the computer answered.

The cruiser was now entering the upper atmosphere and they started to feel the aerodynamic deceleration. While acceleration caused by the engines was automatically compensated, with the generators damaged to such an extent it was impossible to compensate for the deceleration caused by aerodynamic forces. The deceleration quickly increased and little by little the hull started to vibrate. The external hull heated violently and they were almost certainly visible from the ground as a reddish glow. By now they were over the Midwest of the United States, where it was a few hours before dawn, and the cruiser must have shone like a bright star in the black sky.

"I am energizing the shields and altering the trajectory," the computer announced.

"Try not to gain altitude too quickly. The more we stay in the dense layers of the atmosphere the more speed we will lose," Tom answered. The vibrations were getting stronger and stronger and an impressive booming sound reverberated through the hull. Suddenly they heard a great bang and the ship started to rotate. The computer quickly regained control, bringing the attitude back to the optimum orientation. "A piece of the exterior hull over the engine room has detached," it announced.

"Check whether the strength of the hull has been compromised," Tom ordered.

After a few seconds the machine declared that it had been a part that had already been damaged by the explosion and there was no danger. "For now," she concluded in what was meant to be a soothing a tone but which was by no means comforting.

Slowly the noise and the vibration decreased as the ship returned to the highest layers of the atmosphere. "The first phase of the aerodynamic maneuver was performed according to predictions," the computer said with far less concern in its voice. Now the hull was cooling and they were probably invisible from Earth once more.

The second dive into the atmosphere was performed slightly north of the Marquesas Islands. This time the ship sank much deeper into the denser layers of the atmosphere, heating up to the point that the inside temperature became rather too high for comfort. The deceleration was considerable, as was their return bounce, to the point that Tom feared they wouldn't be able to modify their trajectory downwards sufficiently with the low power that was available. The computer reassured him, declaring that the maneuver was proceeding fully in accordance with anticipations.

The following entry in the atmosphere occurred over the deserts of Australia. It was shortly after sunset, but it was not very likely that the light would be noticed by anyone. This phase of the aerobraking was crucial, because the speed would decrease below the orbital velocity and therefore after this rebound they would no longer have any chance of aborting the landing. Initially, the entry went according to plan: by now they were used to the strong vibration and to the glare of the shock wave in front of the ship. All of a sudden they heard a loud bang, followed by a sudden roll and an increase in the vibration.

"Twenty-six, compensate for the roll," Tom exclaimed. Because the computer didn't answer immediately, he immediately said: "Ashkahan, take manual control. Compensate the attitude as soon as possible."

The Aswaqat began to work with the manual-control terminal and the roll speed decreased. Tom saw that also Heiqwahan had started to work with his terminal. "The computer is damaged." He said. "The structural parts we shed contained a part of the engine room's memories. I should be able to restore the higher-level functions very shortly."

Tom was trying to distinguish, amongst all the noise, the hiss of air escaping from the hull. He couldn't, so he interpreted it as sign that the damage was located in the engine room that had been depressurized for two days. He also got his hand terminal and, while Ashkahan manually stabilized the ship, began to orient the hull to get the necessary aerodynamic forces to curve the trajectory upwards. Shortly afterwards the cruiser began to rise again towards space.

"Admiral, I am taking back my functions," the computer said with altered voice.

"Try to hurry up, I don't know for how long we can continue on manual control," Tom answered.

It took no more than two minutes. "I see you can do without me," the computer said trying to dissipate the tense atmosphere. Tom appreciated her effort: it was the second time in two days that the computer had suffered such damage that it had caused what in human terms would have been described as a loss of consciousness, yet it was reacting well.

"Try to determine whether the fragment reached the ground, and if so, where it ended up. As soon as the supply ship arrives, we must recover it; it must to be radioactive like the inside of a generator. Luckily it must have ended in the middle of a desert," Tom decided, happy that they had planned the trajectory in such a way that they crossed mostly seas and deserts.

"It is so radioactive that the rescue team will find it without any problem," the computer stated.

"I never believed that a ship could be controlled manually in such a situation. To do it in real life, I mean, not in a simulation," Ashkahan said, still amazed at her own work.

"Get ready to do it again: we don't know whether the computer will again suffer damage in our next dive into the atmosphere," Tom answered. He had the impression that the computer was fervently praying that it would not be necessary.

In the following thirty minutes they made other two passes through the outer atmosphere, one slightly south of the island of Zanzibar and one over Mauretania. This time the speed was much lower and they had no problems. Shortly afterwards they were north of Bermuda and then in sight of the coast of the United States. At that point they had to begin the final phase of the approach. "Twenty-six, broadcast the transmission, until we need all the power we've got."

The original flight plan included an orbit around Earth, broadcasting a long speech on the main radio and television frequencies. The situation had forced them to change their plans, because for aerobraking they needed all available power. Therefore they had prepared a brief message to be broadcast immediately before the last atmospheric entry.

Tom's image, recorded the day before in the library, appeared on the screen. "Admiral Thomas Taylor, general commander of all the Starfleets of the Galactic Confederation and Plenipotentiary Ambassador to the Solar System sends this message to all the peoples of Earth." As he was listening to his own voice he thought about what was happening at that precise moment in the zone covered by the transmission: certainly at that time there were not many people in front of television screens, but many more were probably listening to him by radio. For a short time nothing would happen: they would simply wonder what might be advertised by such a strange commercial. And then, as Susan told him the day before, while they were making the recording, he was not believable: a Starfleet admiral with that sort of name, a fairly strong western accent and those magniloquent titles. If they wanted to be taken seriously, they ought to have had the words read by Yyrtl, with her four arms, or by a Tteroth, with his wings stretched out.

The message continued for a few minutes with greetings and other diplomatic statements and concluded with the intention of going immediately to the United Nations to present his credentials to the General Secretary and to give a speech in front of the General Assembly. The purpose of the message was really to prevent the civil and military bureaucracies from wasting precious time waiting for someone to decide on the best way to give the news. Doing it this way there were no possible alternatives, other than taking them as soon as possible to the United Nations and broadcasting everything live.

"What do you say, Susan?" he asked as soon as the message ended.

"Even worse than yesterday," she answered.

Now they had to focus on landing. "If we make a mistake and crash, they will show our autopsy live," he said to Susan in a low voice.

He realized that Ashkahan had reacted with a grimace: she couldn't know about the Roswell case and wouldn't understand the joke. "Don't worry, Ashkahan, I was joking. If we crash nothing of us will remain to be the subject of an autopsy, anyway."

"You don't know how much this reassures us," Susan murmured.

He looked at the map projected on the screen: it was by now time to prepare for landing. "Twenty-six, put me in contact with the tower."

The computer established the connection. "This is star cruiser Charlie Hotel 23426. We ask for immediate authorization for landing. We have an emergency. I repeat, we request assistance for an emergency landing."

The answer came immediately. "Received, Charlie Hotel. We don't see you on the screen. Communicate position and type of emergency, and repeat the aircraft type." The flight controller didn't know about the radio message and was thinking of a normal type of emergency, so to speak.

"We are at one hundred and twenty thousand feet, range thirty five miles at one hundred and fifteen degrees. Approach speed is two thousand three hundred knots, vertical speed one thousand two hundred knots. Yes, you understood correctly: this is the flagship of the Starfleet of the Galactic Confederation: Star cruiser Charlie Hotel 23426."

"A moment, please," was the answer. But his tone of voice had changed: Tom would bet he was unsure, between being angry at an idiotic joke and worried about having gone crazy.

"I request a quick answer: we are almost in free fall. An accident has deprived us of seventy-five per cent of our propulsion. Anticipated time to impact six minutes."

After a moment of silence he heard a series of noises and the distant sound of a very agitated but incomprehensible conversation. Then someone took the microphone.

"This is General Woods speaking. We are following you on radar, what kind of assistance do you need?" The general was panting: perhaps through emotion but more likely because he had rushed there.

"Thank you General. We will try to land; give us the coordinates of a point on the airport that can be fenced off with a radius of at least one thousand meters. Even if the maneuver is successful we will release radiation from the engine room and nobody must approach within a distance of a kilometer from the hull. In less than thirty hours a supply ship will arrive. It will recover the ship and decontaminate the surrounding zone, but prior to that nobody must approach."

A quick and muffled conversation followed and then the general gave them the coordinates. Clearly he was aware of their broadcast and was trying to behave more or less normally.

"I have entered the coordinates into our computer. In less than four minutes we will be on the ground. If anything happens, nobody must approach the wreckage or to any part that is detached from it. If anyone is contaminated, put him in hypothermia and bring him to the supply ship as soon as it arrives. We lost a piece of the hull's structure above the Australian desert: it is strongly contaminated too. If we get down safely, we must go immediately to the United Nations."

"The fragment fallen in Australia has already been located. I will communicate with the search team not to get close. Welcome on Earth, Admiral and best wishes."

Tom closed the link and concentrated on the maneuver. At two minutes from the impact the ship was still going down too fast, slowed by the engines powered by the only generator still working. "Twenty-six, get ready to bring the generator to hundred fifty percent of nominal power."

The computer confirmed that the maneuver was proceeding according to their simulations; everything now depended on the chance that the generator would work for at least a minute and a half at such a high power. Thirty seconds passed slowly, then they started hearing the low hum and the vibration transmitted from the generator to the whole structure, a sign that it was working above its normal capacity. By now they could clearly see the coast in front of them and they started to distinguish Kennedy airport not too far from the sea. No aircraft was visible, which meant that the airport had been closed and flights diverted elsewhere. Tom wondered what had been said to the thousands of people flying to New York or waiting for embark. He could just imagine what it was like to tell all those furious people that the airport was closed because the runways were being used by a damaged star cruiser. . .

The vibration increased, but the speed of descent reduced rapidly. Everyone was staring at the screen on which the computer was displaying the altitude and the position relative to the point assigned for landing. At thirty seconds from impact, the altitude had reduced to five hundred meters and they were practically vertically above the landing point.

"Twenty-six, slightly reduce power, let her go down a little faster." It was better to touch down at a slightly higher speed than to try to maintain that power level for too long.

At two hundred meters from the ground they heard bangs coming from the engine room, and the lights went off. The screen also went off, but the generator was still working. After about twenty seconds, at fifty meters from the ground, as Tom guessed, an explosion made the whole hull boom. A few instants later, a second explosion and then a third, much greater, bang followed. The cruiser, no longer sustained by its engines, started falling, slowly rotating to one side. Tom thought: *If we were only at fifty meters, we've made it!* Then the ship hit the ground. The force of the impact was reduced by the way in which the bottom of the hull crumpled. Everyone remained motionless and immediately the emergency lights came on.

"All crew: report on your condition," Tom said relaxing against his seat. One by one everyone confirmed that they were unharmed. "Twenty-six, report on the situation in sickbay."

"Del-Nah is in controlled biostasis. I kept the equipment always under power. However now the air-conditioning systems is out of use," the computer answered.

That wasn't a source of worry, Tom decided. "Let in external air. It will stink of burnt kerosene, but we can breathe it," Tom answered.

"Well, as a triumphal arrival, that was not bad. We shed pieces over half the world and we were almost shattered on the runway," Tom said getting up.

"I am just happy we got down in one piece," Susan answered, getting up too.

"Let's go. The true work starts now," Tom concluded as they all moved towards the exit.

Part II

The Science Behind the Fiction

Humans and Aliens
The Science Behind the Fiction

1 Introduction

The main scientific theme of the novel is extraterrestrial intelligence (ETI). Bioastronomy and SETI are, generally speaking, controversial subjects, because the basic issue of both, the existence of extraterrestrial life, is still an unproved statement. Even if there is a substantial difference of opinion, with astronomers and physicists generally for a positive answer and biologists for a more skeptical approach, many think that simple living organisms, comparable in complexity with our bacteria, are widespread in the Universe. More doubts are cast on the existence of extraterrestrial complex life and above all of intelligent extraterrestrials (ETIs).

The search for ETIs, the so-called SETI, is by now an established research field. Initially, one of its motivations was the belief in the impossibility of interstellar travel. If it is impossible to send probes to extrasolar planetary systems, which at that time were postulated and by now have been discovered in the hundreds,[1] the only possibility for a detailed study of the life forms that exoplanets might host was to hope that a number of them were intelligent. If they had developed technology, they could contact us, directly supplying the information we seek. After more than 50 years, SETI has not yet produced any result. On the one hand, this might be expected, owing to the difficulties of the search, which has frequently been defined as searching for a needle in a haystack of cosmic size. On the other hand, perhaps also due to overoptimistic initial hopes, many look at SETI research with increasing disillusionment.

The so-called Fermi Paradox is often quoted [1, 2, 5, 6]: If the Universe hosts many intelligent aliens, and if long-distance space exploration is not forbidden by some physical laws, how is it possible that we have not yet

---

[1] In March 2015, the total number of exoplanets already discovered was 1821 (plus several thousands of unconfirmed candidates), in more than 1100 planetary systems.

© Springer International Publishing Switzerland 2016
G. Genta, *A Man From Planet Earth*, Science and Fiction,
DOI 10.1007/978-3-319-21115-2_2

entered in contact with them? However simple this question may appear, it is quite difficult to give an answer which doesn't exclude one of the terms of the dilemma: the existence of alien intelligences or the possibility of interstellar travel. The first problem we encounter when speaking about ETIs is that we lack a definition of intelligence and we don't even agree on whether an intelligent being is necessarily self-aware. There is even no agreement whether humans are the only intelligent and self-aware beings living on our planet, or whether other animals, such as some apes, dolphins or even octopuses may be considered as such, at least to a certain extent. I think that none of them can qualify as intelligent beings, and that only members of the *Homo sapiens sapiens* species are endowed with intelligence. To quote Jacques Arnould [7], humans "... are the only beings we know with certainty to have the ability to ask not only about how but also about why things happen, about beings, and about themselves". This implies that only one intelligent species exists on Earth, and that the birth of intelligence must be located in time between 80,000 and 150,000 years ago. In this view, Neanderthal men were possibly intelligent, but this is immaterial in what follows.

In the novel it is assumed that this is a common situation, and a planet on which there is more than one aboriginal intelligent life form is a rare exception. Perhaps this is also because once there is an intelligent species on a planet, it prevents other species from developing intelligence or even tends to cause its extinction. Something of this kind might well have happened to *Homo sapiens neanderthaliensis*, if we accept that it was actually a species distinct from our own.

2   What Kind of Extraterrestrial Intelligences?

Apart from the basic question about the existence of ETIs, another question is often put forward: if such a thing exists, what kind of aliens will we find, if our search does have success? How similar to terrestrial life may the life we will find be, and, in particular, how similar to us (how human) may ETIs be?

This issue is an old one, and the first to take a firm stance on the subject was Galileo Galilei who, in his *Istoria e dimostrazioni intorno alle macchie solari e loro accidenti* ('*History and descriptions of the sunspots and their details*') (1613) states that he ...considered as a false and point of view that was to be condemned to assume the existence of inhabitants on Jupiter, Venus, Saturn and the Moon, intending 'inhabitants' to be understood to mean animals like ours, and particularly human beings. But then he states that ...it is possible to believe that living beings and plants exist on the Moon and the planets, whose characteristics are not only different from those of the beings on Earth, but also from what our wildest imagination can produce. More than half a century later, in 1686, Bernard Le Bovier de Fontenelle published his book *Entretiens sur la pluralité des mondes* ('*Discussions on the plurality of worlds*'), which had a great success, with 33 editions in French and many translations, published before the death of the author. He follows a similar approach, assessing that the living and perhaps intelligent beings existing outside our planet will be widely different from those we know. In the following three centuries the wide variety of the known terrestrial living beings became wider and wider, and Galileo's words assumed a new meaning. Nevertheless, when the idea of extraterrestrial intelligent beings became popular, it was often embodied in humanoid form. The drawings accompanying the articles written in 1835 by Richard Locke to describe the alleged discoveries made by the famous astronomer Sir William Herschel on the Moon, depict winged beings with a humanoid body (Fig. 1). Similarly, the aliens of early science-fiction novels were humanoid, as well as almost all of those starring in science-fiction movies up to the present day. The popular image of aliens is that of humanoids, and most of those who claim to have had an 'encounter of the third type', or to have been abducted by aliens, describe them in the same way (Fig. 2).

Some science-fiction writers tried to describe ETIs not only with different body shapes but also based on a completely different biology. The examples are many, from *The Black Cloud* by Fred Hoyle to the inhabitants of a Kuiper belt asteroid described in *Camelot 30 K* by Bob Forward. Many of these ideas are reported in the book *The Science of Aliens* by C. Pickover [8].

At present most scientists who believe in the existence of ETIs, think that it is highly unlikely that their general body plan is humanoid[2] [9, 10]. Evolution is governed by casual mutations and natural selection, and can operate only by working on the genetic material present in the original life forms: under these conditions evolutionary processes occurring in different places at different times yield different results. Moreover, evolution is the result of random processes, and if we could start the evolution of life on Earth again, the results would probably be widely different.

Much depends on the way life started in the various systems. If the origins are completely unrelated, it might be expected that not only the body shapes would be different but also that the biochemistries may be completely independent from each other, provided that living beings based on a biochemistry different from ours may exist. If life started from organic material (the so-called building-blocks of life) that spread through the galaxy, there will be similarities – at least as far as the biochemical processes are involved.

If the supporters of panspermia are right (and I think it is unlikely), and all living beings derive not only from organic material but from some sort of

---

[2] The discussion about whether intelligent living beings must have an aspect similar, in a general sense, to that of humans is usually referred to as the problem of the *predominance of the humanoids*.

Fig. 1 The Great Moon hoax of 1835. Sketch of an inhabitant of the Moon allegedly seen by Sir John Herschell with his telescope, from the frontispiece of the book *Delle scoperte fatte nella Luna dal Dottor Giovanni Herschel*, (On the Discoveries made on the Moon by Dr. John Herschel), published in Naples in 1836

organised matter containing genetic instructions coming from space, then similarities may be greater, and they will be even larger if what is carried through space are actual viruses or spores. In this case, even convergent evolution might be possible.

It was a fairly common opinion that similar body forms could evolve independently, provided that the environments were not too different, owing to convergent evolution. However, this consideration has also been discounted: convergent evolution can produce similar results only if the genetic material of the species involved derives from an ancestor that already

Fig. 2 Sketches of aliens reconstructed on the basis of the descriptions supplied by abductees (sketches by Brian Mansfield and Michelle Sullivan, reported in [8] and then in [2])

possessed the potential for the common form. This rules out convergence in the case of organisms evolving from distinct roots, even if the environments were identical. The classical example of the ichthyosaur, the shark, the tuna and the dolphin (a reptile, two different types of fish, and a mammal) may be countered by observing that these apparently different creatures are genetically quite close. They all belong to the same phylum (chordata) and their seemingly distant evolutionary places are caused by our viewpoint of their divergence in the evolutionary tree [11].

One hypothetical example of convergent evolution and of the predominance of the humanoids is the result of the study by palaeontologist Dale Russel of the Canadian Museum of Nature. Starting from the hypothesis, that the catastrophe that wiped out the dinosaurs and, as it was then believed, indirectly delivered the Earth to mammals, had never happened, he simulated the evolution of reptiles toward intelligence. The result is a humanoid 137 cm tall, called 'dynoman' by researchers (Fig. 3): the similarity of the body structure with that of *Homo sapiens* is apparent. However, this exercise is about the evolution towards intelligence of terrestrial reptiles and terrestrial mammals which, as said earlier, is a completely different thing from the predominance of humanoids as applied to ETIs.

The very concept of 'humanoid' aliens may mean different things. Sometimes it is interpreted in a literal way, a humanoid being an ETI with two legs and two arms, a body that has an upright posture and a head in which the brain and the most important sensory organs are located. Another interpretation is a living being with any body shape but with an intelligence comparable

Fig. 3 Model of a hypothetical humanoid, deriving from the evolution of dinosaurs, next to his ancestor, the *Stenonychosaurus*, a small dinosaur living between 70 and 80 million years ago in Canada [2]

with human intelligence, both in quantity and quality. As already stated, in humans intelligence goes together with consciousness, and we take usually it for granted that this is the case for all intelligent beings, but this is another point that can be debated [2].

Underlying the hypothesis of the predominance of humanoids there is the implicit assumption that the human anatomy and/or intelligence are a sort of natural masterpiece, the apex of evolution and adaptation, at least in terrestrial-type environments. Most of the supporters of SETI rightly assume that, if we ever contact an ETI, it will be much older than ourselves, by many millions (or more) years (see below). Thinking in terms of terrestrial standards, these are times in which a species can evolve and become extinct, so the very assumption that humanoid ETIs (in any sense) exist means that this is the ultimate form toward which evolution tends—a very anthropomorphic (and perhaps chauvinistic) approach.

However, there is no doubt that the human body and intelligence are very successful products of evolutionary pressures on this planet. Moreover, the very concept of a 'generalist' animal, which can occupy many ecological niches and spread out over the whole planet (and many other planets if the promises of the space age are to be taken seriously) is a much more successful answer than a host of 'specialist' animals, each one suited for a particular task.

It may be argued that physical evolution may stop with the birth of intelligence, and that then only cultural (and technological) evolution takes place. A conscious species that attributes a large value to the life of individuals, tends to oppose natural selection with increasing success as soon as science and technology advance. Medicine is, after all, a long battle against natural selection: humans do not want survival of the fittest to apply to their own species, but that each individual is granted the longest and the best possible life, even those who are the least fit to survive. Medicine has to face the problem of mending the damage caused by an individual's genetic inheritance, by correcting the defects carried by the individuals who could not survive in a natural situation, but only do so thanks to science. In this way, an intelligent species works to maintain its own genetic inheritance unchanged, thus stopping the mechanisms (mutation and natural selection) of evolution.

There is little doubt about the ethical propriety of this way of proceeding and it is a specific duty of medicine to grant a life as healthy and normal as possible to every individual.

Here again we may argue that these statements are too anthropomorphic: this is the result of our own ethics, and perhaps only of our modern ethical stance. In the past, we had cultures in which there were no ethical problems in allowing new-borns that appear unfit to die or directly killing them. (Nowadays we would say that this was by performing artificial selection instead of natural selection). Some ETIs may behave like Spartans!

But already today, and probably more often in the future, medicine could have another task: not only allowing everybody to live, so stopping evolution, but to facilitate it, by producing favourable mutations. On this subject there is general agreement that practices of this type are to be morally condemned. The negative moral judgment, however, may be motivated by two types of reason: the intrinsic illicit character of any practice that aims to modify the human species, and the opportunity of avoiding any intervention for fear of causing damage instead of benefit. In the first case, the ban on these practises is likely remain in force forever, whereas in the second it is only provisional: when the techniques in this field have improved, what now is forbidden might become acceptable.

These considerations apply to all ETIs, who might behave in different ways when faced with this ethical choice. It seems that we might find two types of ETIs: species undergoing little physical changes even over a long period, and others, who design their bodies and, at least partially, their minds, using genetic engineering. The latter may eventually differentiate into many species adapting themselves to different environments, or different tasks, or even just in accordance with personal taste.

3   How Old Are ETIs?

As already stated, most SETI scientists are sure that there is an overwhelming probability that ETIs are much older than we are. The point is that the timescales of the universe are completely different from those of human history. The Universe is about 13.8 billion years old and the Solar System about 4.6 billion. Life is about one billion years younger; a precise dating is unknown, but the figure 3.7 billion years is often put forward. Complex life is about 500 million years old and our species just 100,000 years old. We have had the technology allowing us to communicate over interstellar distances [12] for less than 100 years. These figures are different from each other by several orders of magnitude, so that any intelligent species is likely to be older than us, perhaps even by as much as a billion years, a time larger by several orders of magnitude relative to our historical time, and also to the times involved in biological evolutionary processes.

It seems that the processes of physical, chemical, biological and social evolution are each faster than the preceding stage by a few orders of magnitude.

All that has been said about the assumed great age of ETIs might be false if the duration of an intelligent species is comparable to our historical or biological times. In this case the only result would be that the number of intelligent species in our galaxy would be so small that we are likely to be alone.

So either they don't exist, or they are much older than us.

The point is that this difference in age is so great that we cannot even attempt to describe an alien species, and aliens would necessarily be excluded not only from science fiction but also from any scientific attempt to understand them. That is not to mention the fact that the Fermi Paradox would become really difficult to solve, without excluding either the existence of ETIs or the possibility of interstellar travel.

A number of explanations have been put forward to get us out of this impasse.

Two of them may be mentioned. First, the early stars consisted of just hydrogen with some helium, and they produced heavier nuclei, which were then released in space and entered in the formation processes of subsequent stars and planetary systems. Reasoning in this way, stars with a metallicity high enough to have terrestrial-type planets rich in the elements needed for life (carbon, oxygen, nitrogen, etc.) cannot be extremely old. This probably means that life would be impossible in a planetary system 10 billion years old, but would not apply to systems, say, 6 billion years old. However, expanding this argument, it is possible to exclude systems older than the Solar System.

Another explanation may be that the Universe has only allowed the existence of advanced living organisms quite recently. One reason may be gamma ray bursters, which seem to be much less frequent in recent than in ancient (in a cosmic sense) times. Because each of them may sterilize a large part of a galaxy, the duration of a planetary biosphere may have been too short to allow complex, and above all, intelligent life to evolve, when such bursters were much more frequent than at present. In this case, it would be reasonable to assume that no intelligent species would be much older than ourselves, and it is possible to attempt to describe these older, but not much older, ETIs.

4 Technology

Does an intelligent species need to develop technology? There are a few considerations that suggest a positive answer. On Earth, hominids started developing technology when their consciousness and intelligence were far less advanced than those that characterise *Homo sapiens sapiens*. Any advance on the road to intelligence was marked by a revolution in technology and when a fully conscious and intelligent species appeared, differentiation in culture and the duration of a particular technology began, suggesting that further technological and cultural advances were not driven simply by physical evolution, but also by the different cultural environments which are different in space and which also changed with time.

In general, technology has considerable value in the natural selection process and the value of intelligence is basically expressed in the possibility of developing technology. If intelligent beings are 'generalist' animals, this is mostly because they are able to use objects that increase the potential of their bodies, working like prostheses. Examples are an axe that makes up for the lack of claws or fangs, a skin garment to make up for the lack of fur that would protect from the cold, and so on. What other species accomplish by slowly modifying their bodies, intelligent life forms obtain in an incomparably shorter time by implementing tools and devices that are purposefully designed [13].

Another reason why intelligence and technology are linked to each other is a sort of feedback action: technology is a consequence of the development of intelligence, but itself causes problems which require an increase of the size of the brain and thus a development in intelligence. This may be easily understood by comparing the area of cranial cortex that is devoted to the control of the hands with that used to control the feet.

As a consequence, we should expect an intelligent being to have some organs devoted to the manipulation of objects, some sort of arms and hands in a general sense (at least in the sense used in robotics).

A further consideration about technology: on Earth the use of fire was of paramount importance in the development of technology. A source of chemical energy, readily available and exploitable with simple technology, might be an important requisite in the road towards higher intelligence. In an oxygen-rich atmosphere energy may be obtained easily by burning any type of fuel – with a biochemistry similar to ours many organic materials will serve. Although life uses water as a solvent, the impossibility of using fire is a severe handicap on the road to intelligence for any aquatic life form.

Different conditions might be found in the case of living beings based on a different biochemistry: in a reducing atmosphere, energy may be obtained by burning oxidising materials. It must however be noted that the amount of oxidizer required to burn a certain quantity of fuel is much greater than the quantity of the latter: it is thus easier to burn fuel in an oxidizing atmosphere than to burn an oxidizer in a reducing atmosphere.

As a general result, it is likely that ETIs have some sort of organs for locomotion and manipulation. Manipulatory devices are best derived from locomotion organs like legs, whereas it is more unlikely that they would derive from flippers or other swimming devices. The difficulties linked to intelligent aquatic life suggest that it unlikely that ETIs can live, swimming in a fluid in which they are supported by hydrostatic forces—obviously there is nothing against a lifestyle in which a land life form spends part of time in the water, like otters.

Organs for the manipulation of objects might derive from wings, and we cannot exclude the possibility that an ETI might derive from a flying life form that flies according to aerodynamic forces. However, the large mass of the brain may be a serious drawback to flying, so it might be expected that, as the brain evolves, the ability to fly reduces. An exception may be an ETI that has evolved on a planet with low gravity and perhaps an atmosphere with a high density, where there might be few problems in supporting a heavy brain.

Technology poses a problem linked with the age of ETIs: how is it possible to understand or, far less, to predict, a technology so much more advanced than ours? We are experiencing a phase of such rapid technological progress that predicting a technology more advanced than ours by a few decades is practically impossible, not to mention a technology more advanced by thousands of years (or millions, or even billions of years, given the timescale of the universe).

The result is, again, the complete impossibility of describing aliens. This is even more true if the we accept the hypothesis that technology proceeds following an exponential curve, finally leading to a technological singularity [14]. In this case, however, there may be several alternatives. A first, radical position, states that there are limitations to technological advancement and that we are probably quite close to reaching it. Since the same limitations should also hold for ETIs, their technology may be barely more advanced than ours, even if they are older by billions of years [15]. The supporters of this view have strong doubts about the possibility of interstellar travel, because they state that if we are close to reaching the limits of technology, there will be no time to develop it.

A limit to technology may not, however, be an absolute limitation but, at least up to a point, a limitation linked to our own capabilities. An often quoted example is what happened in our own past: the various species of hominins that preceded *Homo sapiens* had phases of quick technological progress but when they reached a certain level, their progress stopped until a new evolutionary step occurred, usually accompanied by a growth of brain volume. At this point, equipped with new intellectual faculties, humans could resume their path of technological advancement (Fig. 4) [1, 2].

The assumption on which Fig. 4 is based seems to be completely antithetical to the one that is the basis of the theory that suggests development leads to a technological singularity. The latter assumes that technological progress follows an exponential pattern over time, while the curves in the figure are S curves, with a horizontal asymptote. A similar effect is seen in the development of population: the typical exponential growth (Malthusian style) from a certain point may change into an S curve, finally yielding a steady state condition.

This may not be so, because the exponential growth might be the result of the several S curves tending to asymptotes that exponentially grow in value. What grows exponentially would not be the technology of a single species, but the technology reached by the various species following one another. This is in obvious contrast with the assumption that intelligent species may put an end to biological evolution.

The assumption made in the novel is that technology eventually follows a slowly growing linear trend, which may well be the outcome of a sort of S curve, adapted to an open environment, dominated by the expansion of the civilization into space. The older ETIs have a technology much more advanced than ours, including interstellar travel, but their further advancement is slow and this is compatible with a basically stable society.

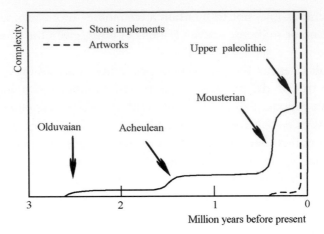

Fig. 4 The increase in complexity and variety of stone tools and in the complexity of artwork in the development of hominins. Note the discontinuities linked with the appearance of *Homo habilis* (Olduvaian culture), *Homo erectus* (Acheulean culture), primitive *Homo sapiens* (Mousterian culture), and finally of the *Homo sapiens sapiens* and *Homo neanderthaliensis* (upper Paleolithic culture)

An even more important point is the consequence of the observation that during most of its development humans developed a technology based on empirical experience, and trial-and-error attempts. Progress attained in this way is not only slow, but in many cases the results may be provisional: what was obtained by a master craftsman could be lost due to the death of the 'owner' of the secret or of a group of adepts. Ancient history and mythology is full of 'lost secrets'.

Everything changed when a technology based on experience was gradually replaced by a technology based on science, a process that started in western civilization in the Middle Ages [16]. The basis for this achievement is essentially the belief that the world is based on rationality, and thus things may be explained by the human mind. Science supplies an explanation of things, and can be put at work to invent new technologies and to rationalize old ones. In turn new technologies allow the construction of new instruments, which allow science to further advance, in a cycle which becomes faster and faster.

Actually, this interaction between science and technology is not essential in gaining some of the highest achievements of modern technology. After all, neolithic farmers created new species of plants and animals without any knowledge of genetic engineering, and ancient blacksmiths created steel without knowing about iron-carbon phase diagrams, even introducing complex treatments like carburizing or nitriding to obtain harder surfaces. Instead of proceeding through scientific knowledge of the properties of the materials they were working with, they discovered the relevant procedures by trial and error, often giving *a posteriori* explanations based on magical or mythological reasoning. After all, the most complex devices existing on Earth, living beings, were not designed, but evolved without any conscious mind designing them on the basis of scientific knowledge! We cannot be sure that all modern technology could be developed without our scientific background, but we can be certain without any doubt that, if this could be done, it would take an incomparably longer time. An alien civilization might develop an advanced technology, perhaps even interstellar travel, without any scientific knowledge of the universe, but this process would be painstakingly slow.

Because, on our planet, science was developed only by one of the many civilizations [16] and only a relatively short time ago, scientifically driven technology may be an exception instead of a rule. Moreover, we are even not sure that on our planet the development of science will continue indefinitely. Alarming symptoms are, for instance, the scientific illiteracy of the majority of the population, even in the most scientifically advanced countries; the widespread extent of pseudo-sciences and of ideologically driven beliefs, even occasionally among the scientific community (such as the acceptance of Lysenko's theories in Soviet times and the arbitrary use of scientific concepts outside their field of application [17]) and the growth of anti-scientific and anti-technological movements.

The phase of quick technological advancement that we are witnessing now is thus an exception caused by a number of particular circumstances. One of the reasons of a slow-down of the pace of technological advancement might be the lengthening of the duration of human life, with a consequent decrease in the birth rate and an increase in the average age of the population. Another one might be expansion into space, with a decrease in the population density and an increase in the distance between human communities, causing an increase in the time needed for the various groups to communicate with each other.

Technology also influences the way intelligent beings get the food and energy required for their vital functions. On Earth a life form may be autotrophic, heterotrophic or a mixture of the two forms. An autotrophic life form obtains its energy from some nonliving source (from the Sun, but in some cases from hot submarine hydrothermal vents, chemical reactions, etc.) and uses it to synthesize the chemicals needed for its life from non-living matter. A heterotrophic living being uses organic material already produced by other life forms to obtain both energy and the substances it needs.

Autotrophic beings may not need any sort of mobility, whereas heterotrophic ones usually need to move to obtain their food. A particular form of the latter are predators. If predation is defined as a procedure '...where the predator strives to reach a final goal (the consumption of the prey) as quickly as possible, and the prey tries to terminate the sequence of events as early as possible. . .' [18] (as reported in [19]), it is clear that it is difficult to imagine a predator having no mobility, although some examples may be found among plants and animals on our planet.

The energy consumed by the human brain is large: it constitutes only 2 percent of the weight of the body, but it consumes 20 percent of the energy. From a purely energetic viewpoint it seems that an intelligent species needs to be heterotrophic, and possibly quite high in the food chain, because it needs food that can be assimilated quickly and supplies a lot of energy. These considerations would indicate that ETIs should be, at least initially, predators.

Perhaps it not just a matter of energy. Many palaeontologists think that the need to coordinate individual efforts for hunting, in absence of a body specialized for this way of living (humans do not run particularly fast, although being able to run for a long time, and are not endowed with fangs, claws or other natural weapons), and the increase of the amount of time that is free from the primary function of feeding, undoubtedly had a positive effect on the development of intelligence [20]. Conversely, an autotrophic species does not need to be particularly smart to obtain energy.

Intelligence however caused what we call the Neolithic Revolution: our species learned to produce its own food by cultivating the land and raising animals instead of gathering vegetables and hunting animals. We are still predators, but of a very peculiar type. This process has not yet been completed: the way we exploit the food from oceans is still mainly consistent with pre-neolithic practices.

However, a second neolithic revolution is predictable: we may, in a not too distant future, produce our food directly, synthesizing it from inorganic matter. It is probable, at least at the beginning, that we will give our synthetic food an appearance, taste and smell similar to that of 'natural' food, but the advantages are huge. By this means the quantity of food that we can produce depends only on the quantity of energy we have at our disposal, so that an energy-rich society will be capable of nourishing a much larger number of humans, relative to what was possible in the past.

Two points are worth mentioning. First, the term 'natural' used above as opposed to 'synthetic' may have meaning if it refers to the process of producing food, but not if it refers to its product. If a molecular structure is reconstructed, atom by atom, the synthetic product cannot be distinguished from the natural one. This may also apply to the process, if the bioreactor producing meat, for instance, uses the same biological reactions as those that take place in the original animal cell.

The second point is that the whole thing may have its own risks, like those that accompanied the original Neolithic revolution. A sparse population of hunters-gatherers was much less susceptible to droughts and other natural catastrophes, than a much more crowded agricultural population. Similarly, an energy crisis might cause havoc in a population feeding on artificial food.

A further observation is that the widespread use of technologies to produce food from non-living matter may cause a more predatory style of living to be considered unfit, barbarian or even unethical. This is the case with the aliens in the novel. At any rate, an alien species which is much older than ourselves is likely to be technologically autotrophic, even if originally it was as heterotrophic (with a predatory lifestyle), as we are.

5   Body Shape

From what we can infer from the situation on Earth, the rise of intelligence and consciousness is linked to the development of the nervous system. Neurogenesis seems to be directly linked to multicellular organisms, and as soon as the nervous system started to become complex, the process of encephalization, i.e., the concentration of most of the nervous system in a central 'brain', began. To reach a complexity sufficient for intelligence and consciousness, the nervous system—the brain, if the encephalization process is a universal trend—must have a minimum size. If the processes of extraterrestrial life are based on carbon chemistry (organic chemistry), this size cannot be much smaller than that of the human brain, because the size of organic molecules is the same throughout the whole universe. This statement must be considered as an estimate of the order of magnitude required, because it seems that structural complexity counts more than sheer size. If, despite the fact that organic molecules and amino-acids have been found on comets and meteorites, and even in interstellar space, ETIs are based on a different chemistry, or on non-chemical processes, they may have different sizes, but at any rate there should be a minimum size.

On Earth, most animals moving on a solid surface use some form of legs. Continuous rotational motion is unknown in nature, except in microscopic organisms [21]. Evolution is characterized by a gradual reduction of the number of legs: from the filaments (parapods) of the *Annelida* (for instance, the millipedes) to the articulated legs of arthropods, there was a continuous reduction of their number (10 in crustacea, 8 in arachnida, 6 in insects). With terrestrial vertebrates, the number of legs reduced to four. Apodal animals, like snakes, are not here considered as candidates to develop intelligence, also because of the lack of manipulatory organs. It is thus difficult to imagine a type of land locomotion not based on legs suitable to intelligent beings.

A large number of legs, together with a low position of the centre of mass (that is a low height of the centre of mass when compared with the 'track' of the legs) allows the animal to remain easily in static equilibrium during all phases of walking. To achieve static equilibrium during walking an animal with less than six legs must have an articulated body. By coordinating the motion of the legs and of the body, even a biped can remain in static equilibrium when walking. In general, the larger the animal, and the lower the planet's gravity, the easier it is to remain in equilibrium on fewer legs, in the sense that the response required from the nervous system to avoid falling down may be slower. From this point of view, low gravity makes the operations linked with motion simpler.

Against this, a maximum walking speed exists for any animal; to go faster the animal must change its gait, performing a transition from walking to running or jumping [22, 23]. This speed is greater for taller animals and under higher gravity. Since high speed is in general an important factor in natural selection, either to run away or to chase food (or anticipating others in the search for food), there is a strong incentive to shift from walking (a sequence of static equilibrium positions) to running (alternation of equilibrium positions with others in which the equilibrium is not guaranteed). Large animals, possibly with a smaller number of legs, may thus have an advantage. Animals, not able of using a biped stance, may use the tail as a support for stabilizing the body while walking on the hind legs: the tyrannosaurus is an example of an animal that adopted an erect posture thanks to this third support, even though it undoubtedly had a primitive brain.

These considerations do not prove that all intelligent species must be biped, but they are good clues that point in this direction.

Humans (and humanoids) are not only biped, but also have an erect posture with the head above the trunk, instead of forward of it, as in most land animals. This is made necessary by the increase in the weight of the brain, reducing bending stresses in the spine. With encephalization, the brain is located in a strong box of bone, close to the organs of hearing and sight. This seems to be reasonable, and might apply to many different configurations and even to different biological processes. Also setting the organs for sight and hearing (generally speaking, the detectors for electromagnetic waves and pressure waves in the surrounding medium) and olfactory organs (again generally speaking, organs able to recognize chemicals suspended in the medium) close to the brain and in a head, located above a more or less articulated neck, allows the environment to be better monitored. It is even more important if the eyes are in a frontal position, allowing binocular vision, which is important in the evaluation of distances.

While the frequency range of the electromagnetic or pressure waves or the sensitivity of the sensory organs may differ widely from one environment to another (or, as terrestrial biology shows, even between organisms evolved in the same environment), it is difficult to imagine a radically different interaction with the environment. Less important seems to be the location of the mouth in the head and close to the brain: there seems to be no functional reason for that, whereas it may be well be a result of 'historical' reasons. The simplest multicellular animals had the mouth at the front end and the excretory opening at the opposite one; this layout was basically maintained later. At any rate, it is likely that any heterotrophic animal ingesting food has taste organs, i.e., provision for recognising the chemical nature of the food, and the most obvious solution is that they are placed at the food intake opening.

Another sense in humans is touch. Any manipulatory organs, at the least, must have some provision for evaluating the force exerted and the characteristics of the surface on which they operate. Similarly, a set of proprioceptive organs must be present, to assess the relative positions of the various parts of the body.

The senses just listed must be interpreted in the widest possible way: any organs, enabling a being living close to a gamma-ray source, emitting little light, to detect gamma rays, may be referred to as sight organs. Similarly, organs able to detect pressure waves in a fluid may be considered hearing organs, even if they have little to do with sound.

Given these broad definitions, it is impossible for us to imagine other senses in addition to ours or substituting for them. We must then conclude that either they do not exist or else we are so blind to a part of the real world that we completely miss some of its features. This is a theory advanced by those who support extrasensory perception (e.g., ETIs with telepathic powers, etc.) but there is no proof, not even a serious clue, that such senses might exist at all.

The sensory organs in animals on Earth agree with strict criteria concerning economy. A number of eyes greater than two, for instance, might be an advantage, but would require an increase of the volume of the brain, since a non-negligible part of the brain of upper animals is devoted to the processing of signals from the eyes. It would be a waste to devote precious 'computing power' to sight, above that essential to analyze images supplied by two eyes, and compare them to obtain stereoscopic vision. As a consequence, few animals have more than two eyes. The same applies to hearing, while taste and smell, which do not need stereoscopic processing, have only one symmetrical organ.

These considerations show that the general body plan of humanoids is in a way optimized for an intelligent being, and that intelligence has been favoured by the challenges created by the humanoid form. In particular, a bipedal form with erect stance seems to be a logical consequence of an intelligent being with manipulatory appendages that have derived from the front legs of quadrupeds, together with carrying a heavy braincase with multiple sensory organs at the forward end of the body. However, such a configuration exerts demands on the equilibrium organs and on the brain to ensure steady and rapid bipedal walking and running. These features were obtained only in *Homo sapiens*, whereas *Australopithecus* was far more apelike, and still earlier members of the *Homo* genus were even more primitive in this respect.

Do these considerations allow us to take a firm stance in favour of the predominance of humanoids? To state it in another way, are these considerations sufficient to conclude that it is likely that ETIs have a body-form resembling humans?

Probably not. One thing that has not been considered up to now, is symmetry. On Earth all terrestrial and many marine animals, starting with the arthropods, and in particular all vertebrates, have bilateral symmetry, at least as far as the exterior body shape is concerned. The distribution of the internal organs may depart from this feature, but symmetry is basically maintained. All animals with bilateral symmetry have an even number of legs and the few that stand on three supports use a strong tail (located in the plane of symmetry) for this purpose. Likewise, they have an even number of organs for sight and hearing and, if they have a single mouth and a single anus located on the plane of symmetry, this is because of the fact that they have a single digestive tract. Yet there are two nostrils, but this seems to be a completely arbitrary fact.

This symmetry is so familiar to us that most of the machines we build have a plane of symmetry and, even when we try to imagine strange shapes for fictional aliens, we end up with something that has similar features. On our planet, evolution also developed living beings with various orders of axial symmetry, such as starfish (mostly of order five) or octopuses (order eight), but none of them has articulated limbs or is adapted to live on land; the mainstream of evolution that led to intelligence took a different path.

The presence of a symmetry plane has some peculiar dynamical characteristics for a moving object and robots with axial symmetry (mostly of order six) did not show any particular advantages as walking machines [24]. However, this does not seem to rule out the possibility that in other biospheres intelligence may develop in animals with completely different characteristic symmetries. There is no doubt that we can imagine (although with difficulty) ETIs with no symmetry at all but anything of this sort may appear very awkward to us. Axial symmetry is a possibility, but other symmetries, far less familiar to us, are also possible. In the mineral world, crystals exhibit many possibilities in this respect; perhaps astrobiology will supply other examples. The fact that we cannot imagine them is immaterial: according to Galileo, this is exactly what we should expect from extra terrestrials.

6   The Aliens in the Novel

In accordance with what we have seen in the previous sections, the aliens appearing in the novel are older than humans from Earth but not so much that their way of living is so alien as to be beyond imagination. The oldest species are about 100 to 150 thousand years older than we are, while planets hosting species which have not yet reached intelligence are mentioned. The older species have a technology that is far more advanced than ours, and reached interstellar travel a long time ago, but the model assumed in assessing their science and technology is an evolutionary model with a slow rate of evolution.

Technological advances occur through slow refinement of the technologies and scientific theories that underlie them and the expanded human lifespan, together with the great distances and relatively slow communications, prevents any such advances from being rapid.

Rapid progress, like the type we are now experiencing, is even considered as a pathological situation, typical of that period one character calls a 'pre-space crisis', characterized by 'rapid development, demographic expansion and perhaps even a good dose of aggressiveness', when technology seems to create more problems than those it is able to solve.

The number of intelligent species is high, but compatible with that obtained from Drake's equation [2], with a moderately optimistic set of parameters. The technologically advanced species (by which the species able to travel among the stars are intended) formed a political entity, called the Galactic Confederation, which extends over the largest part of the Milky Way (Fig. 5). This confederation contains about 400.000 species, with about 10 million inhabited planets. This means that the average distance between the places of origin of intelligent species is something like 500 light years, and that between inhabited planets is about 200 light years. The overwhelming majority of planetary systems host no life and many are what, in the jargon used in the Galactic Confederation, are called type C planets (well developed life, but no intelligent beings).

Planets having no life are assumed to be not directly habitable by humans, since we cannot breathe in their atmosphere, if they have one. This is quite specific, to the point that finding free oxygen in the atmosphere of an exoplanet is considered as a clue of the presence of life. In the novel, most inhabited planets are said to have been terraformed, and terraforming planets seems to be a routine operation in the Galactic Confederation.

Some planets, like Earth, are type B planets, i.e., they host intelligent life forms, which have not yet mastered interstellar travel. The politics of the Confederation is to avoid any contact with these planets: this assumption may

a

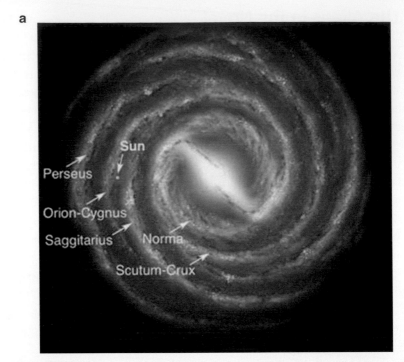

b

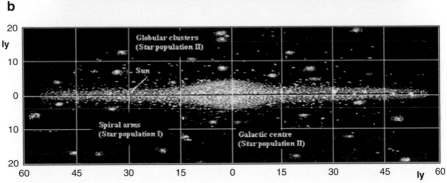

Fig. 5 A 'map' of the Milky Way indicating the position of the Sun with respect to the spiral arms. (a) plan view and spiral arms (b) cross section. Scales in thousands of light years

give one explanation of the Fermi Paradox [5], and was first put forward by Tsiolkowsky [25], even before the Fermi paradox was devised. The hypothesis that advanced civilizations carefully avoid contact with less advanced ones has been put forward in many scientific papers and science-fiction novels, and was taken to its extreme in *Star Trek* with the 'Prime Directive'.

In the novel, the assumption that the biogenic elements are brought from space, possibly by comets, causes the biochemistry of all living species to be quite similar. Genetic information is encoded in some molecule which perform the functions that is performed by DNA in terrestrial life. While nothing is said about the chemical nature of these substances, their information storage, and their function in the transmission of information from generation to generation are similar to those of DNA everywhere. Nothing is said about the possibility that information might be stored in some other way, what is important is that the result is an evolutionary process leading to more and more complex forms of life. Although different, these forms of life are assumed in the novel to be fairly uniform as far as their general characteristics are concerned. Many intelligent species are humanoid in the broadest sense, and in many evolutionary lines it is possible to find common typologies: the terms arthropod, reptile, mammal, etc., may be applied in a broad sense, although this doesn't mean that types that are not present on Earth may not have evolved in places where the environmental pressures lead to them.

The anthropologists (in a broad sense) of the Confederation use a set of 3 numbers (x–y–z) designating the number of arms, wings and legs that characterize the body plan of the various species. What we call humanoids are thus of the 2–0–2 type, and some of the characters of the novel are 2–2–2. At a certain point humans of the 4–4–6 type are mentioned to give an example of the variety of the possible human forms.

The point that is the basis of the novel is that the similarities in the life processes and the parallel evolutionary paths caused the intelligences, which had evolved in the various planetary systems, to be similar enough to allow them to understand each other, to the point of using the term 'human' as synonymous with 'intelligent life form'.

It might be said that all ETIs are thus humanoid, not in the physical, but in the psychological sense. This allowed the various species to understand each other and to form a complex, articulated society.

The society described is thus a multi-species society, even if the great distances between the planets inhabited by the different human species make contact between them infrequent. Actually, the only institution in which the species enter easily in contact with each other is the Confederation Starfleet. The few planets where it is possible to find humans belonging to different species are the capitals of the various regions (the sectors) in which the Confederation is subdivided and above all the Confederation capital itself.

Interbreeding between species is obviously impossible, even if the mechanisms of sexual reproduction are not very dissimilar from species to species, and sexual intercourse between individuals of different species are frowned upon, sometimes with hypocritical prudery. This makes a multi-species society something deeply different from a multiracial society: the difference between races are considered small and irrelevant, whereas those between species are objective and important, to the point that they may make it impossible, or at least difficult, for members of the various species to live in the same environment.

The advanced technology and advanced age of the species doesn't make them 'morally advanced'. Moreover, one of the assumptions at the base of the novel is that there is not such a thing as 'moral progress'. Good and evil co-exist in the soul of human beings of all species, and it is impossible to make a clear cut differentiation between the two. Some of the characters attribute a metaphysical meaning to these considerations.

This society, consisting of different species, all essentially human, at a certain point encounters a species that is radically different from the others, something defined as a non-human intelligence. No understanding is possible with such an entity, and the only relationship between that species and the others rapidly evolves into what we see most frequently in nature when at least one of the species involved is heterotrophic: the relationship between predator and prey [19]. Because the new, really alien, species seems to have a competitive edge over the others, it plays the role of predator.

The possibility that in our galactic neighbourhood there are hostile aliens is usually called the Nasty Aliens Hypothesis (NAH). Many SETI scientists think that, owing to the impossibility of excluding this hypothesis, it is wise not to broadcast any signal aimed at ETIs, i.e., not to perform active SETI. Although it is the general opinion that this kind of danger is quite remote [6], the few messages broadcast in the past have been widely criticized. In the novel, attention is focused on the NAH, and in particular on the difficulties for humans (from Earth and extraterrestrials) to understand an ETI of a completely different type, which is dubbed non-human.

7 Space Travel

The species belonging to the Galactic Confederation described in the novel have all reached the stage of FTL (Faster Than Light) interstellar travel. The very possibility of FTL travel is controversial, ever since, at the beginning of twentieth century, Einstein's theory of special relativity stated that no material object or transmission of information may move with a speed greater than that of light. The theory of relativity has been confirmed by countless experiments and today few physicists have any doubts about it. Science-fiction writers realized that few interesting plots involving contact with ETIs or true interstellar civilizations could be developed if this cosmic speed limitation was taken into account. In general, the attempts to overcome it are based on another result of modern science: the awareness that the Universe is much more complex than it may appear and that apart from the three spatial dimensions of which we are aware, and the time dimension, other dimensions that are beyond our comprehension might exist. If it were possible to warp space-time in a way to cause the points of departure and arrival of the journey to be closer to each other than they are in the normal sense, it would be possible to reach a destination in a time shorter than that taken by light to cover the same distance. This approach to FTL travel does not imply actual motion at a speed greater than the speed of light and thus does not violate special relativity.

Naturally, we must not imagine that modern science represents the ultimate pinnacle of human knowledge. There are, moreover, undoubtedly many things that it is not able to explain. Relativity might in the future be replaced by other scientific theories, which will include it (because it had been confirmed by many experiments), just as Newtonian mechanics is included in relativity as applying under particular circumstances. A deeper knowledge of the Universe might perhaps open up perspectives that are unimaginable today.

However, general relativity has not yet been studied deeply enough. For instance, some solutions of its basic equations yield particular types of black holes, often called wormholes, which, in an impressive way, look like the space-time tunnels of science fiction.

Today scientists are more open-minded on the subject, than they were even in the recent past, and scientific papers on the subject have been published in journals and presented at conferences. However, we are still far from being able to state anything convincing on the matter. Moreover, even if we could prove the theoretical possibility of FTL travel, there would still be enormous problems to be solved before implementing the relevant technology. The huge quantity of energy required for FTL interstellar journeys constitutes a problem that is far from being solved.

The most common approaches usually followed in describing FTL travel are essentially of two types:

- The use of 'space-time tunnels', like the so-called wormholes mentioned above, to connect distant points with a path much shorter than their spatial distance, thus allowing travel over distances of many light years in a short time. In many cases, the term 'hyperspace' is used in science fiction to designate the multi-dimensional space that is crossed by these 'tunnels' (Fig. 6).
- The creation of a sort of spacetime bubble of warped spacetime around a space ship, which moves inside it at speed lower than that of light, but travels between points in a time shorter than that taken by light. Usually this approach is referred to as 'warp drive'.

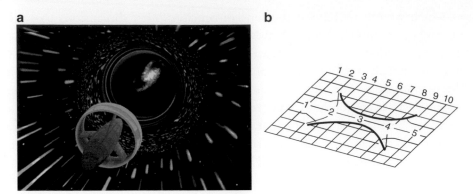

Fig. 6 Wormholes. (**a**) A starship entering a wormhole (NASA image); (**b**) Sketch of a two-dimensional wormhole, connecting two points on a plane. The *numbers* indicate equal distances through normal space and through the wormhole. An actual wormhole is thought to exist in multi-dimensional spacetime. In the first case the tunnel may be a pre-existing anomaly of spacetime or something created by the 'hyperdrive' of the starship and may be either temporary or permanent

In the novel the first of these two approaches is used and little is said about the details, except that there are entry and exit points, called gates, to and from hyperspace in all planetary systems, ready to be used by travellers. The Starfleet had installed and maintains automatic stations at these 'hyperspace gates' to improve the safety of space navigation, in a way not dissimilar to what is done on Earth with lighthouses. Contrary to what is shown in Fig. 6, in the novel the hyperspace gates are said to be invisible from normal space.

From a scientific viewpoint, wormholes and warp drives are just hypothetical ideas, in a field that is largely speculative. It is likely that if humans one day succeed in travelling at speeds faster than that of light they will do so by exploiting physical principles and phenomena that are still completely unknown. The theoretical and technological knowledge are all still to be developed, and the only thing that we can do is to study all unknown phenomena with an open and critical mind.

As a result, we are still far from being able to give a positive answer to the possibility of FTL interstellar travel. However, even if there is no certainty that eventually efforts in this direction will lead to anything serious, even over long or very long time-scales, it is possible to imagine the consequences of the possibility of travelling over interstellar distances in times which may be measured in terms of hours and days and not decades or centuries.

Even if radically new technologies do allow fast interstellar journeys, it is likely that they will remain long and expensive. The trick that allows travel at superluminal speed, for instance, will probably be used only at a distance from celestial bodies, and the portions of the journey that are within the original system and the system that is the final destination will be performed at a lower speed. Moreover, it might be necessary to limit speeds because of the risk of encountering micrometeorites or other, even larger, bodies. Hours, or even days, might be required just to leave the departure system and enter the destination planetary system. Interstellar journeys may not take many years, but almost certainly will require many days or weeks as a minimum.

This situation reminds us of the one that prevailed in the nineteenth century, when all the inhabited zones of our planet could be reached, but long journeys, too expensive for most people, were required. However, these communication difficulties did not prevent institutions from building and maintaining political entities, which included territories located in different continents.

Little is said in the novel about the technical details of FTL journeys: it is a technology so advanced that we have the same chance of understanding it as a Neolithic man would have of understanding a nuclear power station. The main character realizes that he lacks the basic concepts for understanding and decides to postpone any attempt in this direction until he has time to start studying the whole matter from its basic foundations. He adopts the same attitude of those of us who drive a car without wondering about how the engine works, which is even more feasible in the story, thanks to the presence of powerful computers that deal with the technical details.

Other important devices related to space travel are usually described in science fiction: artificial gravity, devices to compensate for acceleration and shields. The first two are related, and an acceleration compensator is absolutely essential if speeds close to the speed of light are to be reached in a reasonable time: at an acceleration as high as 5 g, which cannot be withstood by human beings for anything more than a very short period, about 70 days are required to reach the speed of light. If a technology were to exist that would allow an acceleration compensator to be built, something well beyond our present possibilities, it will also be possible to produce true artificial gravity, which will make space travel much more comfortable without the need for simulating gravity by rotation, a thing we can do at our existing technological level.

Shields, at least those that stop radiation, are really essential for safe space travel and a lot of current work is being done in this area. The basic idea is to produce a strong magnetic field, which acts as a miniature magnetosphere around the spacecraft and, thanks to the use of superconducting magnets, we are not too far from achieving this [26]. The shields used by spacecraft in the novel are however much more advanced than this, being able not only to stop harmful radiation but also to provide protection from weapons. Rather than our magnetic devices they seem to be related to the hypothetical 'force fields' which are omnipresent in the novel to the point of substituting even for pieces of furniture such as chairs and beds. It is no wonder that such advanced technology has changed even some of the common things that we are used to. However, even in this fictional world, shields used against weapons require an awful lot of energy, and therefore cannot withstand hostile action for a very long time: the ancient battle between offensive and defensive weapons has reached equilibrium.

8   Artificial Intelligence

Another field in which the ancient galactic civilization has made great progress is that of artificial intelligence. Most machines are controlled by a computer of some sort. In a way, this is nothing more than an extrapolation, and from this point of view not a very bold one, from what we have now on Earth.

No machine has yet reached an intelligence level similar to that postulated by the supporters of strong artificial intelligence (AI), and computers are said to be neither really intelligent nor conscious. However they are programmed so well that it is probable that the computers controlling starships would at least pass the Turing test, i.e., it would be impossible to tell them from a human being during an interaction based on exchanging messages on any subject, or at least on subjects they had been programmed to deal with.

In addition, little is said here about the details: they may be based on a digital technology, on a form of artificial neural network (ANN) or on some other more advanced technology. They certainly have processors and memories, miniaturized material objects, which may be assembled, disassembled, or even destroyed. In the last case, some information may be extracted by studying their remains even after an accident or the voluntary destruction of the device.

In the novel there are no robots in the usual sense of autonomous machines having a more or less humanoid shape, able to perform many different tasks. Actually, the advantages of building general-purpose autonomous machines instead of a variety of different devices, each one specialized for a specific task, are the subject of considerable controversy [27].

Honda, for instance, has developed a humanoid robot, called Asimo, which is intended by its designers to be a mechanical worker. One of the intended applications for robots of this kind is to operate construction machinery, either autonomously or under teleoperation. Under this scheme, the same machine can be operated directly by a human or a robot, because the latter can use the controls designed for the former, but it is questionable whether this strategy is better than directly supplying the machine with enough artificial intelligence to act autonomously or under teleoperation.

It seems that giving a machine human shape perhaps has advantages only where relationships with humans are its most important task, as in the edutainment industry, health care, home services and similar activities. Apart from these applications, zoomorphic or anthropomorphic configurations seem to have few advantages.

The assumption is that, in a society much more advanced than ours, humans are so used to being attended by machines that even in the mentioned applications there is no need to give a humanoid shape to machines. In a sense, it may be said that the society described has gone beyond the stage of anthropomorphic machines. For instance, in the sick bay of starships there is no robotic nurse or doctor, but the patients are directly treated by the starship's computer, which controls the smaller processors operating various devices such as stretchers, which can recover the injured and move them into the sick bay, and other biomedical apparatus. This doesn't necessarily imply that there are no anthropomorphic devices used by people on some planets, but they are not described because they are not present where the action develops.

Glossary and Acronyms

AI Artificial Intelligence

Astronomical unit Distance unit, equivalent exactly to 149,597,870.7 km (92,955,807.3 miles). It roughly corresponds to the mean Earth–Sun distance.

Drake equation An equation, introduced by the radio-astronomer and SETI specialist Frank Drake, yielding the number of extraterrestrial civilizations that can enter in contact with us. The coefficients entering the equation are highly hypothetical, so that the equation is useful to understand which parameters govern the phenomenon, but is unable to supply a reliable numerical result.

Encephalization The evolutionary process leading to the formation of a single central nervous system in the form of a brain.

ETI ExtraTerrestrial Intelligence

Exoplanet A planet orbiting a star other than the Sun.

FTL Faster than light. FTL travel is considered to be a violation of physical laws because, as a consequence of Relativity, neither matter nor information can move at a speed higher than light speed. However, a better understanding of physics appears to suggest some possibilities in this field.

Gamma ray bursters Object producing flashes of gamma rays and associated with extremely energetic explosions that have been observed in distant galaxies, which may last from ten milliseconds to several minutes.

Hyperspace (as used here) Hypothetical multi-dimensional space through which spacetime tunnels connecting points in the normal space may be obtained, to realize FTL travel.

Light year Distance unit, equivalent to 9.4607×10^{12} km (almost 10 trillion km). It is the distance light travels in one year.

Metallicity The metallicity of a star or of another astronomical object (e.g., a nebula) is the proportion of its matter consisting of chemical elements other than hydrogen and helium. The metallicity of an astronomical object may provide an indication of its age, because older stars have lower metallicities than younger stars such as our Sun.

NAH Nasty Aliens Hypothesis. The hypothesis that hostile aliens exist in our galactic neighborhood, so that it is unadvisable to do 'active SETI', i.e., to broadcast radio transmissions toward space searching for a contact. The final consequence is that, if everybody follows this prudential rule of not betraying his position, SETI becomes impossible.

Neurogenesis (as used here) The evolutionary process leading to the formation of a nervous system.

Parsec Distance unit, equivalent to 30.857×10^{12} km or 3.26156 light years. It is the distance at which the Sun-Earth distance subtends an angle of 1 arcsecond.

SETI Search for ExtraTerrestrial Intelligence

Singularity (technological) Hypothetical moment in the future when artificial intelligence surpasses human intelligence, radically changing civilization, and even human nature. It is a point in the future beyond which no predictions may be made.

Terraforming An astro-engineering enterprise aimed at transforming the physical and environmental characteristics of the surface of a planet to make it suitable for supporting human life. The term terraforming, introduced by Isaac Asimov, has been widely used in science fiction, but now the possibility of terraforming planets, and primarily Mars, has been seriously considered. Terraforming a planet raises heated ethical arguments, in particular if the planet may have indigenous life that is likely to be destroyed in the process. One of the main points is whether it will ever be possible to be absolutely certain that no indigenous life exists on a planet.

References

1. J. Heidman, *Extraterrestrial Intelligence* (Cambridge University Press, Cambridge, 1995)
2. G. Genta, *Lonely Minds in the Universe* (Copernicus, Springer, New York, 2007)
3. P.D. Ward, D. Brownlee, *Rare Earth* (Copernicus, Springer, New York, 2000)
4. R. Shapiro, *Planetary Dreams* (Wiley, New York, 1999)
5. S. Webb, *If the Universe Is Teeming with Aliens. . . Where Is Everybody?* (Copernicus, Springer, New York, 2002)
6. P. Musso, The problem of active SETI: an overview. Acta Astronaut. **78**, 43–54 (2012)
7. J. Arnould, *Impossible Horizons, The Essence of Space Exploration* (to be published)
8. C. Pickover, *The Science of Aliens* (Basic Books, Boulder, CO, 1999)
9. G. Genta, Some Engineering Considerations on the Controversial Issue of Humanoids, in *7th Trieste Conference on Chemical Evolution and the Origin of Life: Life in the Universe*, Trieste, Sept 2003
10. G. Genta, On the controversial issue of humanoids: a mechanical engineer's viewpoint. Acc. Sci. Torino Mem. Sci. Fis. **27**, 35–46 (2003)
11. E.J. Coffey, The improbability of behavioral convergence in aliens – behavioral implications of morphology. J. Br. Interplanet. Soc. **38**, 515–520 (1985)
12. P. Morrison, J. Cocconi, Search for interstellar communications. Nature **184**(4690), 844–846 (1959)
13. E. Righetto, *La scimmia aggiunta* (Paravia, Torino, 2000)
14. R. Kurzweil, *The singularity is near* (Viking Books, New York, 2005)
15. P. Musso, How Advanced is ET? in *7th Trieste Conference on Chemical Evolution and the Origin of Life: Life in the Universe*, Trieste, Sept 2003
16. R. Stark, *How the West Won: The Neglected Story of the Triumph of Modernity* (ISI Books, Wilmington, 2014)
17. A.D. Sokal, J. Bricmont, *Intellectual Impostures* (Editions Odile Jacob, Paris, 1997)
18. J.A. Endler, Interactions Between Predator and Prey, in *Behavioural Ecology: An Evolutionary Approach*, ed. by J.R. Krebs, N.B. Davies (Scientific Publications, Oxford, 1993), pp. 169–196
19. C.S. Cockell, M. Lee, Interstellar Predation. J. Br. Interplanet. Soc. **55**, 8–20 (2002)
20. R. Leakey, *The Origin of Humankind* (Orion Books, London, 2000)

21. A. Azuma, *The Biokinetics of Flying and Swimming* (Springer, Tokyo, 1992)
22. G.A. Cavagna, P.A. Willems, N.C. Heglund, Walking on Mars. Nature **393**, 636 (1998)
23. A.E. Minetti, Invariant aspects of human locomotion in different gravitational environments. Acta Astronaut. **39**(3–10), 191–198 (2001)
24. G. Genta, N. Amati, Walking machines and robots: is present technology able to imitate nature? Atti dell'Accademia delle Scienze di Torino, Memorie Sc. Fis, **26**, 49–75 (2002)
25. V. Lytkin, B. Finney, L. Alepko, Tsiolkovsky, Russian cosmism and extraterrestrial intelligence. Q. J. Roy. Astron. Soc. **36**, 369–376 (1995)
26. P. Spillantini et al., Shielding from cosmic radiation for interplanetary missions: active and passive methods. Acta Astron. **42**(1), 14–23 (2007)
27. G. Genta, N. Amati, Non-zoomorphic versus zoomorphic walking machines and robots: a discussion. Eur. J. Mech. Environ. Eng. **47**(4), 223–237 (2002)

Printed in the United States
By Bookmasters